Smart Grid Infrastructure & Networking

About the Author

Krzysztof (Kris) Iniewski, Ph.D., manages R&D development at Redlen Technologies Inc., a start-up company in British Columbia, and is also Executive Director of CMOS Emerging Technologies, Inc. (www.cmoset.com). From 2004 to 2006, he was Associate Professor in the Electrical Engineering and Computer Engineering Department of the University of Alberta, where he conducted research on low-power wireless circuits and systems. Dr. Iniewski has published more than 100 research papers and holds 18 international patents. He is the editor of *Nanoelectronics* and *CMOS Nanoelectronics*.

Smart Grid Infrastructure & Networking

Krzysztof Iniewski, Ph.D.   Editor

New York   Chicago   San Francisco
Lisbon   London   Madrid   Mexico City
Milan   New Delhi   San Juan
Seoul   Singapore   Sydney   Toronto

McGraw-Hill books are available at special quantity discounts to use as premiums and sales promotions, or for use in corporate training programs. To contact a representative, please e-mail us at bulksales@mcgraw-hill.com.

Smart Grid Infrastructure & Networking

1 2 3 4 5 6 7 8 9 0   QFR/QFR   1 8 7 6 5 4 3 2

ISBN 978-0-07-178774-1
MHID 0-07-178774-7

This book is printed on acid-free paper.

Sponsoring Editor	**Project Manager**	**Indexer**
Michael Penn	Deepti Agarwal, Neuetype	Lars T. Berger
Editing Supervisor		**Art Director, Cover**
Stephen M. Smith	**Copy Editor**	Jeff Weeks
	Lisa McCoy	
Production Supervisor		**Composition**
Pamela A. Pelton	**Proofreader**	Neuetype
	Eina Malik Gupta	
Acquisitions Coordinator		
Bridget L. Thoreson		

Contents

Contributors xiii

Preface xv

1 Demand-Side Energy Management
Albert Molderink 1
Demand-Side Developments 3
    Implications on Consumers 4
    Implications on Transmission/Distribution ... 5
    Implications on Generation 6
    Optimization Potential 6
Technical Challenges 7
    Requirements on Demand-Side
        Management 9
    Pitfalls of Large-Scale Control 10
Control Methodologies 10
    Objectives and Stakeholders 11
    Level of Optimization 12
    Optimization Tool-Chain Approaches 13
    Current Research 14
Conclusion 17
References 18

**2 The Modernization of Distribution Automation
Featuring Intelligent FDIR and Volt-Var Optimization**
*Mietek Glinkowski, Bill Rose, Michael J. Pristas,
David Lawrence, Gary Rackliffe, Wei Huang,
Jonathan Hou* 21
DA Systems Architecture and Communication 23
August 14, 2003: The Day That Changed
    Everything for DA 24
A Smarter Distribution Network Arises 25
What Is Distribution Automation? 26
Information Technology and Communications 26
Fault Detection, Isolation, and Restoration 27
FDIR System Architectures 29
Volt-Var Optimization 34
Conclusion 41
References 41

3   Advanced Asset Management   *Mietek Glinkowski,*
Jonathan Hou, Gary Rackliffe **43**
An Aged and Aging Infrastructure 44
Industry Priorities 45
Asset Health Management 46
Benefits from AHM 46
AHM Technologies 47
Challenges 47
Best Practices 48
Advanced Asset Management Practices 49
Asset Health and Condition
    Assessments 50
Risk-Based Planning 51
Decision Support 51

4   Wide-Area Early Warning Systems
Innocent Kamwa **53**
Introduction 53
Enhancing the PMU Vulnerability
    Assessment Functionalities 57
        Hydro-Québec PMU/C Requirements
            and Performances 58
        Hydro-Québec PMU/C Sample
            Performances 61
PMU Placement for Vulnerability
    Assessment 63
        Background 63
        Disturbance-Based Coherency
            for Network Partitioning 64
        The Fuzzy c-Medoids Algorithm
            for PMU Placement 64
Wide-Area Severity Indices 67
        Reference Network 67
        Time-Domain COI-Referred
            Response Signals 68
        Frequency-Domain COI-Referred
            Response Signals 69
        Interpretation and Data Mining
            of Wide-Area Severity Indices 70
        Transient Energy versus Voltage-Based
            Classification 76
Data Mining–Based Catastrophe Prediction 78
        Background 78
        Data Organization for Training
            and Testing 79

Selected PMU-Based WASI Features   81
Black-Box versus Comprehensible
    Predictors   82
Early Detection of System Oscillations
    and Damping Condition Assessment   88
    Background   88
    Multiband Modal Analysis   89
    Linear SIMO Signal Identification   91
    Illustration of Real-Time Modal
        Analysis of Output-Only Signals   96
Conclusion   101
References   103

5 **The Integration of Renewable Energy Sources
into Smart Grids** *Mietek Glinkowski, Jonathan Hou,
Dennis McKinley, Gary Rackliffe, Bill Rose* **107**
The Smart Grid Connection   109
Advances in Smart Grid–Enabled Wind Power
    Technologies   110
Wind Energy in the Context of Smart Grids   112
Turbine Level Solutions: Intelligent Wind
    Converters   116
Inside the Tower: Medium Voltage Switchgear   120
Grid Interconnection Solutions   121
    AC: AC Wind Interconnection Solution   121
    DC: HVDC Interconnection
        for Offshore Installations   124
Support of the Grid Operation: Wind SVC
    with Energy Storage   126
Wind Power Summary   127
Solar Power in the Context of Smart Grids   128
General Considerations of Photovoltaic Plants   129
Photovoltaic Generation   131
Grid-Connected Plants   131
Intermittence of Generation and Storage
    of the Produced Power   132
Voltage-Current Characteristic
    of the PV Module   133
Expected Energy Production per Year   135
Inclination and Orientation of the Panels   136
Voltages and Currents in a PV Plant   137
Connection to the Grid and Measure
    of the Energy   138
Conclusion   139
References   140

**6   The Microgrid in the Electric System
Transformation**   *Steven Pullins* **141**
What Is a Microgrid?   142
Advantages of a Microgrid   143
Architecture and Design of a Microgrid   145
Barriers   150
Are Microgrids Significant to the
      Electric System?   152
Conclusion   153
References   153

**7   Enhancing the Integration of Renewables
in Radial Distribution Networks Through
Smart Links**   *A. Gómez-Expósito, J. M. Maza-Ortega,
E. Romero-Ramos, A. Marano-Marcolini* **155**
Introduction   155
DC Links in Radial Distribution Networks   157
Back-to-Back VSCs Topology   160
General Optimization Framework   165
      Objective Function   165
      DC Link Model   167
      Network Constraints   168
Results   169
      Use of DC Links for Loss Reduction   169
      Use of DC Links for Increasing
         the Load Level   172
      Use of DC Links for Increasing
         the DG Penetration   174
      Economic Assessment   176
Conclusion   178
Acknowledgment   178
References   179

**8   Voltage-Based Control of DG Units and Active
Loads in Smart Microgrids**   *Tine L. Vandoorn,
Lieven Vandevelde* **181**
Introduction   181
Control Strategies for the DG Units   183
      Communication-Based DG
         Unit Control   184
      DG Unit Controllers without
         Communication   185
Control Strategies for the Active Loads   187
      Communication-Based Active
         Load Control   188

Active Load Control without
Communication 191
Primary versus Secondary Active
Load Control 193
Smart Microgrids 195
Conclusion 197
Acknowledgment 198
References 198

9   Electric Vehicles in a Smart Grid Environment
David Dallinger, Daniel Krampe, Benjamin Pfluger **201**
Introduction 202
Load Shifting Using Electric Vehicles 202
Control Equipment 204
Battery Degradation 207
Regulation Reserve Market 208
German Markets for Ancillary Services 208
Dynamic Simulation Approach 211
Results of the Dynamic Approach 213
Electricity Markets 217
Effects of Fluctuating Electricity
Generation 218
Earning Potential of Smart Grid Devices 220
Conclusion 222
Acknowledgment 224
References 225

10   Low-Voltage, DC Grid–Powered LED Lighting
System with Smart Ambient Sensor Control
for Energy Conservation in Green Building
Yen Kheng Tan, King Jet Tseng **227**
Introduction 227
Low-Voltage DC Microgrid 229
Motivation: AC Grid versus DC
Grid Technology 230
Emerging Standards for LVDC 231
LED Solid-State Lighting System 232
Intelligent Wireless Sensors Network System
for Smart-Induced Green Building 234
Wireless Sensor Network and Its Sensors:
A Technology Overview 235
Case Study: Low-Voltage, DC Grid–Powered
LED Lighting System with Smart Ambient
Sensor Control for Energy Conservation in a
Green Building 238

LED Lighting Simulation 240
Implementation of a Low-Voltage,
    DC Grid–Powered LED Lighting System 243
Smart Ambient Sensor Control for Energy
    Conservation in a Green Building 244
Conclusion 247
References 247

11 **Multiple Distributed Smart Microgrids
    with a Self-Autonomous, Energy Harvesting
    Wireless Sensor Network**   *Josep M. Guerrero,
    Yen Kheng Tan* **249**
Introduction 249
A General Approach of the Hierarchical
    Control of MGs 254
Hierarchical Control of AC-MGs 255
    Inner Control Loops 257
    Primary Control 257
    Secondary Control 260
    Tertiary Control 264
    Results 265
Self-Powered Smart Wireless Sensors
    for Microgrids 272
    Overview of Wireless Sensor
        Networks 272
    Problems in Powering the
        Sensor Nodes 273
    Wind Energy Harvesting for Smart
        Wireless Sensor Networks 273
    Magnetic Energy Harvesting
        for a Smart Wireless Sensor
        Network 281
Conclusion 286
References 287

12 **Wireless Sensor Networks for Consumer
    Applications in the Smart Grid**   *Hussein T. Mouftah,
    Melike Erol-Kantarci* **291**
Introduction 291
Communication Standards for WSNs 295
    Zigbee 295
    Ultra Low-Power Wi-Fi 298
    Z-wave 299
    WirelessHART 300
    ISA-100.11a 301

WSN-Enabled Consumer Applications
in the Smart Grid 301
    WSN-Enabled Demand Management
        for Residential Consumers 301
    Coordination of PHEV
        Charging/Discharging Cycles 308
Security and Privacy of WSN-Based
    Consumer Applications 313
Summary and Open Issues 315
References 317

**13 Zigbee-Based Wireless Monitoring and Control
System for Smart Grids** *Abiodun Iwayemi,
Chi Zhou, Peizhong Yi* **321**
Introduction 321
Zigbee-Based Building Energy Management
    Demonstration System 323
Zigbee/IEEE 802.15.4 and Wi-Fi/
    IEEE 802.11b Overview 324
        Zigbee/IEEE 802.15.4 324
        Wi-Fi/IEEE 802.11b 325
        Main Interference Sources
        of IEEE 802.15.4 326
Performance Analysis of Zigbee under Wi-Fi 327
        BER Analysis of Zigbee under Wi-Fi 327
        PER Analysis of Zigbee under
        Wi-Fi Interference 328
Frequency Agility–Based Interference
    Avoidance Scheme 329
        Interference Detection 330
        Interference Avoidance 331
Simulation and Implementation Results 333
        Simulation Results 333
        Implementation 335
Conclusion 340
References 341

Index **343**

Contributors

David Dallinger   *Fraunhofer Institute for Systems and Innovation Research* (Chap. 9)

Melike Erol-Kantarci   *School of Electrical Engineering and Computer Science, University of Ottawa, Ottawa, Ontario, Canada* (Chap. 12)

Mietek Glinkowski   *ABB, Inc.* (Chaps. 2, 3, 5)

A. Gómez-Expósito   *Department of Electrical Engineering, Universidad de Sevilla, Seville, Spain* (Chap. 7)

Josep M. Guerrero   *Aalborg University, Denmark* (Chap. 11)

Jonathan Hou   *ABB, Inc.* (Chaps. 2, 3, 5)

Wei Huang   *ABB, Inc.* (Chap. 2)

Abiodun Iwayemi   *Illinois Institute of Technology* (Chap. 13)

Innocent Kamwa   *Hydro-Québec Research Institute* (Chap. 4)

Daniel Krampe   *Fraunhofer Institute for Systems and Innovation Research* (Chap. 9)

David Lawrence   *ABB, Inc.* (Chap. 2)

A. Marano-Marcolini   *Department of Electrical Engineering, Universidad de Sevilla, Seville, Spain* (Chap. 7)

J. M. Maza-Ortega   *Department of Electrical Engineering, Universidad de Sevilla, Seville, Spain* (Chap. 7)

Dennis McKinley   *ABB, Inc.* (Chap. 5)

Albert Molderink   *University of Twente, Enschede, Netherlands* (Chap. 1)

Hussein T. Mouftah   *School of Electrical Engineering and Computer Science, University of Ottawa, Ottawa, Ontario, Canada* (Chap. 12)

Benjamin Pfluger   *Fraunhofer Institute for Systems and Innovation Research* (Chap. 9)

Michael J. Pristas   *ABB, Inc.* (Chap. 2)

Steven Pullins   *President, Horizon Energy Group* (Chap. 6)

Gary Rackliffe   *ABB, Inc.* (Chaps. 2, 3, 5)

E. Romero-Ramos   *Department of Electrical Engineering, Universidad de Sevilla, Seville, Spain* (Chap. 7)

Bill Rose   *ABB, Inc.* (Chaps. 2, 5)

Yen Kheng Tan   *Energy Research Institute @ Nanyang Technological University, Singapore* (Chaps. 10, 11)

King Jet Tseng   *Nanyang Technological University, Singapore* (Chap. 10)

Lieven Vandevelde   *Electrical Energy Laboratory, Department of Electrical Energy, Systems and Automation, Ghent University* (Chap. 8)

Tine L. Vandoorn   *Electrical Energy Laboratory, Department of Electrical Energy, Systems and Automation, Ghent University* (Chap. 8)

Peizhong Yi   *Illinois Institute of Technology* (Chap. 13)

Chi Zhou   *Illinois Institute of Technology* (Chap. 13)

Preface

Smart grids, for many the biggest technological revolution since the invention of the Internet, will play an important role in tomorrow's societies. Governments around the world are currently pumping large sums of money into smart grid research, development, and deployment. Their aims are many and diverse.

Smart grids have the potential to reduce carbon dioxide emissions through the integration of distributed renewable energy resources, energy storage, and plug-in hybrid electric vehicles. Moreover, they can increase the reliability of the electricity supply (reduced blackout rate) in real-time measurement by monitoring and controlling the generation, transmission, and distribution of electrical networks. Furthermore, they can render the utilization of power generation stations and electricity transport infrastructure more efficient, deploying dynamic pricing and demand-response strategies. In addition to these benefits, at this stage in the game we cannot even imagine the products and services that will evolve as the smart grid takes hold. Nevertheless, we are confident that the engine of innovation will find new ways to exploit the potential of the smart grids, creating many more jobs than they eliminate through automation.

Besides updating power electronics sensing, monitoring, and control technology, in the last decade smart grids have made a key advance in the area of telecommunications. The infrastructure and networking aspects of smart grids are at the heart of this book, bringing together leading experts from academia as well as from the power electronics and telecommunications industries.

Today's power grid is a system that supports electricity generation, transmission, and distribution operations. It is composed of a few central generating stations and electromechanical power delivery systems operated from control centers. Power flows mainly from the central generating stations (high-voltage producers) toward the medium- and low-voltage customers. One prerequisite for grid stability is the balance between energy consumption and generation. Presently, energy generation follows the time variant consumption. With the integration of more and more small-scale renewable energy generation facilities, e.g., exploiting solar and wind power, the energy flow in some scenarios is reversed. Energy may flow from the customer (owning a renewable energy generation facility) into the grid, leading to a more complex grid structure. However, many renewable resources reveal an intermittent and unpredictable nature of supply. This makes their integration a challenging task and requires an upgrade to the aging eclectic energy infrastructure.

In the coming decade it is expected that the world's energy infrastructure will undergo a transformation in scale similar to that which recently has happened in telecom and media industries. The evolving smart grid combines the electrical power infrastructure with modern digital distributed computing facilities and communications networks. It is a collection of complex, interdependent systems whose key functions include reliable and efficient power delivery. This is facilitated through wide-area situational awareness, peak curtailment through demand-response schemes, widespread integration of intermittent renewable energy resources through real-time control and electric storage, and a shift from a largely fossil fuel–driven transport system to electric transportation.

This book contains 13 carefully selected chapters. They cover the areas of smart grid infrastructure, security, and networking. Particular emphasis is placed on the integration of renewable energy and electric vehicles with the smart grid concept. The authors are well-recognized experts in their fields and come from both academia and industry.

With such a wide variety of topics covered, I hope that the reader will find something stimulating to read and will discover the field of embedded systems to be both exciting and useful in science and everyday life. Books like this one would not be possible without many creative individuals meeting together in one place to exchange thoughts and ideas in a relaxed atmosphere. I would like to invite you to attend the CMOS Emerging Technologies events that are held annually in beautiful British Columbia, Canada, where many topics covered in this book are discussed. See www.cmoset.com for presentation slides from the previous meeting and announcements about future ones.

I would love to hear from you about this book. Please e-mail me at kris.iniewski@gmail.com.

Let the smart grids in the world prosper and benefit us all!

Krzysztof (Kris) Iniewski, Ph.D.

Smart Grid Infrastructure & Networking

CHAPTER 1

Demand-Side Energy Management

Albert Molderink
University of Twente, Enschede, Netherlands

E nergy efficiency, electricity supply, and sustainability are important research topics in society [1]. The provisioning of energy is subject to increasing resource usage, scarcity, and environmental concerns. Next to a slightly increasing energy usage in the Western world, developing countries like China and India show growth figures up to 25%. Since the production capacity of fossil energy sources, especially crude oil, cannot keep up with this growth, the oil reserves are diminishing and energy is becoming scarce and expensive. Furthermore, a lot of fossil energy sources are obtained from politically less stable regions. This, in combination with the growing awareness of the greenhouse effect, drives the search to renewable energy sources.

Other effects of the previously mentioned problems can already be seen in the *electricity supply chain*. The electricity supply chain consists of electricity generation, transportation, distribution, and consumption. In this supply chain a lot of changes are ongoing or expected. Within these changes four trends can be identified:

1. *An electrification of the energy distribution:* A growing part of the consumed energy is transported and consumed as electricity.

2. *An increase in energy consumption:* The energy consumption, and the electricity consumption in particular, increases.

3. *More dynamic electrical loads are occurring:* Electricity consumption does not only increase, it also becomes more fluctuating and sometimes even uncontrollable.

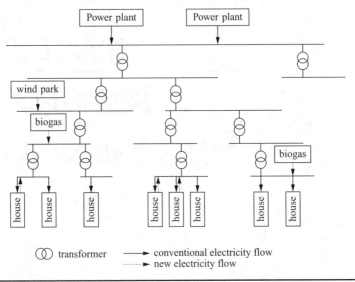

FIGURE 1.1   Sketch of the (changes in the) electricity grid structure.

4. *An increase in distributed electricity generation:* Nowadays more and more electricity is generated on a relatively small scale lower in the grid, whereas in the past all electricity was generated in a few large power plants and was transported via the grid to the consumers.

The most important change in the electricity supply chain is a shift from centrally produced electricity flowing downward through the grid to the consumers, toward a distributed electricity generation on different levels in the grid. These trends and changes result in challenges to maintain a reliable and stable supply, but they also open opportunities. Using widespread information and communication technology (ICT) to monitor and manage this distributed generation and other (domestic) technologies allows the grid to become more efficient and more sustainable.

The electricity supply chain is sketched in Fig. 1.1. The shift toward sustainable generation and the addition of small-scale distributed electricity generation may look rather harmless at first, but it has severe consequences on the electricity supply chain.

The electricity grid is no longer a matter of one-way traffic, for which it was built and designed for many years. This changes the process of decision making in electricity management completely. New concepts for managing/controlling the electricity supply chain are needed. Demand-side management (DMS) in the broad sense can be seen as such a concept to treat the supply chain management problem.

In this chapter we give an overview of demand-side developments, forthcoming challenges to integrate these new technologies in the concept of a smart grid, and control methodologies that are proposed for DSM.

Demand-Side Developments

In the last decades, more and more stress is put on the electricity supply and infrastructure. As shown in Fig. 1.2, the still ongoing electrification of the energy demand caused a significant increase in electricity demand. Furthermore, the electricity demand became very fluctuating, caused by the stochastic nature of demand; people switch on the dishwasher or washing machine whenever they like. The electricity supply chain was designed decades ago and is completely demand driven; consumers just switch on devices when they want to and the generation side has to deal with this fluctuating and hard-to-predict demand. In the "old-fashioned" supply chain the base load of the total system is supplied by large-scale, inflexible but quite efficient generation, whereas fast-reacting (and relatively inefficient) peak capacity has to be reserved to serve the peak loads. Demand peaks result in peaks in generation and transmission, which define the requirements in the supply chain. Thus, due to the fluctuating demand, grid requirements have increased. When electricity demand rises and becomes more fluctuating, e.g., with a large-scale introduction of electrical cars without charge time optimization, the efficiency of conventional power plants drops [2] and large investments in grid capacity are required to be able to transport all electricity (peaks) from the power plants to the consumers.

On the other hand, the reduction in the CO_2 emissions and the introduction of generation based on renewable sources become important topics today. The current rate of natural resources consumption will lead to a depletion of these resources, urging for alternative methods to provide the required energy demand. However, renewable resources are mainly "fueled" by very fluctuating and uncontrollable sun, water, and wind power. To maintain grid stability, all generated electricity must be consumed. Therefore, the peaks in renewable generation should be lower than the electricity consumption. A consequence is that within the current demand–supply philosophy only a limited percentage of the

FIGURE 1.2
Energy usage per household for Dutch households (CBS).

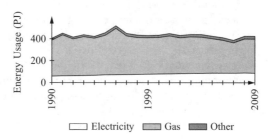

conventional generation can be replaced with renewable generation. Supplemental peak generation capacity is required to keep the demand and supply in balance, resulting in an even more fluctuating generation pattern for the conventional power plants.

So, the demand side of the original supply chain has to deal with *more fluctuating and increasing consumption* and it faces the introduction of *(renewable) small-scale generation* at the demand side (or between the original consumption and generation).

While a lot of research is ongoing to enable the possibility to supply our energy needs with renewable sources, still a lot of improvements can be achieved on the efficiency of current systems as long as not all energy needs can be supplied sustainably, e.g., by preventing the usage of peak power plants.

Therefore, the challenges we face are (1) to increase the efficiency of current power plants, (2) to reduce the stress on the grid resulting from higher demand peaks and prevent investments in grid capacity, and (3) to facilitate a large percentage of renewable sources for electricity generation in the grid while maintaining a stable grid and a reliable supply.

In the next three subsections the implications of the current trends on consumption, the transmission, and the generation are discussed. And in the last subsection of this section the optimization potential of new technologies is introduced.

Implications on Consumers

The current trends in the energy supply decrease the flexibility on the generation side, urging for more flexibility on the consumer side of the supply chain. At the moment, especially increasing energy prices and growing awareness of the greenhouse effect drive consumers to adopt new domestic technologies to save money and energy.

An example of these technologies is a *microgenerator*. Microgenerators generate electricity at kilowatt level in or nearby buildings, resulting in less transport losses. Often microgenerators are more energy efficient than conventional power plants and some are based on renewable energy sources [2, 3].

Other new technologies are energy buffers and smart devices. Energy buffers can (temporarily) store energy. Heat buffers are already common in current buildings, but more and more electricity buffers are introduced. These energy buffers make it possible to shift electricity consumption in time, e.g., shift consumption to earlier times by filling the buffer and supply the demand with the stored energy. Smart devices are defined as devices with the ability to temporarily switch off (parts of) the device or devices that can shift the demand in time. A smart fridge is an example of a device that can shift load in time; the cooling temperature of the fridge should stay between certain bounds; within these bounds there is freedom to start cooling earlier. This is sketched in Fig. 1.3.

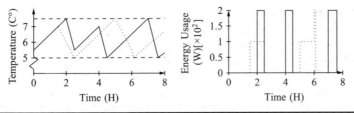

Figure 1.3   Shift the load of a fridge in time.

Unfortunately, some of these technologies may introduce more fluctuations on the electricity grid. For example, if due to human behavior all microgenerators start producing simultaneously, the generated energy must be consumed locally, more likely by the grid. However, these new domestic technologies also introduce freedom in the electricity consumption patterns. These devices can be monitored and managed to change their consumption profile resulting in more flexibility of the consumers.

Implications on Transmission/Distribution

The increasing demand and peaks lead to an increase of required grid capacity and therefore large investments. However, to reduce fluctuations and to incorporate more renewable sources, generation and consumption need to be matched. To make this possible, significant improvements in the grid infrastructure and more intelligence in the grid are required.

The foreseen changes in production and consumption will increase the stress on the grid while at the same time stability, reliability, and fault tolerance of the grid become more important since the community more and more depends on electricity. Therefore, the streams through the grid should be monitored and managed.

Demand-side management plays a key role in allowing a larger share of renewable electricity generation. However, there is also a geographical aspect on the possibilities for this generation. Large-scale sustainable electricity generation is often only possible on remote places with a low density of population and therefore a low electricity demand (e.g., large wind power farms offshore or solar panels in the desert). It is expected that the renewable potential in Europe is large enough to supply all electricity [4]. But this electricity needs to be transported to the customers, requiring a large transmission capacity. The large distance between areas with high potential for generating renewable electricity and areas where the electricity is consumed puts additional pressure on the grid. To transport the sustainable electricity from the generation site toward the customers in Europe, e.g., a European-wide, interconnected, high-capacity electricity grid is required, in combination with a European-wide electricity market.

Implications on Generation

A shift toward a sustainable energy supply has large consequences for the electricity generation. Today, coal is the main source of electricity generation. A future without electricity generation using coal is almost unthinkable since coal is cheap, abundant, and can be harvested in more stable countries [5]. Unfortunately, coal is one of the most polluting fossil fuels concerning the amount of CO_2 emission. A better option is sustainable electricity generation using renewable sources (sun, wind, tides, etc). However, this requires profound changes and improvements of the electricity grid and the electricity supply.

Large-scale sustainable electricity generation has large differences from conventional power plants, both in generation capacity and controllability. In particular, sun and wind are very fluctuating in strength; consider a cloud blocking the sun, which needs to be handled properly to ensure a stable supply of electricity. It is, in general, agreed that it is both desirable and necessary to manage this new type of generation and adapt the rest of the grid infrastructure to facilitate the sustainable, unmanageable generation.

So, next to high-capacity lines for long-distance transportation of electricity, a sustainable electricity supply also requires more and better monitoring and control capabilities of all types of generation on different levels of the grid.

Optimization Potential

The current trends in the electricity supply chain have implications for all parts of the grid. Furthermore, a number of challenges mentioned in the introduction of this section should be solved to maintain a reliable and dependable supply. A solution for these challenges may be to transform customers from static consumers into active participants in the production/consumption process. Consumers can exploit the potential of the new technologies by shifting load and/or generation to the most beneficial times, where beneficial depends on the optimization objective. More can be reached when a large group of consumers together work toward objectives, but this requires coordination and consumers lose (a part of) the control over their devices.

The low flexibility at the generation side of the electricity supply chain is compensated by an increase of flexibility on the consumer side. However, this transformation of consumers into active participants also requires a change in the state of mind of people, i.e., consumers for whom the availability of energy was always evident should cooperate in keeping the quality and reliability of supply at a high level. A start of this transformation is awareness of the energy consumption. Just this awareness already leads to a decrease of electricity usage of up to 20% [6]. Furthermore, it requires readiness from politicians and policy makers.

Technical Challenges

In this section the foreseen transition toward a better monitored and managed grid, a so-called *smart grid*, is studied in more detail. First, a definition of the concept of smart grids is given and the technical challenges are derived. These comprise the technical requirements that are necessary elements in the development of management methodologies for the future electricity supply chain. The technical challenges for the smart grid boil down to requirements for demand-side management.

The improved version of the grid is often called a smart grid. It is hard to give a definition of a smart grid; different parties have their own definition. In [7] is stated that the smart grid is not a "thing" but rather a "vision": "*The smart grid vision generally describes a power system that is more intelligent, more decentralized and resilient, more controllable, and better protected than today's grid.*" The definition given in [8] is rather common:

> A smart grid generates and distributes electricity more effectively, economically, securely, and sustainably. It integrates innovative tools and technologies, products and services, from generation, transmission and distribution all the way to customer devices and equipment using advanced sensing, communication, and control technologies. It enables a two-way exchange with customers, providing greater information and choice, power export capability, demand participation, and enhanced energy efficiency.

For a successful introduction of a smart grid we face a number of technical challenges. In [9] five key technologies required for the smart grid are identified:

1. Sensing and measurement
2. Integrated communications
3. Advanced components
4. Improved interfaces and decision support
5. Advanced control

Since you can only manage what you can measure, sensing and measuring are an important part of the smart grid. The health parameters of the transmission lines and substations should be monitored to protect the grid from outages. Monitoring and forecasting of the weather can be used for forecasting load and potential output of renewable sources. This can subsequently be correlated with transmission line capacity. Next to the grid, also the generation, storage, and consumption sites and devices need to be monitored to be capable of balancing generation and usage and respecting transmission limitations.

To transport all information, a high-speed communication infrastructure is required. This integrated communications infrastructure moves the information between sensing and measurement devices toward the operators and management information back to the actuators. Creating a homogeneous infrastructure requires standards be respected by all stakeholders, from home networks and all devices connected to it via the smart meters and the distribution companies to the overall network operators. The National Institute for Standardization and Technology (NIST) addressed this problem and is working together with the Institute of Electrical and Electronics Engineers (IEEE) to create smart grid standards [10]. The integrated communications infrastructure should be designed with the future in mind, meaning that capacity, security, and performance should be sufficient to facilitate future applications as well. A fast, reliable, and well-designed integrated communications infrastructure glues all the parts of the smart grid together.

A smart grid is built up by a network of advanced components. The grid itself should consist of efficient transmission elements connected by advanced flow control devices. On a domestic level, a lot of technologies are in development. The technologies can be subdivided into three groups:

1. *Distributed generation* (DG): The local electricity production
2. *Distributed storage* (DS): The local energy storage
3. *Demand-side (load) management* (DSM) *in the narrow sense:* The control of the load of specific appliances (c.q. flexible fridges)

New tools are required to assist the grid operators. The grid operators' job became much more challenging in the last years, from response times of minutes some years ago, they now have to react in seconds. To have enough information to make decisions, data mining is very important. An improved interface is required to visualize the large amount of data in such a way that it can be understood at a glance. Furthermore, decision support tools help in making decisions, e.g., fast simulations to forecast consequences of decisions.

To make use of all control capabilities and to exploit all optimization potential, advanced control systems need to be developed. Advanced protection systems can adjust relay settings in time for better protection of the grid and even increased power flows in some cases [9]. Controlling flows can, e.g., increase stability, increase damping of oscillations, operate transmission networks as efficiently as possible, and assure maximum utilization of transmission assets. The growing share of technologies on a lower voltage level that can influence real and reactive flow can enhance operators' ability to influence grid conditions significantly. Furthermore, coordination of (renewable) generation, storage, and consumption is fundamental to reach all targets of a smart grid.

Requirements on Demand-Side Management

As shown in the previous subsection, DSM is already incorporated in some emerging advanced components in the smart grid. However, the intelligence of these components mainly focuses on the component itself and extends its focus at most to its local environment. We want to refer to this kind of DSM as demand-side management in the narrow sense. Demand-side management in the broad sense asks for management tools (on prediction, on planning, on control) that, besides focusing on DSM in the narrow sense, also include intelligence to reliably integrate all parts of the smart grid.

The optimization objective can differ, depending on the stakeholder of the control systems, the system state, and the rest of the electricity infrastructure. Therefore, a control methodology for DSM in the broad sense should be able to work toward different objectives. Next to different objectives, control methodologies can have different scopes for optimization: a local scope (within the building), a scope of a group of buildings, e.g., a neighborhood (microgrid), or a global scope (virtual power plant). Finally, there are a lot of different (future) domestic technologies and building configurations and it should be possible to incorporate new technologies. As a consequence, the control method-ology needs to be very *flexible and generic*.

The goal of the control methodology is to monitor, control, and optimize the import/export pattern of electricity and to reach objectives that may incorporate *local but also global goals*. In this context, local objectives concern energy streams within the building, e.g., lowering electricity import peaks and using locally (in or around the building) produced electricity in the building. Global objectives on the other hand concern energy streams of multiple buildings, e.g., on a neighborhood, city, or even (parts of) a country level. These objectives can be on different levels, e.g., on a neighborhood level to consume local generated electricity locally or on a national level to optimize production patterns of large power plants. Thus, the control methodology optimizes the runtime of individual devices to work toward local and global objectives.

Furthermore, the control methodology should be able to optimize for a single building up to a large group of buildings. Thus, the algorithms used in the control system should be *scalable* and the *amount of required communication should be limited*. The control methodology should try to exploit the potential of the devices as much as possible while *respecting the comfort constraints of the residents* and the technical constraints of the devices. Furthermore, the control system should consume significant less electricity than it saves.

Furthermore, *limitations on the communication links* should be taken into account. Due to the latency of communication links, sending information between system elements about system state and decisions made require a certain amount of time. However, deciding whether it is profitable/required to switch on a large consuming device (e.g., a washing machine) or reacting on fluctuation in generation needs to be done virtually instantaneously. Thus, a local control system has to be able to make these *realtime decisions* or these decisions need to be made beforehand.

Pitfalls of Large-Scale Control

The technical challenges lead to several requirements for DSM as mentioned in the previous section. The complexity of the methods used in DSM is of more importance in the smart grid. Especially for realtime control low-complexity management tools should be available, since the possible decisions to choose from have increased significantly. This problem is noticed in the requirement for scalable methods, but we would like to stress that scalability should not be achieved by neglecting the cooperation between all elements of the smart grid. For example, *turn-based* decision making, in which each level (domestic, neighborhood, city) makes its own decisions based on global information from the past, could lead to an unstable electricity grid, so coordination between different components should not be neglected.

The control methodology should prevent oscillating behavior caused by over-steering and large fluctuations (peaks), e.g., when a lot of buildings react on the same steering signal. This is called *damage control*. Damage is often caused by prediction errors and/or using more potential then available (e.g., maximum electricity import is too low) or synchronous behavior (all buildings reacting at the same time).

Control Methodologies

In this section the characteristics of control methodologies for DSM are discussed.

Control methodologies for DSM in the broad sense can work toward objectives on different levels. On a high level, a large group of buildings is combined to improve efficiency of power plants by reducing fluctuations in demand, or the flexibility is used to compensate for fluctuating renewable generation to allow a higher penetration rate of renewable energy. On a medium level, the electricity streams through the grid are managed to optimally use the available grid capacity. On a low level, the locally generated electricity is kept within the neighborhood and peaks in consumption are lowered (*peak shaving*). We have mentioned before that a control methodology should focus on:

1. Improving the efficiency of existing power plants

2. Facilitating the large-scale introduction of renewable generation

3. Allowing large-scale introduction of new domestic technologies, both producing and consuming, using the current grid capacity

while at the same time maintaining grid stability and reliability of supply.

Based on the way the control methodology is used, extra requirements may arise. One possible application of the control methodology is to act actively on an electricity market for a group of buildings. To trade on such a market, an electricity profile must be specified one day in advance. Therefore, it should be possible to determine a forecast of the net electricity profile of the managed group of buildings one day in advance. Another application can be to react on fluctuations in the grid, e.g., caused by renewable generation, asking for a realtime management. Reacting on fluctuations requires a realtime control and the availability of sufficient generation capacity at every moment in time to be able to increase or decrease the consumption. To achieve sufficient capacity, again planning must be determined in advance, in combination with realtime control, to react on the fluctuations. Thus, a combination of *prediction* of demand and generation of devices, a *planning* of the use of these devices, and *realtime control* are needed.

To create a successful smart grid solution and exploit all optimization potential the introduced technologies need to be monitored and synchronized to each other. Based on the measured data during monitoring, prediction and trends can be generated, which can be used during planning and the realtime control.

Objectives and Stakeholders

An important issue is the large number of stakeholders involved in the transition toward a smart grid: governments, regulators, consumers, generators, traders, power exchanges, transmission companies, distribution companies, power equipment manufacturers, and ICT providers [5]. These stakeholders need an incentive to cooperate, because it seems to be unattractive for some stakeholders to change their view on the supply chain. However, distribution companies can decrease operating and maintenance costs and reduce capital costs. Production companies can introduce new types of generation and increase generation with relatively cheap base-load plants [11]. The consumers can reduce their costs and increase power quality, and finally society will benefit from a stimulated economy and improved environmental conditions [11]. For the electricity retailers, demand-side developments open new possibilities to act on the electricity market. Based on the (partial) control over local (renewable) generation capacity, retailers can remodel their strategies on the electricity market, forcing the original generation and transportation side to adapt to the emerging technologies.

Both [8] and [12] indicate that commercial attainability and legislation are important issues for the success of the introduction of DG. The opinions for the investments and profits differ strongly. On the one hand, the European Climate Forum states that large investments are required while it is unknown what the actual benefits and profits are [4]. On the other hand, the U.S. Department of Energy states that the transition toward a smart grid has already started and that profits are higher than the investments [11]. They even claim that due to all benefits (e.g., improved safety and efficiency, better use of existing assets) the transition toward a smart grid will be market driven.

Level of Optimization

The optimizations can be performed on different levels in the grid, all with their advantages and disadvantages. Roughly the levels can be divided into three groups:

Local Scope

On a local scope the import from and export into the grid can be optimized, without cooperation with other buildings. Possible optimization objectives are shifting electricity demand to more beneficial periods (e.g., nights) and peak shaving. The ultimate goal can be to create an independent building. This can be done in two forms: *energy neutral* or *islanded*. Energy neutral implies that there is no net import from or net export into the grid. A building that is physically isolated from the grid is called an islanded building.

The advantages of a local scope is that it is, besides the technical challenge, relatively easy to realize, there is no communication with others (less privacy intrusion) and there is no external entity deciding which devices are switched on or off (better social acceptance). The disadvantage is that it might result in high investment costs, e.g., in storage capacity and microgeneration.

Microgrid

In a microgrid a group of buildings together optimize their combined import from and export into the grid, optionally combined with larger scale DG (e.g., wind turbines). The objectives of a microgrid can be shifting loads and shaving peaks such that demand and supply can be matched better internally. The ultimate goal can be perfect matching within the microgrid, resulting in a neutral or islanded microgrid. The advantage of a group of buildings is that their joint optimization potential is higher than that of individual buildings since the load profile is less dynamic (e.g., startup peaks of devices disappear in the combined load). Furthermore, multiple microgenerators working together can match more demand than individual microgenerators since better distribution at the time of the production is possible [13]. Finally, within a microgrid the locally produced electricity can be used locally, saving transmission costs and preventing streams from lower to higher voltage levels. However, for a microgrid a more complex control methodology is required.

Virtual Power Plant (VPP)

The original VPP concept is to manage a large group of microgenerators with a total capacity comparable to a conventional power plant. Such a VPP can replace a power plant while having a higher efficiency, and moreover, it is much more flexible than a normal power plant. This last point is especially interesting since it expresses the usability to react on fluctuations. This original idea of a VPP can of course be extended to other domestic technologies. Again, for a VPP a complex control methodology is required. Furthermore, communication with every individual building is required and privacy and acceptance issues may occur.

The three previously mentioned groups mainly differ where the decisions are made. This is tightly coupled with who is responsible for both the control systems and the techniques used. For example, in case of a local scope, homeowners can invest in their own house, reaping the profits made locally. In the case of a VPP, retailers or utilities might invest in domestic generators to be placed in houses, using them to make money on an energy market.

Optimization Tool-Chain Approaches

There are many research projects investigating energy efficiency optimization. From the studied research, simulations, and field tests it can be concluded that the efficiency can be improved significantly, especially when all three types of technologies (consuming, buffering, and generating) are combined.

Several *control methodologies* for DG, energy storage, demand-side load management, or a combination of these can be found in the literature. Roughly these control methodologies can be divided into two groups: (1) *agent-based market mechanisms* and (2) *discrete mathematical optimizations*. The advantage of agent-based market mechanisms is that no knowledge of the local situation is required on higher levels, only (aggregated) biddings for generation/consumption are communicated. The advantage of mathematical optimizations is that the steering is more direct and transparent; the effect of steering signals is better predictable. Another important difference is that in an agent-based approach often every building works toward its own objectives whereas in a mathematical approach the buildings can work together to reach a global objective.

To overcome the scalability and communication issues the structure of the control system is important. A hierarchical structure with data aggregation on the different levels is an often proposed scheme. Such a structure is scalable while the amount of communication can be limited. However, when data is aggregated, information gets lost, so it is a trade-off between precision and the amount of communication. An example of such a hierarchical structure is shown in Fig. 1.4.

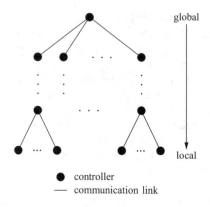

FIGURE 1.4
Hierarchical control
architecture.

controller
— communication link

All control methodologies split the control into a local and a global part, most of them using a hierarchical structure for scalability. Furthermore, most control methodologies use an online algorithm deciding on device level, and some control methodologies use prediction to adapt the production and demand patterns. However, this prediction data is often only used on a local level and, therefore, on a global level hardly any prediction knowledge is available. What is lacking in most control methodologies is the predictability on a global level, which is required for electricity market trading, i.e., insight in the effect of choices. This is also related to dependability.

Several control methodologies are based on cost functions for every devices, which is a nice abstraction mechanism from device-specific characteristics. These cost functions define the normal behavior of the device and express options to deviate from the normal behavior with the desirability (costs to deviate). The cost functions for devices are a very flexible way to express the status of the device and desirability of different options. Since the cost functions are similar for every type of device, new devices can be incorporated in this approach. Furthermore, the control methodologies act on an homogeneous set of cost functions that keeps the algorithms much easier and less computationally intensive. Finally, cost functions of multiple devices can be combined into one cost function to study the effects of a single steering signal on a group of (different) devices.

Current Research

Most of the current research considers agent-based control methodologies. These agent-based control methodologies propose an agent per device [14]. The agents give their price for energy production (switching an appliance off is seen as production); via a market principle it is decided which agents are allowed to produce. Since there are a lot of agents, the information is aggregated on different levels in a hierarchical way. The research described in [15] combines domestic generation, consumption, and buffering of both heat and electricity.

They propose an agent-based system where buildings are divided into groups (microgrids), which are loosely connected to the conventional large-scale power grid. In first instance the goal is to maintain balance within the microgrid without using the large-scale power grid. Furthermore, agents use predictions to determine their cost function. Their field studies show that 50% of the domestic electricity demand can potentially follow a planned schedule (within certain boundaries). To reach this potential, there have to be incentives for the residents to allow some discomfort.

The PowerMatcher described in [16] and [17] additionally takes the network capacities into account. This control methodology is rather mature; it is a product capable of being used in field tests [18]. In these field tests, a peak reduction of 30% is reached when a temperature deviation of 1° of the thermostat in the buildings is allowed. To be able to reach objectives, business agents can be added that influence the biddings in the auction market.

In [19] the authors compare the results of individual (local) and overall (global) optimizations. They conclude that global optimizations lead to better results. Next, they claim that agent-based control methodologies outperform non-agent-based control methodologies since agent-based control methodologies take more (domestic) information into account.

In the literature, some mathematical control methodologies are also proposed. The research described in [20] proposes a methodology that is capable of aiming for different objectives. For every device a cost function is determined for both heat and electricity. Using a nonlinear problem definition the optimal on/off switch pattern is found. The authors of [21] address the problems of both agent and non-agent-based solutions: non-agent-based solution are less scalable and agent-based solutions need local intelligence and are not transparent. Therefore, they propose a combination: aggregate data on multiple levels, while these levels contain some intelligence. The aggregation is done with a database; the control methodology is rule based. In [22] a control methodology is proposed using stochastic dynamic programming (SDP). The stochastic part of the control methodology considers the uncertainty in predictions and the stochastic nature of (renewable) production and demand. The authors of [23] propose a control methodology based on time-of-use (TOU) pricing, where electricity is cheaper during off-peak periods. They combine this approach with a domestic wireless sensor network: when a smart appliance would like to switch on, it has to send a request to a controller. This controller decides, based on the electricity price and the status of the other devices, whether the appliance is allowed to switch on. The TOU pricing can be seen as global steering signals; however, it is a rough steering signal which is equal for a large group of buildings. Furthermore, it is not known in advance what the impact of the steering signals is.

In [24] a combination of existing tools together with a new developed platform is used. The electricity consumption and production per device is forecasted and using genetic algorithm the best runtime for every device is determined. The platform consists of two levels: a global level for global optimizations sends steering signals to the local level and a local level control which uses the global steering signals as input and determines the runtimes based on the steering signals while respecting local constraints.

Most control methodologies use some sort of prediction of demand and/or production. These predictions can be made rather well with neural networks, as described in [25] and [26]. The predictions follow the trend rather well.

At the University of Twente, a methodology has been developed that uses mathematical optimization techniques and a combination of prediction, offline global planning based on the predictions, and online real-time control based on the global planning [27]. The base of the control methodology is (1) using local information, (2) communication using multiple levels, and (3) scalability. The goal of the control methodology is to work toward (global) objectives and the performance of the control methodology is measured by the extent to which the objectives are reached. The methodology uses (1) predictions on a device level to be able to predict the overall result, (2) planning to estimate the energy streams in the building and the grid, and (3) real-time control to respond to changes (e.g., fluctuations in renewable generation) and work around predictions errors.

Based on the previously mentioned considerations, the control methodology uses three steps and is split up into a local and a global part: (1) local offline prediction, (2) global offline planning, and (3) local online control. Because of scalability reasons, the global planning has a hierarchical structure and can aggregate data and plannings on different levels, e.g., within a neighborhood or city.

Due to the predictions and planning beforehand, the predictability of the global electricity streams is improved. The combination of planning (aggregated knowledge on higher levels) and mathematical optimization results in better dependability, and the combination of planning and realtime control improves the damage control. Furthermore, the amount of communication can be limited due to the hierarchical structure. Finally, the requirements on the communication medium are low since the local controller can work independently and a lot of information can be sent on beforehand without high latency requirements.

The combination of prediction, planning, and real-time control exploits the potential of the overall system at the most beneficial times. The hierarchical structure with intelligence on the different levels ensures scalability, reduces the amount of communication, and decreases the computation time of the planning.

Conclusion

Concerns about climate change, increasing energy prices, and dependability of energy supply ask for drastic changes in the energy supply chain, but also in the current demand–supply philosophy. Current trends in energy consumptions result in an increasing and more fluctuating electricity usage, causing a decreasing efficiency of conventional power plants and increasing requirements on the grid and generation capacity. Furthermore, in order to meet the CO_2 emission reductions aimed for in the 20-20-20 agreements, at least a large part of the electricity should be generated by renewable sources, which are to a large extent uncontrollable. This introduces even more challenges to maintaining a reliable, dependable, and affordable electricity supply. Therefore, new ways (1) to achieve a more efficient use of the generated electricity of existing power plants, (2) to facilitate the large-scale introduction of renewable sources, and (3) to allow a large-scale introduction of new technologies for consumption and storage of energy are required while maintaining grid stability and ensuring a reliable and affordable supply.

The current grid is developed based on a demand–supply philosophy in which all electricity is generated in a few large central power plants and is transported top-down and one-way to the consumers. The consumers' side of the supply chain is static; consumers switch on devices and the generation side has to supply the demand. However, to increase the efficiency of current power plants and to allow the introduction of uncontrollable renewable sources, the consumer side of the supply chain should become more flexible; i.e., consumption should be adjusted to generation. To achieve this, the current electricity grid should be transformed into a smart grid and domestic customers should be transformed from static consumers into active participants in the energy supply chain. The main goals of a smart grid are to support the introduction of renewable generation and to keep up with the growing electricity demand and at the same time maintain a stable, reliable, and affordable electricity supply. The consumption can be adjusted to the generation: the decrease in flexibility on the generation side can be compensated by a more flexible electricity grid and a more flexible consumer side. Essential in a smart grid is a system that monitors and manages all parts of the grid. The emergence of smartening the grid and updating the electricity supply chain is emphasized by the numerous initiatives worldwide from the European Union, from governments, and from industries as well as from the academic world. However, to reach a smarter grid, a number of technical (e.g., scalability and dependability), economical (e.g., who has to invest/profit), political (e.g., is it allowed), and ethical (e.g., privacy issues) challenges have to be addressed. To tackle the technical challenges and to realize a monitoring and management system, ICT is seen as one of the key enabling technologies.

An important component of monitoring and management systems for smart grids is, next to sensors and actuators, a control methodology consisting of algorithms to gather information, process this information, and optimize the overall electricity streams. Such a control methodology, capable of exploiting all potentials in a reliable and dependable way, should meet a number of requirements. The control methodology should work with both local and global objectives and should be very generic and flexible. Furthermore, since a large number of buildings is involved, the control methodology needs to be scalable. To be acceptable for the residents, it should also respect their comfort level. Furthermore, to get a dependable and reliable control methodology capable of damage control, a combination of prediction, planning, and control is required. Finally, the requirements on the communication links should be limited and in case of failing communication links the local controller needs to be capable of working independently.

A hierarchical tree structure ensures scalability and limits the required communication. Furthermore, the optimizations based on cost functions result in a flexible and generic control methodology. The separation in local and global controllers distributes the required computational power and ensures the comfort and privacy of the end users. The combination of offline prediction, offline planning, and online control results in a flexible, dependable, and predictable solution.

References

[1]   E. Commission, "Energy 2020, a strategy for competitive, sustainable and secure energy," European Union, Tech. Rep., 2010.
[2]   A. de Jong, E.-J. Bakker, J. Dam, and H. van Wolferen, "Technisch energieen CO_2-besparingspotentieel in Nederland (2010–2030)," *Platform Nieuw Gas*, p. 45, Juli 2006.
[3]   United States Department of Energy, "The micro-CHP technologies roadmap," *Results of the Micro-CHP Technologies Roadmap Workshop*, Dec. 2003.
[4]   A. Battaglini, J. Lilliestam, C. Bals, and A. Haas, "The supersmart grid," European Climate Forum, Tech. Rep., 2008.
[5]   E. S. T. Platform, "Vision and strategy for europe's electricity networks of the future," European SmartGrids Technology Platform, Tech. Rep., 2006.
[6]   S. Darby, "The effectiveness of feedback on energy consumption," Environmental Change Institute, University of Oxford, Tech. Rep., 2005.
[7]   N. E. T. Laboratory, "A vision for the smart grid," U.S. Department of Energy, Tech. Rep., 2009.
[8]   J. Scott, P. Vaessen, and F. Verheij, "Reections on smart grids for the future," Dutch Ministry of Economic Affairs, Apr. 2008.
[9]   N. E. T. Laboratory, "The transmission smart grid imperative," U.S. Department of Energy, Tech. Rep., 2009.
[10] N. I. of Standards and Technology, "Nist framework and roadmap for smart grid interoperability standards, release 1.0," National Institute of Standards and Technology, Tech. Rep., 2010.
[11] K. Dodrill, "Understanding the benefits of the smart grid," U.S. Department of Energy, Tech. Rep., 2010.

[12] P. Fraser, "Distributed generation in liberalised electricity markets," International Energy Agency, Tech. Rep., 2002.

[13] S. Abu-sharkh, R. Arnold, J. Kohler, R. Li, T. Markvart, J. Ross, K. Steemers, P. Wilson, and R. Yao, "Can microgrids make a major contribution to UK energy supply?" *Renewable and Sustainable Energy Reviews*, vol. 10, no. 2, pp. 78–127, Sept. 2004.

[14] J. Oyarzabal, J. Jimeno, J. Ruela, A. Englar, and C. Hardt, "Agent based micro grid management systems," in *International conference on Future Power Systems 2005*. IEEE, Nov. 2005, pp. 6–11.

[15] C. Block, D. Neumann, and C.Weinhardt, "A market mechanism for energy allocation in micro-chp grids," in *41st Hawaii International Conference on System Sciences*, Jan. 2008, pp. 172–180.

[16] J. Kok, C. Warmer, and I. Kamphuis, "Powermatcher: Multiagent control in the electricity infrastructure," in *4th international joint conference on Autonomous agents and multiagent systems*. ACM, July 2005, pp. 75–82.

[17] M. Hommelberg, B. van der Velde, C. Warmer, I. Kamphuis, and J. Kok, "A novel architecture for real-time operation of multi-agent based coordination of demand and supply, " in *Power and Energy Society General Meeting —Conversion and Delivery of Electrical Energy in the 21st Century, 2008 IEEE*, July 2008, pp. 1–5.

[18] C. Warmer, M. Hommelberg, B. Roossien, J. Kok, and J. Turkstra, "A field test using agents for coordination of residential micro-chp," in *Intelligent Systems Applications to Power Systems, 2007. ISAP 2007. International Conference on*, Nov. 2007, pp. 1–4.

[19] A. Dimeas and N. Hatziargyriou, "Agent based control of virtual power plants," in *Intelligent Systems Applications to Power Systems, 2007. ISAP 2007. International Conference on*, Nov. 2007, pp. 1–6.

[20] R. Caldon, A. Patria, and R. Turri, "Optimisation algorithm for a virtual power plant operation," in *Universities Power Engineering Conference, 2004. UPEC 2004. 39th International*, vol. 3, Sept. 2004, pp. 1058–1062 vol. 2.

[21] E. Handschin and F. Uphaus, "Simulation system for the coordination of decentralized energy conversion plants on basis of a distributed data base system," in *Power Tech, 2005 IEEE Russia*, June 2005, pp. 1–6.

[22] L. Costa and G. Kariniotakis, "A stochastic dynamic programming model for optimal use of local energy resources in a market environment," in *Power Tech, 2007 IEEE Lausanne*, July 2007, pp. 449–454.

[23] M. Erol-Kantarci and H. T. Mouftah, "Tou-aware energy management and wireless sensor networks for reducing peak load in smart grids," in *Proceedings of the IEEE Vehicular Technology Conference Fall*, 2010.

[24] S. Bertolini, M. Giacomini, S. Grillo, S. Massucco, and F. Silvestro, "Coordinated micro-generation and load management for energy saving policies," in *Proceedings of the first IEEE Innovative Smart Grid Technologies Europe Conference*, 2010.

[25] J. V. Ringwood, D. Bofelli, and F. T. Murray, "Forecasting electricity demand on short, medium and long time scales using neural networks," *Journal of Intelligent and Robotic Systems*, vol. 31, no. 1-3, pp. 129–147, Dec. 2004.

[26] V. Bakker, A. Molderink, J. Hurink, and G. Smit, "Domestic heat demand prediction using neural networks," in *19th International Conference on System Engineering*. IEEE, 2008, pp. 389–403.

[27] A. Molderink, V. Bakker, M. G. C. Bosman, J. L. Hurink, and G. J. M. Smit, "Management and control of domestic smart grid technology," *IEEE Transactions on Smart Grid*, 2010.

CHAPTER 2

The Modernization of Distribution Automation Featuring Intelligent FDIR and Volt-Var Optimization

Mietek Glinkowski, Bill Rose, Michael J. Pristas, David Lawrence, Gary Rackliffe, Wei Huang, Jonathan Hou
ABB, Inc.

The history of the distribution network on the power grid goes back to the days of the early "current wars" between Thomas Edison and Nicolai Tesla. The first distribution automation product was introduced in 1940 when oil circuit reclosers, designed to provide fault detection on rural networks, were installed on distribution lines. In 1941, the first single-phase hydraulic recloser with dual timing was launched, and the first three-phase hydraulic recloser was introduced in 1946. The first reclosers placed outside of the substation were installed on lines in the 1950s. More advanced automation of the distribution networks emerged in the 1960s with solid-state technology. Microprocessors were introduced in the mid-1970s but were not widely accepted by utilities because they were considered too costly and risky to deploy. It wasn't until the 1990s when microprocessor technology blossomed and the costs of remote communications media were acceptable that automation made its way into the utility infrastructure.

The role of the distribution grid for years was to step down the power from high-voltage transmission lines to a medium-voltage level and distribute it as needed to homes and businesses with minimal incidents. When there were disturbances or faults, such as damage caused by storms or the occasional substation fire, utilities would ideally dispatch their personnel to the trouble spot and manually repair or replace the damaged equipment as quickly as possible. In reality, however, utilities often did not even know where the fault occurred, why it occurred, or where the failed equipment or line was located. In fact, utilities would often have to wait for customers to call them to even know if there was an issue on the grid. Fortunately, the occurrence of power outages was manageable in the first several decades of electrification in the United States.

Utilities in general are motivated by market conditions. For years the primary driver for distribution automation (DA) has been the utility's desire to provide reliable and clean power to its customers at the lowest cost. As the market has matured, other drivers for DA have emerged, such as the need for real-time information of critical power parameters and remote assets throughout the network. Beyond the substation, however, utilities knew little about the condition of their distribution network until a disruptive "event" occurred. These remote assets of the utility's distribution system in hindsight could be characterized as the "final automation frontier."

For the past two decades, the deployments of new power generation and transmission lines have not kept pace with market growth. Environmental laws and the emergence of data centers that support the Internet as a large power consumption entity have put a strain on the power generation capacity in the United States. Power delivery systems in many parts of the United States have become saturated and highly congested. Thus, the occurrence of power disruption events has increased with the added load to these systems, and the ability to identify and rectify outages has increased in importance. Utilities were now being measured on the "uptime" of their total power delivery system, especially in these highly saturated and congested networks.

One of the solutions developed to address this growing concern was technology that would automatically identify, isolate, communicate, and mitigate fault conditions to safely and quickly restore power. As automation and communication technologies evolved, utilities have been able to systematically "IT (information technology) enable" these remote assets as part of their integrated infrastructure management systems. To "IT enable" a product or equipment is defined as providing monitoring and control intelligence plus realtime communications capabilities to previously isolated and stand-alone remote distribution assets.

Asset management is another market driver that has impacted utility behavior. The ability to analyze past service call information and to correlate this data with remotely monitored distribution assets in real time can reduce maintenance costs, prolong equipment life, and prevent premature failures of equipment and distribution lines.

More recently, the term "network optimization" has been used to define the ability to maximize the utilization of remote assets, such as capacitor banks and voltage regulators, to improve system reliability and efficiency, avoid the investment of additional generation, and improve overall demand response.

Safety is another critical element of distribution automation, particularly in the medium-voltage environment. When you evolve from a manual recloser product and power restoration practice to an automated remote recloser system, safeguards, rules, and procedures must be well-designed, documented, communicated, and understood within the utility service community. Service personnel can now monitor local conditions and the switching status through the user interface of the automated recloser and coordinate their activities with remote system personnel to assure safe and effective procedures and practices.

Technology is now also available that monitors and records numerous power data points from the distribution network, issues predefined alarms to remote users, and maintains historical data. The enabling core technology is a high-accuracy current and voltage line sensor system that provides numerous realtime power parameters and information on the health of the network to the end user. This technology also contributes to the prevention of injuries due to faulty equipment or distribution lines. Distribution line service personnel know ahead of time the location and nature of network problems. They can also now safely and efficiently correct emerging problems before they contribute to more serious outages.

Thus, safety procedures for automated recloser and distribution network monitoring systems rank as the highest priority for distribution automation technology while at the same time maintaining the importance of uptime efficiency initiatives for the utility.

DA Systems Architecture and Communication

When it comes to system architecture and communications within DA, there is no "one size fits all." Historically, it started as a decentralized stand alone application due to a lack of communication technology, poor reliability, and high cost of implementation. The distribution systems in North America began primarily as radial networks and therefore, the early DA systems evolved based on this design. In Europe and in parts of Asia, ring mains and matrixed networks have dominated the utility landscape. DA schemes are now different depending on the power network topology and operating conditions.

	Positives	Negatives
Stand-Alone	Fast Simple	Views only itself Results in longer duration of outages
Peer-to-Peer	Better decision-making More efficiency Handles more events Can handle a wider variety of events	Requires some communication Higher costs Complexity level is higher
Centralized	Command-centered model "All knowing"	Cost Complex Large
Intelligent Decentralized	Lower cost Chunking the problem into manageable pieces Optimum hybrid between peer-to-peer and centralized	Vendor-to-vendor compatibility between "chunks"

TABLE 2.1   The Four Major Distribution Automation System Architectures

There are today four major DA system architectures, each with unique advantages and disadvantages, and each with their own contribution to the smart grid, as demonstrated in Table 2.1.

August 14, 2003: The Day That Changed Everything for DA

Most of the apparatus on the distribution lines operated untouched for decades. In fact, much of the original equipment remains today. Most utilities were historically silo-oriented in their strategic planning and infrastructure design. However, everything changed one summer day—August 14, 2003—when a widespread power outage occurred throughout parts of the northeastern and midwestern United States as well as Ontario, Canada. In fact, it was during this Blackout of 2003 that the term "smart grid" was used for the first time, by a reporter writing about the need for a modernized, highly integrated, and technology-driven power grid [1].

The New York ISO (independent system operator) that was responsible for managing the New York State power grid reported that a 3500 MW power surge affected the transmission grid at 4:10 P.M. on August 14. Outages were then reported over the next half hour throughout Ohio, New York, Michigan, and parts of New Jersey. These were followed by other areas including Vermont, Connecticut, and most of Ontario. When it was over, a large area that stretched from Lansing, Michigan, to Sault Ste. Marie, Ontario, from the shore of James Bay to Ottawa, throughout New York, and into northern Ohio was without power.

The official analysis of the blackout reported that more than 508 generating units at 265 power plants in the United States and Canada shut down during the outage. At full capacity, the NYISO-managed power system was carrying 28,700 MW of load. During the peak of the outage, the load had dropped to 5716 MW, a reduction of 80%.

It was apparent from the investigation of the event that the U.S. and Canadian grids were operating independently (both internally and externally) of the other, and a series of badly timed power events resulted in the second-largest power outage in the history of electrification. ISOs had a solid reputation for managing events within their own service territory. However, they were ill prepared for a series of major events that originated outside of their own domain.

In retrospect, this event was a blessing in disguise. The electric utility infrastructure was in dire need of modernization on every level. The silo mentality that permeated throughout the utility structure was experiencing a transformation. Heretofore stand-alone equipment and remote assets were becoming "IT enabled," as stated earlier, a term that references the introduction of microprocessor based, smart user interfaces, resident memory, and remote communications capabilities. This information was now accessible by utility personnel "24/7" through dedicated software platforms designed specifically for the application. Supervisory control and data acquisition (SCADA) and other discreet utility network management systems have been in place and refined over the past several decades. However, there had been a large information gap beyond the substations and throughout the transmission and distribution networks. More resources and information about this momentous event may be found on Wikipedia ("Northeast Blackout of 2003") [2].

A Smarter Distribution Network Arises

Historically, existing distribution networks have been neglected from the vantage point of design, architecture, technological sophistication, and functionality of the devices. Only since the 2003 blackout have electric utilities started to publicly recognize that distribution networks have to be on par with technological advancements in the transmission and generation portions of the grid in order to provide better end-user reliability. What used to be a simple radial overhead line with a fuse or oil recloser can now be modernized to utilize the newest technologically advanced devices and algorithms that can detect, isolate, and restore systems in the event of faults. This includes technologies such as FDIR (fault detection, isolation, and restoration). From an efficiency point of view, distribution networks could now be optimized to lower the losses, improve voltage regulation, and manage control of reactive power via a variety of volt-var optimization (VVO) techniques.

In the information technology domain, the distribution system—being the "middle man" between the low-voltage service network, electricity meter, and the high-voltage bulk transmission system—has to play an increasingly important role connecting, managing, and enabling the flow of clean, reliable power and information from the level of individual electricity customers to the system-wide energy control centers.

This chapter will look briefly at a few of the key technologies and network innovations that are already moving the distribution network to a truly modernized, smart grid status.

What Is Distribution Automation?

The IEEE defines distribution automation as "a system that enables an electric utility to remotely monitor, coordinate, and operate distribution components in a real-time mode from remote locations"[3]. We similarly define distribution automation as "any operation of device (or devices) outside the power substation that is triggered by an automatic response to any change in a power distribution network." DA can function as centralized or decentralized intelligence as auto-operations in normal (load) and abnormal (fault, fault restoration) situations, with load transfers and two-way power flow for the distributed grid.

Information Technology and Communications

A key enabler for distribution automation has been the advent of IT, including the development of reliable, faster, and cost-reduced microprocessor and communications technologies. It has only been since 1971 that the Intel 4004 4-bit microprocessor was introduced. Since then, Moore's Law has generally held true, where microprocessor complexity has doubled every 24 months. Eight-bit microprocessors led the way through the 1970s and 1980s with the 8080, 6502, and 6809 designs. Sixteen-bit microwars were ever present in the 1980s with the Intel 80×86 and Motorola 68000 families. Thirty-two-bit designs dominated the 1990s. And 64-bit designs from Intel, AMD, and PowerPC led the way over the last decade.

Parallel to the microprocessor wars was the evolution of digital radio frequency (RF) communications. GSM (Global System for Mobile Communications) 2G digital cellular networks were first proposed to replace 1G analog cellular networks in 1987. GSM could move 9.6 kbps. CDMA (multiple access) followed with speeds of 9.6 to 14.4 kbps. GPRS (packet service) was introduced next with 9.6 to 115 kbps of throughput. In 2003 EDGE (enhanced GSM) obtained 385 kbps. Qualcomm first designed EV-DO (evolution data only) in 1999; this technology was considered 3G and increased the bandwidth to 1.5 Mbps to 2.4 Mbps. HSPA (high-speed packet access) 4G was first introduced in 2008. It achieves speeds of 10 Mbps to 50 Mbps.

This brief summary of the evolution of microprocessor and RF communications serves as a strong reminder that the technologies for monitoring and control of the distribution automation network exist today. Today's analog-to-digital sensing and communications products are typically based on reliable and inexpensive 8- and 16-bit microprocessor designs incorporating a variety of communications media.

Fault Detection, Isolation, and Restoration

Customer service, grid reliability, asset optimization, and sustained revenue are key drivers for investor-owned utilities, municipalities, and cooperatives. These drivers are measured by various parameters, but two that address outage management and stand out as key performance indicators (KPIs) are SAIDI (system average interruption duration index) and CAIDI (customer average interruption duration index). These two KPIs are used in the energy industry to measure operational effectiveness.

In the past, equipment was deployed as stand-alone assets. Fuses would open and then require manual intervention to replace. Circuit breakers would require a close signal to reenergize. More recently, reclosers were programmed to open and then reclose the circuit in an attempt to "burn-off" the cause of the fault. After multiple attempts, they would stop the reclosing activity and lock out, requiring a manual reset. More recently, reclosers have evolved to work with semi-intelligent auto-sectionalizers to limit the spread of the outage and to sectionalize the system into smaller pieces.

Today, intelligent electronic devices (IEDs) are installed in reclosers. Their activities are coordinated with other feeder reclosers, tie-point switches, substation control equipment, and network operation center algorithms to facilitate the fastest possible response to an outage. The most recent improvement to this technology is known as automated fault detection, isolation, and restoration, better known as FDIR.

FDIR has emerged as an integrated application due to the advancement in microprocessor and communication technologies. Speed, performance, and bandwidth have increased almost exponentially, while the cost of these technologies has decreased.

In the timeline above (see Fig. 2.1), the fault occurs at (t0), the fault is detected within subcycles (t1), and the circuit is opened. At this point a circuit breaker or recloser will typically make three attempts to restore power. If the fault is permanent, then it is isolated at (t2) and unfaulted portions of the feeder may be reenergized through an alternative path. Later, when the fault is cleared, the circuit is restored at (t3) and reconfigured back to its original configuration.

Fault Detection

A fault can be detected by a fuse or a breaker. The problem with a fuse or breaker is that utility personnel must find the fault, repair it

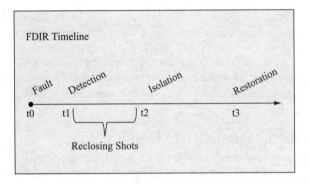

FIGURE 2.1   Fault, detection, isolation, and restoration timeline.

and then replace the fuse or reset the breaker. Today, protection relays, reclosers, and switches include IEDs, microprocessor-based devices with analog-to-digital conversion functionality that can measure line voltages, currents, and current waveforms, and make decisions by analyzing these measurements. A common IED decision is to open the circuit. Another is to try three recloser operations prior to opening the circuit. More recently, the decision process includes coordination with other IEDs to isolate the fault and to automatically pick up unfaulted sections of the feeder.

Recloser Shots

A recloser shot happens after the device detects a downstream fault, opens the circuit, and still sees upstream (source-side) voltage. The recloser will close the circuit in an attempt to burn off the fault. If the fault remains, then the recloser will open the circuit. The recloser tries to restore power, typically three times at time intervals of 0.5 seconds, 15 seconds, and 45 seconds. If the fault is permanent, then the recloser locks out.

Isolating a Fault

Fault isolation is performed by coordinating-source (substation) circuit breakers, reclosers, and tie-point IED actions. Isolation and pickup of unfaulted feeder sections can be accomplished with stand-alone IEDs that use coordinated timers, or with IEDs that coordinate through communications. The pros and cons of these architectures are described later.

The utility determines the location of the fault to repair via phone calls from customers (worst case), by visually inspecting the feeder (driving and observing fault indicators or downed lines), or by receiving messages from field IEDs through the SCADA system telling them the approximate GPS position of the fault (best case).

Restoring Service

Once the fault is repaired, the feeder is restored to its original configuration. This can be accomplished manually or automatically through the system of coordinated IEDs. Great care must be taken when restoring power and returning the feeder to its original configuration.

FDIR System Architectures

There are three major system architectures for addressing FDIR: loop control scheme, peer-to-peer messaging, and decentralized. These are described next.

The loop control scheme is implemented with IEDs, but does not require communications. In Fig. 2.2, the source is a substation feed and the circuit breaker includes an IED, VTs (voltage transformers), and CTs (current transformers). The circuit breaker supplies power downstream. If a fault occurs between the Source 1 circuit breaker and the sectionalizing recloser, the circuit breaker opens the circuit. The circuit breaker performs the *recloser shots* algorithm. If the algorithm fails to burn off the fault, then the circuit breaker locks out in the open position. The recloser and the tie-point recloser see the voltage "not present" condition on the upstream side and they both know the *recloser shots* algorithm. When the Source 1 circuit breaker locks out, the recloser also opens, isolating the fault. The tie-point recloser back-feeds the unfaulted section by closing and restores service to this portion of the system.

Figure 2.3 is a simple example. Normally a feeder will have a few sectionalizing reclosers. The behavior of the system is made possible through IEDs that all have knowledge of the reclosing shots algorithm. The IEDs coordinate their retry and lock-out behavior with timers and counters.

If the permanent fault occurs between the last feeder sectionalizing recloser and the tie-point, then it is possible for the tie-point recloser to back-feed Source 2 power into a faulted section.

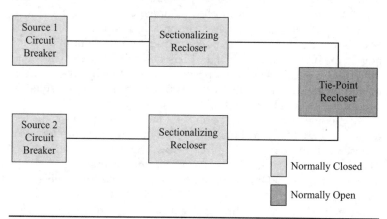

FIGURE 2.2 FDIR using a loop control scheme.

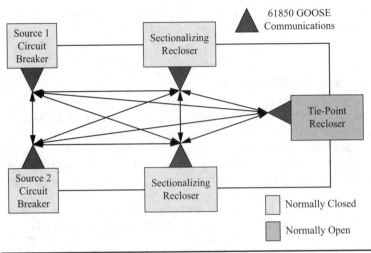

FIGURE 2.3 FDIR using peer-to-peer messaging.

The peer-to-peer messaging FDIR scheme is implemented with IEDs and peer-to-peer communications. The communications medium can be fiber-optic cable or copper wire, but is normally based on RF technology. A communications protocol that supports peer-to-peer addressing and coordination, like IEC 61850 GOOSE, must be used in order to implement this architecture. Figure 2.3 shows each IED with a point-to-point connection with every other IED. Each IED must have knowledge of the feeder network topology. It is likely not necessary for the IEDs on Feeder 1 to communicate with IEDs on Feeder 2, so these links may be removed from the protocol addressing mechanism. The chain of IEDs on the feeder must be known to each IED, and the tie-point recloser must be able to communicate with both feeders. In a more complex arrangement with many tie-points, this structure can become complex to implement.

When a fault occurs somewhere on a feeder, the section of the fault is immediately known through peer-to-peer communications among the IEDs. The actions of the IEDs are similar to those described in the loop control scheme, except coordination with hardware timers and counters is not needed. The *recloser shots* strategy and results are communicated to all affected IEDs.

In this architecture, if the permanent fault occurs between the last feeder sectionalizing recloser and the tie-point, then the tie-point recloser will not back-feed Source 2 power into the faulted section. The section of the fault is known to all communicating IEDs.

Decentralized FDIR is normally implemented on a substation computer with multiple feeders (see Fig. 2.4). It can also be implemented across multiple substations with an interconnected network.

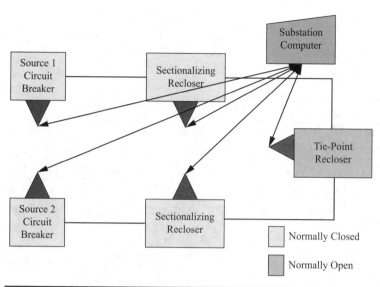

FIGURE 2.4   Decentralized FDIR.

Decentralized FDIR uses a connectivity matrix to represent the distribution network scheme, which includes three types of components: sources, switching devices, and loads. It reads the status of all devices and updates the connectivity matrix if there is a change in the network from opening or closing actions. FDIR logic reads the load current of each device and updates realtime loading for each load. Capacity of sources is updated frequently, especially when alternate power sources are in the network and can potentially be switched in to back-feed power to unfaulted sections.

A lock-out signal from a recloser or IED triggers the FDIR logic (see Fig. 2.5). It uses the connectivity matrix to find IEDs and switches next to the fault. It communicates with these IEDs to isolate the fault by opening the switches around it. Next, it attempts to restore power to unfaulted healthy loads. FDIR logic searches all possible paths from the sources to the normally open tie-point reclosers. If a connected path or multiple paths are found, and if the source or sources have enough capacity to power the additional load, then FDIR calculates the expected load on the new path, sends a load change message to the protection relay to change its settings, and sends a close command to the tie-point recloser.

Once the fault is cleared, FDIR logic is able to restore the network to the prefault condition by reversing the FDIR actions and safety-checking all conditions.

FDIR can execute in manual, automatic, or test modes. When FDIR is first deployed, utility operators can select manual mode in order to gain familiarity and confidence with the system. In manual mode, each switching action is confirmed by the operator through the

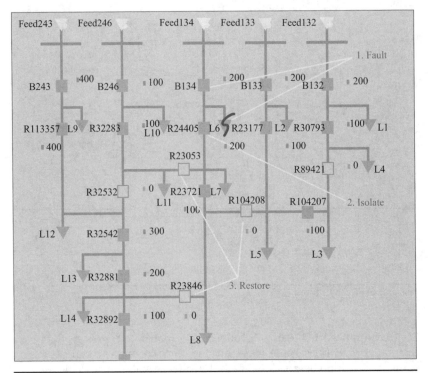

FIGURE 2.5   A decentralized FDIR network.

FDIR human machine interface (HMI) interface on the substation computer. In automatic mode, isolation and restoration actions are executed automatically; switches are opened and closed, the fault is isolated, and the load is restored without human intervention. The operator can run FDIR in test mode to test the behavior of FDIR logic. In test mode, FDIR logic will simulate a fault on any load, generate the isolation and restoration actions, and record them to an operator log without sending actual commands to IEDs.

Decentralized FDIR functionality is described using the network in Fig. 2.5. The network includes one substation with two feeders and a second substation with three feeders. Normally open tie-point reclosers connect the feeders together. When a fault occurs at load (L6), then the recloser (B134) locks out. FDIR logic sends an open command to (R24406) to isolate the fault. FDIR logic then checks the network and finds a path for load (L7) and (L8) to other sources: (Feeder243), (Feeder246), and/or (Feeder133). If (Feeder246) has enough capacity to power extra loads (L7) and (L8) through (R233053), then FDIR logic updates protection relay settings (B246), (B32283), and (R23721) and sends a command to close switch (R233053) to restore power to loads (L7) and (L8).

Figure 2.6 provides a summary of the various types of FDIR.

FDIR Summary	Comm	Scalability	Impact on Customer Satisfaction	Comment
Loop Control Scheme	No	Small	Good	• Good for small deployments • No communications infrastructure required • Simple IEDs measure fault on downstream, voltage on upstream, coordination with counters and timers • Can connect alternate source to a faulted section • Longer response time to full restoration
Peer-to-Peer Messaging	Yes	Medium	Better	• Improved automated handling of small deployments • Faster fault detection and isolation • More accurate knowledge of where the fault occurred • Improved response time to full restoration • IED configuration complexity increases greatly as the size and capabilities of the network increase
Decentralized	Yes	Large	Best	• Best for large deployments • Even though it is called "decentralize," it is more accurately described as substation or multiple substation centralized • Has network connectivity map in 1 location—the substation computer • Accurate fault location determination • Circuit loading information makes switching activities more safe • Best response time to full restoration • Full implementation yields best impact on O&M costs • Best customer satisfaction

FIGURE 2.6 Summary of different types of FDIR.

33

Volt-Var Optimization

Have you ever wondered how much electric energy the world consumes and how much of this energy is lost on its way from the power plants to the end users that could be saved or how much greenhouse gas emissions could be cut if such energy losses were reduced by even a small amount? The industry is now producing new technologies and integrated systems to help reduce electric energy losses and the demands made on electric distribution systems. There is a wide spectrum of solutions available today that can increase energy efficiency and optimize demand management.

Volt-var optimization (VVO) is the latest evolution of these applications. Differing from the traditional approach using uncoordinated local controls, VVO uses realtime information and online system modeling to provide optimized and coordinated control for unbalanced distribution networks with discrete controls. Electric distribution companies can achieve huge savings in the new frontier of energy-efficiency improvement by maximizing energy delivery efficiency and optimizing peak demand. VVO helps achieve these objectives by optimizing reactive resources and voltage control capabilities continuously.

The active power (P) delivered to the loads is always accompanied by the reactive power (Q). Historically the flow of reactive power has been viewed as the necessary evil. Loads and overhead (OH) lines mostly consume reactive power (inductive character), whereas cables mostly generate power (capacitive character). However, the balance between the two is less than perfect.

The entire system has to carry extra current to support Q, which has several consequences, among them: (a) increased I^2R losses in lines, cables, buswork, transformers, switches, etc., due to Q (see Fig. 2.7); and (b) uneven voltage profile along the lines due to the $I * Z_L$ voltage drop, where Z_L is the inductive impedance of the line.

When the loads and OH lines need reactive power, the power adds in quadrature to the active (real) power (P), and therefore the total power (S) at a given system voltage (V) results in the increase of the current (I). See Fig. 2.8.

$$S\,[kVA] = \sqrt{P\,[kW]^2 + Q\,[kvar]^2} = V * I$$

Both of these factors have been recognized as potential for great improvements. It is estimated that up to 5% of the system losses can be improved by compensating for (or eliminating) the unwanted vars. In comparison, a mere 1% reduction of the peak load demand in the United States would save the country over 8 GW (8000 million watts) of electricity generation [4] [5].

These improvements have been historically initiated in the form of installing fixed (lumped) capacitors (in all three phases) connected along

The energy loss is due to the resistance in the conductor. The amount of loss is proportional to the product of the resistance and the square of the current magnitude. Losses can be reduced, therefore, either by reducing resistance or the current magnitude or both. The resistance of a conductor is determined by the resistivity of the material used to make it, by its cross-sectional area, and by its length, none of which can be changed easily in existing distribution networks. However, the current magnitude can be reduced by eliminating unnecessary current flows in the distribution network.

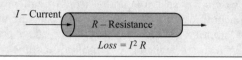

$$Loss = I^2 R$$

FIGURE 2.7   Energy losses.

The voltage and current waveforms on an AC power line are typically sine-shaped. In an "ideal" circuit, the two are perfectly synchronized. In the real azworld, however, there is often a time lag between them. This lag is caused by the capacitative and inductive properties of attached equipment (and of the lines themselves).

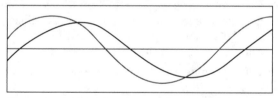

The momentary flow of power at any time is the product of the momentary current and voltage. The average value of this power is lower than it would be without the time lag (for unchanged magnitudes of voltage and current). In fact the power even briefly flows in the "wrong" direction.

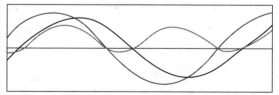

The greater the time lag between the curves, the lower the energy delivery. This lag (expressed as phase angle) should thus be minimized. The average energy delivery per time unit is called active power (measured in W). Reactive power (measured in VAr) is a measure of the additional power that is flowing on the line but cannot be put to effective use.

FIGURE 2.8   Active and reactive power.

the distribution lines. One feeder line of certain length might have one or two capacitor banks installed at 40% and 80% of the line length. These capacitors would act as var generators (or var sources) boosting the voltage of the feeder line around them as well as lowering the current that the line would have to carry to support the same loads. Although this provides the first step to more optimal operation, it is not enough.

The loads and the resulting line currents are very dynamic. Changes and fluctuations are daily (day versus night loads) as well as weekly (workdays versus weekends) and seasonal (winter versus summer). Fixed capacitors cannot change. The next step in the quest for an optimized, smarter system is a voltage and reactive power compensation in the form of switchable capacitors (simple volt-var control, VVC). A capacitor bank would have its own switch (or a contactor in some cases), which would be controlled by a dedicated controller. The controller would switch the capacitors on or off depending on a number of parameters:

- Time of day (i.e., on at 7:00 A.M., off at 7:00 P.M.). This requires a clock that is part of the controller electronics.

- Ambient temperature (i.e., on when T > 90°F , off when T < 80°F). This requires that a temperature sensor be included.

- Voltage magnitude (i.e., on when voltage drops below 0.97 pu, off when voltage rises above 1.03 pu). This requires that a voltage signal be available from the line at a utilization level of 110 to 220V; often a distribution transformer powering the controller is also used as a voltage sensor.

- Current magnitude (i.e., on when current is greater than x amp and off when current is dropping below y amp). This requires a current sensor or CT (current transformer).

- Var flow. This requires the controller to have V and I signals available (for one or for all three phases), perform the vector multiplication of the two with the correct phasing, and compare these calculations with the preset values of Q_{on} and Q_{off}.

Most controllers today provide all of these functionalities. In addition many are also equipped with a remote communications function. Also, few of them have the opportunity of the feedback messaging (two-way communication), i.e., confirmation that the switch is operational or the cap bank is performing as expected, etc.

Although capacitor banks on the distribution feeders boost the voltage of the line by generating vars, voltage regulators (VR) is yet another technique of regulating the voltage on the line. They are, in essence, load tap changing (LTC) transformers that can move their taps up and down to keep the voltage within the desired limits.

At this point smartness of the grid becomes a reality. The switchable capacitors with controllers as well as voltage regulators

Tier	Scope	Communication Pattern	Decision-Making Model	Suitable Applications
Tier 1	Device	No (or yes for notification only)	Local, autonomous	Protection, equipment monitoring, diagnostics
Tier 2	Feeder or substation	Hierarchical (master/slave)	Central	Restoration, VVC (w/o model), VVO (w/model)
Tier 2 P2P	Feeder or substation	Peer to peer	Collaborative	Restoration, VVC
Tier 3	System	Hierarchical (master/slave)	Central	VVO (w/model)

TABLE 2.2   Levels of Volt-Var Efficiency

able to communicate two ways can become part of a bigger picture, i.e., a system that can collect data from the number of controllers and VR locations of a substation, process the data for all the lines, calculate the optimum var flow, and perform the switching of capacitor banks and/or adjusting the taps to achieve the optimum operation. This approach is often referred to as a comprehensive VVO.

As the ultimate step of the volt and var management for a smart grid one can see the emergence of grid-wide VVOs where all solutions described earlier are utilized, combined at the distribution system level, and married with the unbalanced load flow analysis so the single-phase loading and unbalances between phases can also be minimized.

Table 2.2 summarizes the applications, benefits, and required equipment for the different levels of the volt-var efficiency.

In Table 2.2 the different schemes of VVC and VVO are also described from the communication and decision-making process by tiers (see Fig. 2.9).

Another program closely related to smart grid distribution automation and volt-var efficiency improvements is often acronymed as CVR, conservation voltage reduction. The concept is simple. Most electrical loads consume power that is proportional to the supply voltage, i.e., the higher the voltage, the more active power (P) consumed. The specific relationship differs depending on the type of load. The different load models include:

- *kZ load*   The most popular and historically dominating, it is defined as c(k)onstant impedance (Z) load, since the impedance of the load does not change, every drop of voltage (V) will result in a drop of current (I) and therefore, the square of the

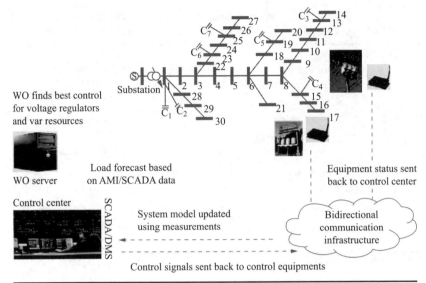

WO finds best control
for voltage regulators
and var resources

WO server

Load forecast based
on AMI/SCADA data

Control center

System model updated
using measurements

Equipment status sent
back to control center

Bidirectional
communication
infrastructure

Control signals sent back to control equipments

Figure 2.9   A schematic showing how VVO works [6].

drop of power (S = V * I) consumed by the load. For example,
a voltage drop of 3% will result in 6% of power consumption.

- *kI load*   In this type of load the current stays constant regardless
  of the voltage applied to the load, so every drop in voltage
  results in an equal drop of power consumed—for example, a
  voltage drop of 3% results in 3% of power consumption.

- *kP load*   This type of load delivers the constant power to the
  process or system it is designed for—for example, a constant
  mechanical power of the motor shaft or constant power of the
  plasma TV or constant brightness of the PC screen, regardless
  of the voltage magnitude (within certain limits of course). In
  this load the power consumed by the load would not change
  or might change only slightly (increase) when the voltage is
  lowered since at lower voltages the current has to increase (to
  maintain the constant power output of the load) and therefore,
  the internal losses of the load will slightly increase.

If one aggregates all the loads connected to one distribution line, the
aggregated equivalent load would typically behave as a mix of kZ,
kP, and kI loads. Therefore, even a small decrease in the voltage
magnitude of the feeder could lower the power consumption of the
load without impacting the loads themselves and yet save the utility
a lot of power. This becomes critically important during high power
demands, when the systems are peaking and the generation resources
are exhausted. CVR is a program by which the utility would lower
the voltage of the distribution feeders by a small amount and therefore
lower the power consumption for this feeder. There is, however, an

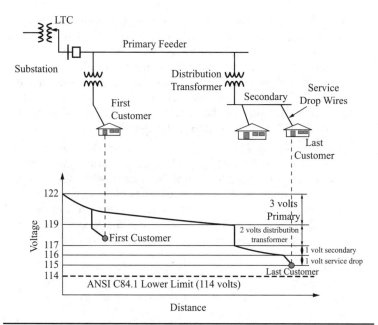

Figure 2.10 Voltage profile distribution.

important constraint. The voltage as supplied to *all the loads* has to stay within the limits prescribed by the regulation (utilities are a regulated industry). It cannot go below or above certain values. Without any VVC or VVO, the voltage profile along the line would resemble the curve in Fig. 2.10.

If one were to drop the voltage of the line at the substation (left side of the graph) even by 1% it would result in an unacceptably low voltage at the end of the line (last house on the line). The solution again is an intelligent VVO scheme. By flattening the voltage profile, greater voltage reduction can be accomplished and therefore more effective power demand management.

The curve in Figure 2.11(a) is for the non-VVO feeder lines; Fig. 2.11(b) is with the implementation of the intelligent VVO. One can clearly see that with a flatter profile there is greater room for adjusting the feeder voltage up or down and still stay within the regulatory limits prescribed. The range (UM + LM) of Figure 2.11(b) is much greater than in Figure 2.11(a).

Recently a new trend has been emerging where customers try to integrate or combine different automation programs for even better results. When FDIR and VVC are both applied on the same network, there is potential for great savings but at the same time some problems need to be addressed (see Fig. 2.11). These are

- *FDIR affecting VVC* When the FDIR function modifies the network, the resulting new network will likely have suboptimized power factor and voltage profile. VVC should correct this and bring the new network to optimum operating condition. For example, when a capacitor bank was on to lift the feeder line voltage before the fault occurred, after the fault isolation and restoration, new loads are added to a healthy feeder and some loads cut out in the faulted one. According to the old network scheme, conservation voltage reduction and volt-var optimization are no longer valid. A "fixed" VVC algorithm will generate wrong outputs.

- *VVC affecting FDIR* If VVC runs "first," its algorithm will modify the network (tap changer positions and capacitor bank energization) to control the voltage on the feeder, which also changes feeder capacities. If FDIR is not aware of the changes, it will make the wrong decision during fault restoration when it checks if the healthy feeder has enough capacity to take over the load from the faulty feeder.

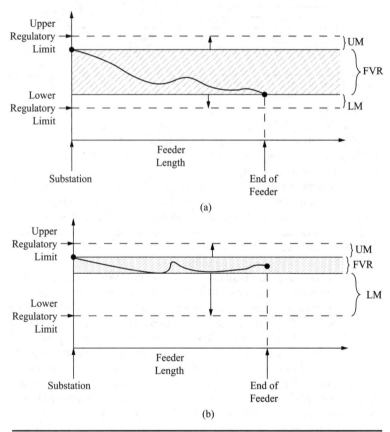

Figure 2.11 Feeder voltage profiles within the allowed regulatory range (a) before VVO and (b) after VVO. FVR—feeder voltage range, UM—upper margin, LM—lower margin, (UM, LM) defines how much the feeder voltage range can be adjusted up and down.

Therefore, VVC and FDIR affect each other and should be coordinated and integrated. This new integrated approach is sometimes called *integrated feeder automation control* (IFAC). This implementation of the two smart grid functions also reduces duplication and overall cost by utilizing common communication channels, network model knowledge, network status model, and HMI (human machine interface) and event reporting.

Conclusion

In summary, all different volt-var and FDIR management solutions can be implemented with different levels of benefits. The solutions can be three phase or, at a more advanced level, single phase. More advanced solutions require more sensing, measurement, communications, and software but they provide greater benefits. Depending on the level of sophistication of the system operators, a solution can be developed to suit the needs of the consumer and the system owners at the investment level that makes the most sense for a specific system. The road to smart grid distribution automation is gradual, one smart technology at a time.

The primary objective here is not necessarily to build new generation, but rather, to reduce the carbon footprint, reduce dependency on oil and other fossil fuels, and optimize network performance—which means identifying and removing bottlenecks (weak points) and maximizing asset utilization (e.g., a circuit breaker for 600 amp, using only 200 amp). There are action items behind each of these acronyms. As stated in the introduction, distribution automation is indeed the "final automation frontier" of the power grid.

References

[1] *"What caused the Power Blackout of 2003 to spread so widely and so fast?"* Genscape. [Online]. Available: http://genscape.com/pages.php?uid=4&sid=10&nid=34&act=news.

[2] *"Northeast Blackout of 2003"* Wikipedia. [Online]. Available: http://en.wikipedia.org/wiki/Northeast_blackout_of_2003.

[3] *"NIST Interim Smart Grid Standards Interoperability Roadmap Workshop,"* contribution from IEEE PES Distribution Automation Working Group. [Online] [PDF]. Available: http://grouper.ieee.org/groups/td/dist/da/NIST%20INTERIM%20SMART%20GRID%20WORKSHOP%20IEEE%20PES%20DAWG%20CONTRIBUTIONMay1st09.pdf.

[4] US Energy Information Administration, International Energy Annual 2006. [Online]. Available: http://205.254.135.7/iea/overview.html.

[5] World Net Electric Power Generation 1990-2030. Energy Information Administration (EIA), International Energy Annual 2005 (June-October 2007), website: www.eia.doe.gov/iea. [Online]. Also available: http://www.scribd.com/sabitavabi/d/59845714-World-Electric-Power-Situation

[6] Xiaoming Feng *et al.* *"Smarter Grids are More Efficient: Voltage and Var Optimization reduces energy losses and peak demands."* ABB Review. 3/2009.

CHAPTER 3

Advanced Asset Management

Mietek Glinkowski, Jonathan Hou, Gary Rackliffe
ABB, Inc.

Smart grid technologies in North America are often associated with automated metering infrastructure (AMI) as a means of facilitating real-time pricing, demand response programs, and other customer-facing applications. More recently, AMI communications—especially wide area wireless—has enabled distribution management systems (DMS) and distribution automation (DA) applications with the goal of creating "self-healing" networks. This vision involves much more than smart meters and extends to protection and control equipment, communication systems, and a wide variety of sensing devices and monitoring and control systems.

With more data—and more detailed data—available from across the distribution system, it becomes possible for a greater level of automation to be introduced to the day-to-day management of the grid. Potential disturbances can be detected sooner and can be addressed without the need for human operator intervention.

As more intelligence is built into the transmission and distribution (T&D) system, even down to the device level, utilities are beginning to explore the potential for leveraging the vast amount of data these devices generate. The purpose is to gain a better understanding of the actual condition of grid assets so that operations and maintenance activities can be optimized along with the performance of the network as a whole.

This is broadly known as asset management—the practice of maintaining the reliability of transformers, circuit breakers, and other equipment in a cost-effective, programmatic manner. Historically, good asset management meant following manufacturer guidelines for regular maintenance at specific time intervals. Now, the advance of technology on several fronts has enabled a more sophisticated approach to come to the fore. This chapter will explore what has become known as asset health management (AHM).

We begin with an assessment of the status quo both in terms of technology and the priorities of utility management. We then move into a discussion of the rationale for AHM, its technological underpinnings, and challenges to its wider adoption. Finally, we explore some best practices in implementing AHM programs.

An Aged and Aging Infrastructure

It has been well-documented that infrastructure investment in North America suffered a long period of decline that has only recently been reversed. The result of this trend is especially visible in the power delivery system where many devices currently on the grid are operating well beyond their designed lifespan. The average age of a power transformer, for example, is over 40 years.

While large-scale (i.e., transmission-level) blackouts draw media and public attention, the fact remains that the vast majority of the outages that occur every year happen in smaller areas, usually at the distribution level. Reports indicate they account for anywhere from $80 billion to $150 billion per year in lost economic activity. Still, according to a study of leading utilities conducted in 2011 by The McDonnell Group, utility executives indicated that it is easier to gain regulatory approval for new capital spending than it is to fund repair or replacement of existing assets.

This speaks to an important distinction. Repairs often fall under operations and maintenance (O&M) budgets, whereas new assets are characterized as capital expenditures. The former are based on risk of failure or establishing that the equipment has reached the end of its useful life. The latter are driven by capacity or reliability requirements. Both types of spending undergo regulatory review, but the criteria applied to one is different from that applied to the other.

Meanwhile, demand for electricity continues to rise—in some parts of the world nearly twice as fast as the overall demand for energy. In North America, the proliferation of electronic devices, the trend toward larger homes, and even the growth of large data centers (which already consume 1.5% of all electricity) have all contributed to the ever-increasing appetite for electric power. In addition to this trend, wholesale power trading has created an ongoing need for large amounts of power to be shipped over long distances, placing even more demands on the T&D infrastructure.

The North American power grid as it exists today was not designed to support the volume and complexity of our power requirements. It evolved from isolated grids that grew into the interconnected system we know today, but this evolution was piecemeal and far from master-planned. The result is a patchwork system with 10 to 15 voltages used at the transmission level alone and total T&D system losses of between 8% and 10%. Compare that to the more recently built Korean system, which utilizes three transmission voltages and boasts losses of just 4%.

This is not to say that our T&D systems are entirely inadequate. On the contrary, they are extremely reliable—so much so that most of us take electricity for granted. The reliability utilities are able to deliver has a lot to do with the people supporting the grid, and that experience base is also aging.

The shifting demographics of utility staff has been discussed for many years, but the fact remains that the industry is poised to lose close to half of its technical workforce in just the next few years. A recent report by GridWise Alliance indicated that "utilities will need to replace 46% of skilled technician positions by 2015 because of retirement or attrition." The importance of the institutional memory these individuals carry with them cannot be overstated.

Industry Priorities

Given all of this, it is perhaps not surprising that utilities are very much focused on preserving reliability, and specifically on the rising cost of maintaining their systems. In the McDonnell Group survey mentioned earlier, 53% of respondents indicated ensuring reliability was their most important overall strategic priority, and nearly all (94%) placed it as one of their top three. In terms of maintenance costs, 74% ranked the increasing cost of maintaining substation assets as a top priority. By comparison, regulatory compliance was rated as a top concern by 35% of the utility executives surveyed.

The McDonnell Group survey also examined the confluence of information technology (IT) with operations technology (OT), which is becoming increasingly prevalent in the context of smart grid deployments. Of the utility executives surveyed, 83% saw better IT/OT integration as very important or critical, but only 29% saw their own company's current level of integration as very good or excellent.

This survey paints a vivid picture of a North American power industry in transition. The challenge is to manage the shift from legacy grid to modern grid while ensuring or improving the reliability that customers have come to expect. It's clear that utilities understand the importance of technology in effecting this change, but in a final observation the McDonnell study authors noted that respondents saw IT as a driver for engineering-based OT solutions and not the other way round. Smart grid technology, then, is the means to a variety of ends, one of which is a comprehensive approach to asset health.

Asset Health Management

AHM, like the broader term "asset management," can take on a range of meanings and connotations depending on who is describing it. At this point in time, true end-to-end AHM has yet to be adopted by most utilities. There are reasons for this, as we will see in a moment, but the fact remains that most utilities, especially smaller organizations, still rely largely on time-based maintenance practices.

The potential returns associated with AHM, however, are compelling. The data captured within AHM solutions can also play an important role in management decisions, and can provide solid backing for utilities seeking to justify expenditures with regulators. Likewise, the data can also offer a rationale for deferring investments. In both cases, the key consideration is the impact on reliability, and that is precisely what AHM seeks to address in a more accurate and thus more cost-effective way.

This represents a shift from time-based maintenance practices toward those that are condition-based, and from an emphasis on historical performance of assets to an emphasis on future performance and risk.

Benefits from AHM

The central benefit derived from implementing end-to-end asset health management is improved reliability. This has implications for utilities, customers, and the wider society as capital resources are allocated more efficiently and harm is reduced when the supply of electricity is protected from interruption. For utilities, the benefits extend further.

One example is that additional value utilities can derive from their AMI deployments when smart meters are used to support AHM with a more detailed picture of distribution system conditions. The data gathered by AHM systems can also become very useful in support of cost recovery (i.e., rate cases) and regulatory compliance. Limited O&M budgets can also be allocated more effectively if the utility knows which assets are most likely to fail and what the impacts would be in the event of a failure. AHM provides the information needed for utility management to reach that level of understanding and to provide substantiation for their decisions.

Another area that stands to benefit from better asset management is safety. Put simply, the less that field personnel have to physically interact with equipment, the less likely a dangerous incident is to occur. More information about the nature of a given device's condition and how it is likely to fail will also help utility staff to approach that device with appropriate caution.

AHM Technologies

Basing critical maintenance decisions on the actual condition of grid assets implies the application of a wide range of technologies to measure, collect, communicate, analyze, and store data from the field. Some of this equipment is already in place. Digital relays, for example, are frequently capable of delivering far more data than what they are used for. They may include digital fault recorders, sequence-of-event recorders, synchrophasor measurements, and breaker contact status and timing information, all of which can be utilized by AHM systems.

There are also several new technologies in various stages of development that promise to deliver still more data at a cost that will allow widespread deployment. One example of this is a gas-in-oil sensor being developed by the Electric Power Research Institute (EPRI) for use in transformer condition monitoring. When fully commercialized at scale, this device would replace current technology costing thousands of dollars with an alternative that EPRI estimates would come in at around $1,000.

Of course, as the cost of collecting field data comes down, the volume of data collected tends to go up, which in turn introduces new challenges. Indeed, the full value of smart grid investments is often confined by limitations in the capacity of communication systems. This is changing, however, and data backhaul throughput is increasing along with reliability, with the industry's focus now moving to data storage.

With a proliferation of data streams, the utility is faced with two interrelated challenges: How to manage the influx of so much data from various sources, and how to make it available to a variety of applications. The latter is especially important since the value of a given piece of information is often amplified when combined with others.

There is an important distinction to be made here as well in terms of "data" versus "information." AHM is data-driven, but simply flooding asset managers with raw numbers from assets across their network is not useful and will not lead to better decision making or lower costs. In order to best leverage the increasing volumes of data coming in from the field, AHM systems must include advanced tools for data visualization so that users can extract meaningful information from them.

Challenges

As the previous section implies, perhaps the greatest hurdle AHM must overcome to realize its full potential is to marshal an ever-increasing amount of data into actionable information. This will only become more difficult as the costs associated with sensor technology, communications, and data storage continue to fall. Utilities face an ongoing question as to what data is important and how it can (and should) be used within the context of optimizing asset management practices.

Complicating matters for utilities is the fact that currently relevant data may be stored in a variety of different systems. Supervisory control and data acquisition (SCADA), energy management systems (EMS), enterprise asset management, mobile workforce management—the list goes on, and each of these has its own unique data format. Integration of these various systems is therefore vital to the ability of utilities to implement comprehensive AHM. Standards are particularly important. The common information model (CIM) is one example of a data standard that seeks to allow information to be shared between systems.

Standards are also important to facilitate interoperability between equipment and systems from different manufacturers. There is inherent risk in being locked into any proprietary data standard, mainly due to the fact that the utility is then tied to a single vendor. Open standards such as IEC61850 for substation communications provide an alternative that allows components from different suppliers to work together.

Of course, much of a given utility's historical data may not even exist in electronic form. Given the age of many T&D assets, this presents a special challenge but one that at least has a certain lifespan.

Finally, a longer-term challenge utilities face is the convergence of IT and OT. As noted earlier, this process is already well underway, but it is likely to play out in numerous ways in the coming years.

Best Practices

The most widely adopted set of best practices applied in the utility asset management arena is PAS 55, which is supported by the Institute of Asset Management (http://theiam.org). PAS 55 has been adopted by many utilities in the United Kingdom, Australia, New Zealand, and Canada in recent years with many successes. National Grid, for example, adopted PAS 55 practices in 2005 and expanded it to all of the company's UK and U.S. operations.

The International Standards Organization (ISO) is starting an International Standard for Asset Management that is in turn based on the PAS 55. This ISO standard will be published in three parts:

- ISO 55000 will provide the overview, concepts, and terminology in asset management.
- ISO 55001 will specify the requirements for good asset management practices, the "asset management system."
- ISO 55002 will provide interpretation and implementation guidance.

Aside from the more rigorous PAS 55, most leading asset managers use a four-step process: plan, implement, evaluate, and act:

1. *Plan* The plan is driven by an established asset management policy that provides a framework for asset management strategy as well as objectives and plans to be developed and controlled. This policy is supported by the company's senior management team and should have an executive sponsor.

2. *Implement* Business unit managers develop organizational asset management strategies and objectives based on the asset management policy. The strategy addresses asset management direction, goals, initiatives, and system performance metrics. Asset management objectives with quantified measurements must be communicated in the strategy with consideration of related risks. Specific asset management plans are then developed to support each of the aspects of the strategy's implementation. These address specific asset life-cycle activities covering planning, design, engineering, operation, maintenance, repair, refurbishment, replacement, and retirement.

3. *Evaluate* Asset management decisions are based on asset conditions, performance status, test results, past performance, system loading, etc. A complex asset health decision process should also evaluate system risk impacts to the environment, customers, employees, and business to determine the optimal solutions. A routine audit and validation of performance improvement is required to ensure proper execution.

4. *Act* Execution of the asset management plan is the most critical step in the advanced asset management practice.

Advanced Asset Management Practices

Traditional maintenance practices used by most of North American utilities focus on time-based maintenance. This practice is easy to set up and expensive to execute. In a way, utilities are performing maintenance on a set schedule, regardless of the condition of the equipment (e.g., an oil sample test every 6 months). This approach is a kind of lowest-common-denominator approach in which the maintenance of all equipment is driven by those units in the worst condition.

A more cost-effective approach emphasizes condition-based maintenance, which seeks to minimize the total ownership costs of the assets from planning, design, engineering, acquisition, and installation through operation, maintenance, repair, refurbishment, and eventually retirement. It accounts for the complete life cycle of the equipment with the objective of managing cost and risk.

For example, an oil sample test once a year might be adequate for newer transformers while more frequent tests are conducted for older units or those with a history of problems.

An advanced asset management system optimizes all assets' reliability, availability, and flexibility to provide the maximum system benefits to the asset owners.

- *Reliability* is a measure of equipment condition. For example, is a transformer reliable to operate at a specific operating requirement? The equipment reliability indicator could change based on its operating history, current health condition, life-cycle status, loading requirement, etc. System and unit reliability are mostly impacted by asset condition.

- *Availability* is a measure of uptime. A reliable transformer may not be available for the next 6 months if the substation is undergoing a major renovation project. Thus, the transformer unit is not available to use, regardless of its condition. Or, the system capacity is limited to 50% due to a cooling system limitation. Availability is mostly impacted by business operation and schedule.

- *Flexibility* is a measure of service availability of a system or subsystem. For example, a double-bus substation layout provides more operational flexibility compared to a single-bus layout. Smart grid technology also adds operational flexibility to allow utilities to manage network assets and to isolate faults. So, flexibility is impacted most by system layout and control schemes.

Asset Health and Condition Assessments

The core of condition-based maintenance is individual equipment/ asset condition assessments or analytics. These utilize past performance historical data, test records, maintenance records, and peer group performance statistics to project equipment condition into the future. Some analytics can also pull information from real-time monitors to provide up-to-date conditions. For example, transformer asset analytics might consider measurements from dissolved gas analysis (DGA) results, electrical tests, corrective maintenance records, SCADA relay data, etc. The result provides a quantifiable and repeatable evaluation method for all similar equipment.

Subsystem or system-wide analytics then consolidate the individual equipment analytics to provide a system-level view of condition evaluation. For example, a substation condition assessment can be developed based on equipment analytic inputs from transformers, circuit breakers, switches, etc.

Risk-Based Planning

A risk evaluation of the impact of asset failure is a critical component in an advanced asset management operation. For example, a network serving a downtown hospital has much higher impact comparing to a network in a rural residential area. This comparison is sometimes referred to as "life or light."

In a typical risk-planning process, asset managers should consider the potential risk/threat likelihood of occurrence, the potential impact severity level, and required risk mitigation efforts. An advanced asset management practice is essentially a risk management practice to minimize overall operational risk exposures along with overall asset life-cycle costs. Asset managers must carefully evaluate overall system risk impact, as the result of this planning can provide a framework for decision and resource prioritization.

Decision Support

Decision support is the most important part of the advanced asset health management practice. The data from smart grid devices is converted into information that has meaning for asset owners in terms of benefits and risks and how action should be taken. There are millions of data points from the network and historical data strings. The decision support system must process all the data from smart grid sensors and operations to provide a best course of action based on set objectives and operation parameters.

This repeatable decision support process provides utilities with a documented method for process improvement and a sound, justifiable basis to support their decisions.

CHAPTER 4

Wide-Area Early Warning Systems

Innocent Kamwa
Hydro-Québec Research Institute

Introduction

It is now a well-established fact that the existing power grid is being operated closer to its stability limits as a result, for example, of increasing renewable generation integration without adequate transmission expansion. Managing this increased exposure to wide-area blackouts requires a trustworthy online monitoring of these limits. It is in this context that wide-area measurement systems (WAMS) were mandated in the Energy Act (Epact05) by the U.S. House of Representatives for the purpose of real-time monitoring of North American inter-connections [1]. Should such a system be implemented, it could provide the following benefits to enhance the situational awareness of system operators and their reaction capability:

- Provide early warning of deteriorating system conditions, so operators can take corrective actions.

- Provide more diagnostic tools than are currently available.

- Allow for the more effective use of automatic controls for self-correction, such as automatic switching or controlling the flow of power.

Although it is widely acknowledged that the technology now exists for deploying a real-time transmission monitoring system [2–4], how to translate the captured data into actionable information for system operators remains a matter requiring further research. While wide-area measurement-based power system stabilizers (PSS) and special protection systems (SPS) have been shown to be feasible and hugely beneficial to system capability and reliability [5-6], the only online use of phasor measurements in control centers reported so far is to improve the state estimator. Since the state estimator will stop functioning anyway during emergency conditions [7], improving its accuracy with phasor measurment unit (PMU) data will be of no value at such critical moments. What is really needed is a flexible and resilient alternative to the present supervisory control and data acquisition (SCADA)–energy management system (EMS) scheme, one that could remain alive in the most adverse conditions.

In this context, the PMU-based WAMS technology appears as a powerful paradigm, complementing and modernizing the state estimator while addressing independent system operator (ISO) security monitoring functions in real time and providing operations personnel with the fast decision-supporting tools they need. Building on this emerging WAMS technology, the wide-area situational awareness system (WASAS) was identified as one of four priorities in the Federal Energy Regulatory Commission (FERC) smart grid policy [8]. It is aimed at providing a visual display of interconnection-wide system conditions in near real time at the reliability coordinator level and above. The wide-area situational awareness efforts, with appropriate cybersecurity protections, can rely on the NASPInet work undertaken by the North American SynchroPhasor Initiative (NASPI) [2] and will require substantial communications and coordination across the regional transmission operator (RTO) and utility interfaces.

However, any successful WASAS would require at its core model predictive analytics to convert PMU data into information in a timely manner for supporting real-time decisions and actions during emergency conditions. Such an analytical engine, which is as new to system operations as the PMU technology, will be loosely defined in this chapter as an integrated early-warning system (EWS). The EWS is a key enabler of a successful WASAS, allowing safe system operation closer to its stability edge under a high penetration of renewable generation and high power transfers across the merchant corridors.

The overall design concept of the EWS is summarized in Fig. 4.1. The design starts with consideration of the various technology options available. The selection of PMU and phasor data concentrator (PDC) brands should take into account the dynamic-performance aspect, since the data will be analyzed online for fast decision making in order to prevent blackouts while maximizing the power transfer capability of the grid corridors. Next, the PMUs should be properly sited throughout the grid so as to maximize the information content

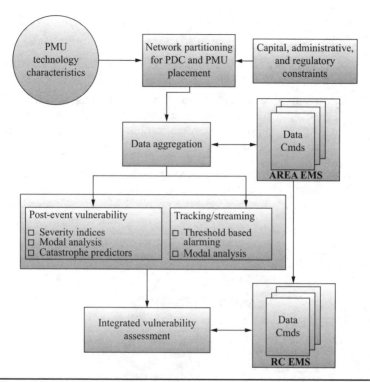

Figure 4.1 Structured design of an integrated PMU-based wide-area early warning system.

of the post-disturbance data records. This is best done by first partitioning the network in electrical coherent areas and then assigning PMUs to the medoid bus of the area [9]. Additional PMUs could be sited at other busses for redundancy purposes or in order to fulfill practical business or regulatory constraints [10]. In a large interconnected system, one can safely assume that each control area represents a reasonable subdivision, equipped with a control center ("Area EMS" in Fig. 4.1), having authority over one or many electrically coherent areas depending on the geographical size of the control area. Next, several control areas could be consolidated into a reliability coordinator (RC) jurisdiction (RC EMS). Finally, the RC could report to the national reliability organization (e.g., North American Electric Reliability Council [NERC]) in real time.

On the information flow side, the PMU data should first be aggregated by electrical zones in order to reduce the noise and communication bandwidth. This is best done by reducing all the phasors in an area into a single pilot phasor whose angle and frequency are then projected in the center-of-inertia reference of the control area [11]. Analysis of the aggregated data could next proceed following two different paths reflecting two complementary needs in vulnerability assessment [12]:

1. Disturbance-triggered vulnerability analysis aimed at assessing the grid vulnerability following an incident

2. Streaming or tracking-mode vulnerability analysis, aimed at assessing the oscillatory impact of slowly varying random load and intermittent generation in order to detect any unstable trend at an early stage of construction

Figure 4.2 presents another view of the hierarchical relationships between a set of area PMUs connected to a regional transmission level PDC, which in turn reports to a reliability coordinator-level

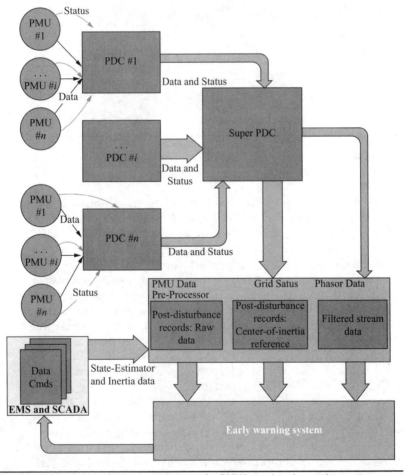

FIGURE 4.2   Upper-level data architecture of a PMU-based early warning system.

(or interconnection-level) PDC. Interestingly, regional PDCs collect not only phasor data but also network status data as perceived by the subject PMU. By status, we mean fault condition, open line, or any other binary information describing the network topology that may be necessary to properly interpret and use the phasor data [3–5]. We also note that the EMS of the control area is responsible for providing additional information required for efficient data aggregation in the context of dynamic-performance assessment. This includes inertia and any other state-estimator information deemed useful at the data fusion and aggregation stage [10]. However, Fig. 4.2 makes it clear that center-of-inertia referencing is not absolutely required in the EWS (although it could improve the analytics). This is especially true for streaming modal analysis of system ambient responses [12].

Enhancing the PMU Vulnerability Assessment Functionalities

Discussions are still ongoing about the best characteristics expected from a PMU with embedded wide-area control and protection capabilities [3–6]. As shown in Fig. 4.2, in the context of vulnerability assessment, in addition to a good compromise between harmonic and interharmonic filtering and speed of response, the PMU should provide grid topology signatures at its own node. Recently, Hydro-Québec has developed a new PMU concept, the so-called PMU/C, which combines all the features required for its new generation of response-based wide-area control systems (WACS) and wide-area monitoring protection and control systems (WAMPACS). These features are essentially the same for an integrated EWS.

Hydro-Québec has a very mature wide-area monitoring system, which has gone through several phases of development from 1976 to now [6]. As illustrated in Fig. 4.3, the system consists of eight PMUs and a SCADA linked directly to the EMS and located adjacent to the TransEnergy control center in downtown Montreal. This represents about 25% of the 735-kV buses in the system. Since 2004, modern PMUs customized to include some algorithmic features specific to Hydro-Québec have been used quite successfully. These new features, which have migrated from the previous generation of Hydro-Québec Research Institute (IREQ)-made PMUs, are voltage harmonic distortion up to the 10th component and voltage asymmetry. These variables are required by Hydro-Québec's preventive control schemes against geomagnetic storm-induced contingencies.

In 1995, wide-area technology was rapidly improving and Hydro-Québec was considering it as a viable alternative for developing more secure special protection systems. However, its planning department expressed serious concerns about the reliability of this not-yet-proven technology. There were also some concerns that existing discrete Fourier transform (DFT)-based voltage magnitude and frequency tracking alone could not provide an acceptable signal quality for fast wide-area

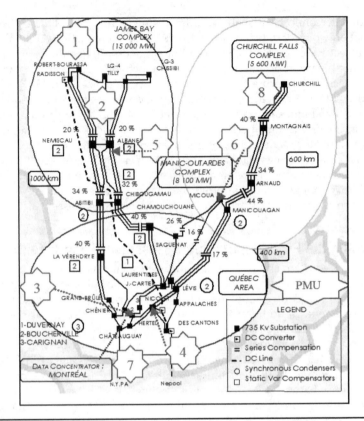

Figure 4.3 Hydro-Québec's transmission grid with the eight GPS-based PMUs constituting its existing angle-shift monitoring system. The circles represent three coherent electrical areas on the 735-kV bulk system.

applications [6, 13–15], given that the heavily series-compensated Hydro-Québec grid shows significant electromagnetic resonances in a frequency range starting from 4 Hz up to harmonic frequencies [13].

Hydro-Québec PMU/C Requirements and Performances

Figure 4.4 presents the overall capabilities of the modern PMUs specified by Hydro-Québec for its wide-area control systems. These capabilities broadly fall into two categories: performance and interoperability requirements. From the interoperability within and outside the substation perspectives, the PMUs should be IEC61850 and C37.118 compliant, respectively. In terms of performance, the PMUs should meet the level 1 static requirements of C37.118 as well as the following criteria:

1. Provide three levels of filtering defined in terms of the speed response achieved as fast, medium, and slow. Higher noise rejection corresponds to slower response filters.

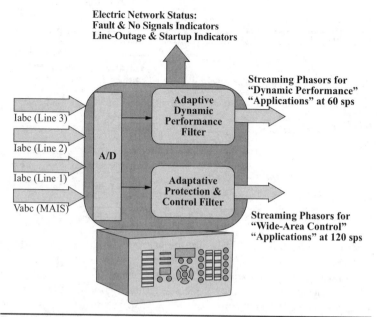

Electric Network Status:
Fault & No Signals Indicators
Line-Outage & Startup Indicators

Streaming Phasors for
"Dynamic Performance"
"Applications" at 60 sps

Iabc (Line 3)

Iabc (Line 2)

A/D

Iabc (Line 1)

Vabc (MAIS)

Adaptive
Dynamic
Performance
Filter

Adaptative
Protection &
Control Filter

Streaming Phasors for
"Wide-Area Control"
"Applications" at 120 sps

Figure 4.4   Enhanced PMU configuration for online dynamic performance assessment and stability control.

2. Accompany the phasor package with a network status binary word concisely describing the measurement context considering, for instance, the important issue of phasor distortion during fault and switching events [3–5]. In the extreme case, the voltage could drop to zero during a three-phase-to-ground fault leaving the phase angle temporarily undefined. In view of control and protection applications, the phasor should therefore be tagged at the PMU level with respect to the transmission line status (fault, outage, switching transient, etc.). Otherwise, the user could mistakenly take strong action based on the wrong information [3–5].

3. Provide real-imaginary sequence quantities in addition to frequency and rate of change of frequency at least 60 samples/s or above without alias. A single unit should be able to process at least one bus voltage and three line currents, but three voltages are preferred in order to monitor a complete corridor using a single device.

Table 4.1 illustrates a typical dynamic performance from Bonneville Power Administration (BPA) and Hydro-Québec. Since there is no standard for dynamic performance yet, comparison between the two sets of requirements can only be done qualitatively. For instance, during faults, the frequency measurements should be blocked using

BPA (NASPI, February 2010)		Hydro-Québec Requirements for WACS		
Criterion	Value	Criterion	Value for Voltage Filter	Value for Frequency Filter
Max. overshoot	10%	Max. overshoot	0%	10%
90% rise time	50 ms	95% rise time	70 ms	75 ms
Settling time to 2%	3 Tp	Settling time to 0.5%	120 ms	235 ms

NASPI: North American SynchroPhasor Initiative

TABLE 4.1   Typical Requirements for PMU Dynamic Performances

a scheme that is intrinsically nonlinear and varies from vendor to vendor.

Table 4.2 presents the step response performance metrics of a PMU/C designed to meet the three-stage filtering requirements set by Hydro-Québec for its automation relays [15]. The filters are adaptive with respect to the fundamental frequency and, in addition [5], can estimate a single subsynchronous oscillation frequency and cancel its contribution to the one-cycle phasor. Although this result is a more accurate phasor overall, Table 4.3 shows that the time response of the fastest amplitude filter is higher than one cycle. However, all the amplitude filters have no overshoot by design. By contrast, there is a 10% overshoot on all three frequency filters, which is basically induced by the fundamental frequency adaptation mechanism.

HQ Voltage Filters	Tr (5%)	Tr (66.7%)	Tr (95%)	Ts (99.5%)	Overshoot
Fast	3.34 ms	15.7 ms	25 ms	39.3 ms	0%
Medium	8.45 ms	48.3 ms	68 ms	118.4 ms	0%
Slow	26.6 ms	81.5 ms	115 ms	159.6 ms	0%
HQ Frequency Filters					
Fast	8.23 ms	24.8 ms	35 ms	197.7 ms	9.7%
Medium	15 ms	55.6 ms	73.2 ms	234.1 ms	9.03%
Slow	34.2 ms	88.6 ms	115 ms	277.1 ms	8.3%

Tr: response time; Ts: settling time.
PMU/C: PMU for Control

TABLE 4.2   Hydro-Québec—PMU/C Performance Specifications

Hydro-Québec PMU/C Sample Performances

The first set of networksignals analyzed result from EMTP simulation of a sudden three-phase fault at the Nemiscau substation in the Hydro-Québec grid (Fig. 4.3), with the fault cleared by a line outage. The results obtained with the one-cycle phasor tracking scheme are shown in Fig. 4.5 for PMUs without and with fault (FTE) blocking features.

The angle shifts at five substations with respect to Boucherville show high temporary values during the fault. This discontinuous behavior translates into spikes on the rate of change of angle. By contrast, the results with variables blocking during fault switching are cleaner and the rate of change of the angle is twice lower during a fault. Figure 4.6 illustrates the corresponding local variables posted from the PMU to a PDC. The fault flag shown in the left plots in dashed yellow was determined separately by pattern analysis of the low-level signals in a way similar to [3–5]. It is very helpful to tag the phasor estimates as suspect during a temporary voltage dip. The lower panel of the same figure illustrates the difference between a local frequency estimate, which uses such a flag to remove fault-induced spikes (right), and a frequency estimate without such a scheme. The raw estimate unrealistically exceeds 2 Hz but, by using the fault flag

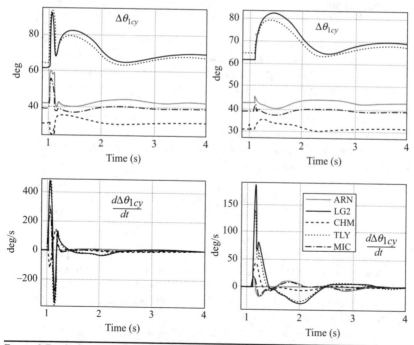

Figure 4.5 Global variables (angle shift and its derivative) with respect to Boucherville substation following a three-phase fault at Nemiskau substation cleared in six cycles by a line outage.

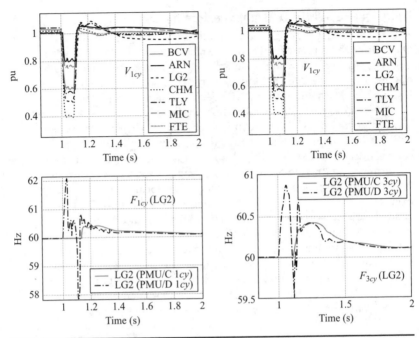

FIGURE 4.6 Local variables estimated by the PMU following a three-phase fault at Nemiskau substation cleared by a line outage (corresponds to the case in Fig. 4.5).

to block the spikes, the frequency signal is considerably smoothened (i.e., orange and dashed orange vs. black). In fact, any action taken for protection or control purposes based on the spiky estimate will be worse than doing nothing. Therefore, qualifying the phasor and frequency using a status description flag (such as Fault, Transient, Open, etc.) should really be a full component of the PMU data collection and dissemination process [5].

To check the PMU/C solution against the C37.118 specification, we can use the following test signal:

$$V = a_1 \sin(2\pi F_c t + F_a(t) + \phi_1) + \sum_{n=2}^{11} a_n \sin\left(n\left[2\pi F_c t + F_a(t)\right] + \phi_n\right)$$
$$+ a_{ss} \sin(F_{ss}(t) + \phi_{ss})$$
(4.1)

A three-phase voltage set is created from the single-phase template. The resulting set is in positive sequence at fundamental, but the changing harmonics are generated to mix positive- and negative-sequence harmonics. The initial phase of all sinusoidal functions has constant randomly assigned values.

Figure 4.7 shows that the fast filter exceeds C37.118 level 1 by a large margin over the 45 to 75 Hz range. All harmonics $n = 2$ to 11 are included simultaneously with a 10% magnitude each. For the fast Hydro-Québec PMU/C filter, the maximum error during a 1 Hz/s frequency sweep is

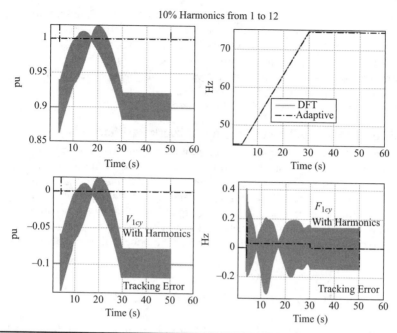

FIGURE 4.7 Stress test consisting of changing harmonics at 1 Hz/s. Amplitude (left) and frequency (right).

0.08%. Under a 45 Hz or 75 Hz steady-state condition with harmonics, the static error is reduced to about 0.003%. The classical DFT behavior is shown on the plot for benchmarking purposes.

Although no direct assessment was made, the PMU/C filtering scheme presented here is capable of matching or exceeding the Western Electricity Coordinating Council (WECC)/NASPI requirements for wide-area monitoring applications [2]. Numerous parametric stress test signals help to prove the effectiveness of the filters, even those with single-cycle filtering, under changing harmonics with the fundamental frequency varying from 40 to 80 Hz and phase, amplitude, and frequency step responses. However, experiments with simulated and actual system responses under fault conditions underscore the need for tagging the phasor when it is unreliable, as during fault-switching events.

PMU Placement for Vulnerability Assessment
Background

As demonstrated in [9], the first step in planning a PMU-based dynamic monitoring system consists of identifying the coherent electrical areas forming the grid. The coherency method is founded on the empirical observation that, following a disturbance (line or generating-unit outage or sudden change in load), certain groups of generators "swing together"; in other words, the generators in each group maintain an almost constant angular difference from each other. Each such coherent group can then be replaced by a single bus. This aggregation procedure applied to all PMUs from the same electrically coherent area greatly simplifies data analysis for dynamic vulnerability assessment (DVA) shortly after a contingency is detected and the resulting wide-area signal responses are recorded. DVA is a key input to EWS.

Disturbance-Based Coherency for Network Partitioning

A group of busses is said to be coherent with respect to angle stability when any network disturbance (bus bar or line short-circuit, single- or multiple-line outage, load or generation switching, etc.) applied outside the group results in all incremental time variations $\Delta\theta(t)$ and $\Delta\omega(t)$ of the bus angle and frequency, respectively, that have the same sign within the group [2]. Consequently, two busses (k, l) are coherent for disturbance d when the following rms-coherency criterion is below a specified threshold:

$$\alpha_{kl}^d = \sqrt{\frac{1}{T}\int_0^T [\Delta\theta_{kd}(t) - \Delta\theta_{ld}(t)]^2 + [\Delta\omega_{kd}(t) - \Delta\omega_{ld}(t)]^2 dt} \qquad (4.2)$$

for $k = 1, ..., n_b$; $l = k + 1, ..., n_b$, with n_b, the number of candidate busses, and T, the observation time window. The angle coherency matrix is formed as:

$$X = \begin{bmatrix} \alpha_{11} & \alpha_{12} & \cdots & \alpha_{1n_b} \\ \alpha_{21} & \alpha_{22} & \cdots & \alpha_{2n_b} \\ \vdots & & \ddots & \vdots \\ & & \alpha_{ii} & \\ \alpha_{n_b1} & \alpha_{n_b2} & \cdots & \alpha_{n_bn_b} \end{bmatrix} \qquad (4.3)$$

where $\alpha_{lk}^d = \alpha_{kl}^d$. It may be observed that all the diagonal elements of this matrix are zero, meaning that the coherency of a given bus with respect to itself is exactly zero. Generalizing this result to $i \neq j$, it appears that the coherency α_{ij} is "large" when busses i and j pertain to two distinct coherent groups and "small" when both i and j busses are members of the same coherent group. The matrix in Eq. 4.3 therefore matches the definition of a dissimilarity matrix. Likewise, the disturbance-based voltage coherency criterion can be set using the relation:

$$\beta_{kl}^d = \sqrt{\frac{1}{T}\int_0^T [\Delta V_{kd}(t) - \Delta V_{ld}(t)]^2 dt} \qquad (4.4)$$

and the corresponding coherency matrix constructed as in Eq. 4.3.

The Fuzzy c-Medoids Algorithm for PMU Placement

Once appropriate measurements of the distance or coherency between busses are obtained, clustering analysis can be applied for region identification [16]. For this task, we adopted the fuzzy c-medoids algorithm (FCMdd) originally developed in [17] for large-scale web mining. A medoid is a representative bus whose average dissimilarity to all busses in the cluster is minimal. Initialization of the medoids is done by the sequential PMU placement technique as in [10], which uses the principle of maximum additional information (or, equivalently, maximum dissimilarity) to add the next bus in the list. Unlike FCM, the proposed method proceeds by searching the medoid within each column of the coherency matrix. Thus, it is able to pinpoint the representative bus of a group, which is important information in the PMU configuration.

The Hydro-Québec grid in Fig. 4.3 is used to illustrate the FCMdd-based PMU placement procedure. The locations of the eight existing PMU locations are marked in the figure. With wind development in the Gaspé Peninsula, another PMU will be added at the Rimouski 315-kV substation, located 100 km east of the Lévis substation. Basically the Hydro-Québec grid is shaped in a reversed triangular form with the apex in the load center located in the Montreal area not far from the U.S. border. Engineers usually use their intuition to break the grid into three relatively obvious coherent areas hemmed in by ellipses, as shown in Fig. 4.3. However, even experienced engineers cannot tell for sure where the area boundaries are located, not to mention which of the busses can be considered the area medoid bus.

For the purpose of generating a database for a probabilistic coherency matrix assessment [9], we started with a base case generated by the control center state estimator on a normal day in 2003. Overall, the database has 288 stability simulations and 7 network configurations for a total of 2016 cases. During each simulation run, the recursive coherency matrix was updated step by step and the value at the end of the simulation was stored as the coherency for that particular disturbance. This leaves us with 288 such matrices per configuration, which are further averaged to obtain the probabilistic coherency matrix for that specific configuration, assuming for simplicity that the 32 contingencies are equally probable.

The clustering methods are initially tested on the 735-kV bus subset of the Hydro-Québec system. Since there are only 65 of these busses, the results are easier to present and interpret. Table 4.3 shows the results when assuming three clusters. The boundaries of the three areas are shown on the Hydro-Québec map in Fig. 4.3 where they are seen to follow quite closely the intuitive solution of system engineers. However, the FCMdd clustering results in area medoid busses, which are natural PMU locations, in addition to the area boundaries. The medoid bus of each group is marked in bold. It is the last bus, and its mates are sorted according to a decreasing order of dissimilarity.

A graphical procedure called CLUSPLOT [18] can be used to represent, as in Fig. 4.8, the partitioned list of busses resulting from the three-cluster partitioning of the 735-kV busses using the FCMdd

Group 1: Northeast (5 busses)	Group 2: South (20 busses)		Group 3: Northwest (40 busses)	
CHU73552	831CHMSTQ75	718SAG73554	783CHI73557	713ABI73557
706MIC73553	731CHOUAN57	814LVDSTQ75	883CHISTQ75	880NEMSTQ75
710MTG73553	714LVDRYE57	804LTDSTQ75	780NEM73557	782ALBNEL57
705MAN73553	703LEV73555	704LTD73554	749LG273557	
709ARN73553	770BRU73554	717JCA73554	720RAD73557	
	790APA73555	755CAN73555	764LG473557	
	715CHE73554	707NIC73555	724TILLY 57	
	700EQUILI90	702DUV73556	723LMOYNE57	
	719CHA73555	708HER73555	750LG373557	
	701BCV73555	**730CAR73555**	722CHISSI57	

TABLE 4.3   FCMdd-3 Clusters (Medoid in bold)

method. Basically, after carrying out the principal component analysis of the coherency data matrix, the first two principal components are plotted along with the percentage of the total variance explained by them. The spanning ellipse of each cluster is then drawn around the objects within the cluster, i.e., the smallest ellipse that covers all objects within the cluster. Interestingly, not only are the three ellipses clearly separated, but their spatial location closely mimics the circles

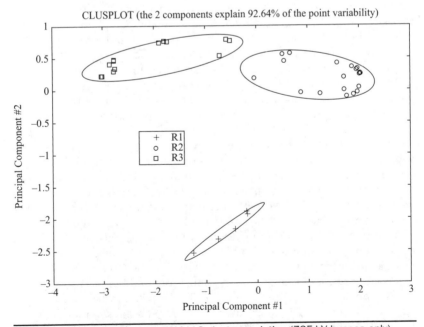

FIGURE 4.8   CLUSTPLOT of the FCMdd-3 clusters solution (735-kV busses only).

shown on the Hydro-Québec map in Fig. 4.3, with two ellipses in the top and one in the bottom of Fig. 4.8. This clear-cut distinction between the areas demonstrates that the FCMdd approach has the advantage of finding the cut-sets while partitioning the network into clusters.

Wide-Area Severity Indices

Reference Network

To define and illustrate the concept of wide-area severity index (WASI) introduced a decade ago [11], we will again consider the Hydro-Québec grid model used by this utility for operations planning, as presented in the previous section. Figure 4.9 illustrates a more detailed partitioning of the same network into nine areas interconnected by "weak" ties, which at the same time define the system cut-set lines [19]. Each area was associated with a medoid bus, which is the location of a primary and essential PMU. However, other

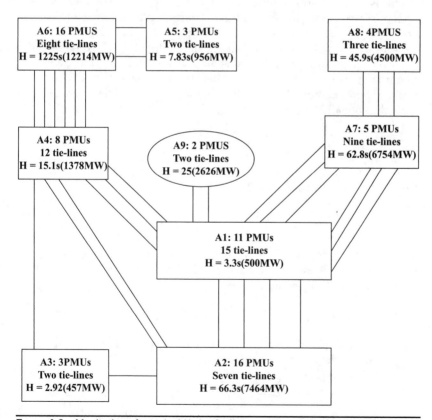

Figure 4.9 Monitoring of a typical Hydro-Québec system configuration with intra-area and intertie PMUs distributed over nine electrically coherent areas. Inertia (H)/generation (MW) data are typical values given for illustrative purposes.

PMUs were added as required for the purposes of redundancy and monitoring of area boundaries, for a total of 68 PMUs in a transmission system with about 619 busses, from 120 to 735 V. During the simulation of a given contingency in the stability program, only the busses with a PMU will be monitored. But the method is designed such that in real time [9, 16], simulation will be smoothly replaced with actual measurements assuming the same PMU configuration.

Time-Domain COI-Referred Response Signals

The severity indices are defined with respect to Fig. 4.9 by associating with each area of the power grid an equivalent inertia representing the total inertia of the generation located in that area. Assuming that each area is coherent following a disturbance, it is reasonable to assimilate its behavior to that of a single large machine with the same inertia and generation. Although this assumption is not perfect, it offers a straightforward means of deriving the center of inertia (COI), which is very useful information for tracking the stability of interconnected areas [43]. In real time, a defense plan or an SPS [15] could readily derive these inertia constants through low-speed communication with the control center state estimator, which holds the actual load and generation dispatch. Before defining the COI quantities, we first need to introduce the area pilot angle concept. For area number k, equipped with N_k PMUs, the area angle is the average angle through all N_k measurements:

$$\bar{\theta} = \frac{1}{N_k} \sum_{i=1}^{N_k} \theta_i \tag{4.5}$$

Then, assuming a total of r areas in the network ($r = 9$ in the present case), we can define the COI of the system as follows:

$$\bar{\theta}_{COI}(t) = \frac{1}{M_T} \sum_{j=1}^{r} M_j \bar{\theta}_j \qquad M_T = \sum_{j=1}^{r} M_j \tag{4.6}$$

where M_T is the total inertia of the grid and M_j is the j-th area inertia, as given in Fig. 4.9. From here on, the area pilot angle and frequency expressed in the COI frame are given by:

$$\theta_i^{coi} = \bar{\theta}_i - \bar{\theta}_{COI} \qquad \omega_i^{coi} = \frac{d}{dt}(\bar{\theta}_i - \bar{\theta}_{COI}) \tag{4.7}$$

Another useful quantity for contingency severity assessment is the dot product of the frequency and the angle [9]:

$$v(t) = \sum_{i=1}^{r} \bar{\omega}_i(t) \left[\bar{\theta}_i(t) - \bar{\theta}_i(0^+) \right] \tag{4.8}$$

where $\bar{\omega}_i$ is the speed deviation in the COI frame and $\bar{\theta}_i(t) - \bar{\theta}_i(0^+)$ is the area pilot-angle deviation from its post-fault value immediately at the clearing instant.

Frequency-Domain COI-Referred Response Signals

It is well-known that sequential spectral monitoring of electrome-chanical oscillations provides important information about the behavior characteristics of system transients. The reason is essentially that the spectral density of the oscillatory signal responses is closely related to the excess kinetic energy injected into the system by the fault [20]. Considering more specifically the deviation of the i-th area pilot-frequency signal $\overline{\omega}_i$ from its immediate post-fault value, $\overline{\omega}_i(t) - \overline{\omega}_i$ (0^+), the excess kinetic energy accumulated by the i-th area equivalent machine over a period of time is proportional to the spectral density of $\overline{\omega}_i$, which justifies using the latter as an indirect measurement of the contingency severity.

Figure 4.10 shows a block diagram for computing frequency-domain severity indices. At the entry, we have the dot product ($x = v$) for each area. The second processing step then computes the spectral density (X_{li}) of the input signal with a prescribed decimation factor N_s; that is, based on an N-point FFT [42], a spectral estimate is provided only every N_s input samples. For each area, the peak of the power spectral density (PSD) is taken over the N spectral lines (X_i) [area-wide, i.e., across the area WASI] and, finally, the largest value among the r areas is selected as the frequency-domain severity index (I_X) [system-wide WASI]. Hence, a short-time fast fourier transform (STFT)-based severity index I_ω can be computed sample by sample (subtracting the post-fault value $x(0^+)$ when the contingency is triggered by a fault), thus yielding a time-varying frequency-domain severity index proportional to the excess kinetic energy injected by the contingency in the most disturbed area.

The scale and time behavior of the WASI patterns in Fig. 4.16 are closely related to the spectral window built into the STFT engine [21]. For transient-stability assessment, short window lengths are the most suitable. We have been using quite effectively for this purpose a 16-point Hamming window whose frequency responses are shown

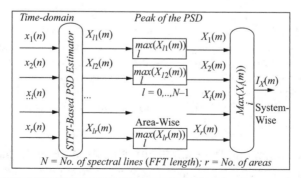

FIGURE 4.10 Computation of frequency-domain-based WASI ($r = 1, \ldots, 9$).

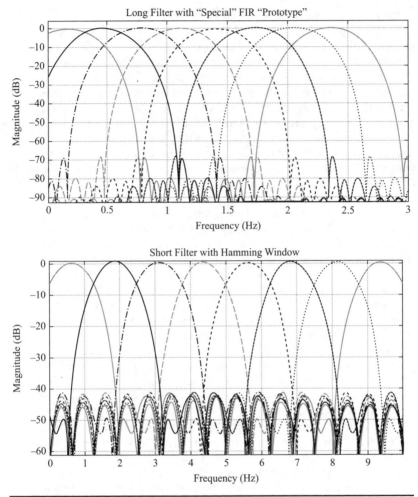

Figure 4.11   Frequency responses of the first eight channels of filter banks used in WASI assessment. Sampling rates: Fs = 20 Hz (top) and Fs = 60 Hz (bottom).

in Fig. 4.11 for the first eight channels. Using the two first bins alone, the filter bank completely covers the electromechanical frequency band, providing about a 1-s response time. By contrast, the long window with a time response near 4 s is based on a customized low-pass filter. It is sharper with, for instance, the second channel centered at 0.47 Hz, a 3-dB bandwidth of about 0.20 Hz and a 70-dB side-lobe.

Interpretation and Data Mining of Wide-Area Severity Indices

More than 60,000 simulations have recently been performed on the Hydro-Québec system in the course of development and actual deployment of a new DSA package to support operations planning and control center activities [21]. To illustrate the relationships between WASI and the actual severity of dynamic events, 120 cases have been selected covering a wide range of phenomena found in actual operations. Figures 4.12 to 4.14 illustrate typical area-wide signals, with WASI patterns shown on the first two strip charts in each figure. The slow system-wise WASI shown in the title are values sampled at 4 and 5 s while the fast WASI are values sampled at 1 and 2 s. Also shown is the end value of slow system-wise WASI, termed post-event WASI. The two last plots of each figure illustrate the area-wide pilot voltage and COI-referred angles. The minimum voltage over a 2-s sliding window is used as the voltage severity index, while the post-fault angle offset with respect to the pre-fault value is used to measure the topological stress induced by the contingency.

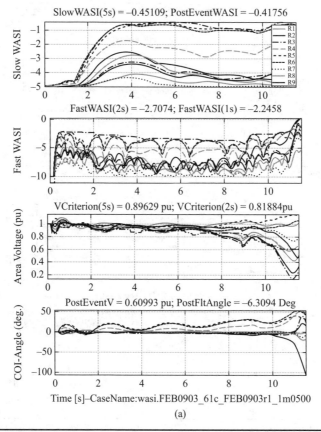

FIGURE 4.12 (a) C1: Severe contingency leading to multiswing instability. (b) C2: Weak contingency leaving the grid in a very secure state.

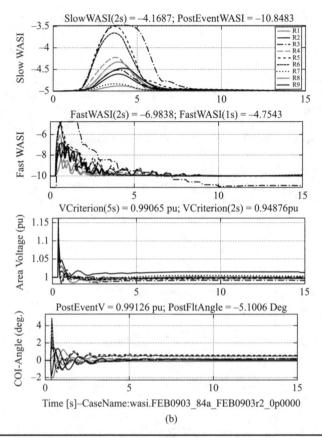

Time [s]–CaseName:wasi.FEB0903_84a_FEB0903r2_0p0000

(b)

Figure 4.12 (Continued).

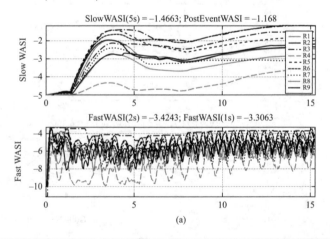

(a)

Figure 4.13 (a) C3: Severe contingency leading to poorly damped voltage swings. (b) C4: Stable contingency inducing a large topological stress (large post-fault angle shift in last strip chart). Legend: R1, ..., R9 represent areas 1 to 9.

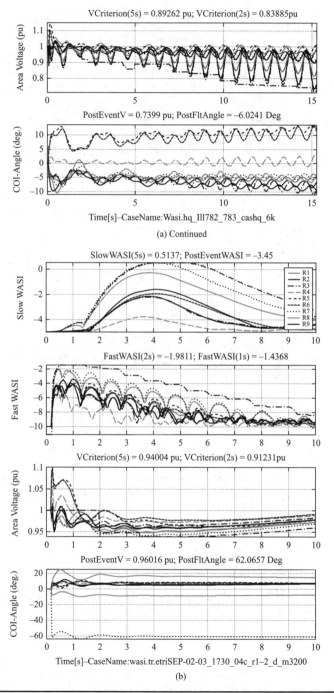

FIGURE 4.13 (Continued).

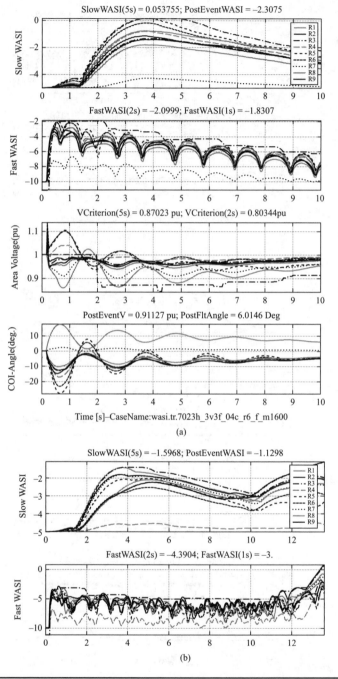

FIGURE 4.14 (a) C5: Stable contingency with large albeit damped transient voltages. (b) C6: Low-energy contingency resulting in damped oscillations and subsequent voltage collapse. Legend: R1, ..., R9 represent areas 1 to 9.

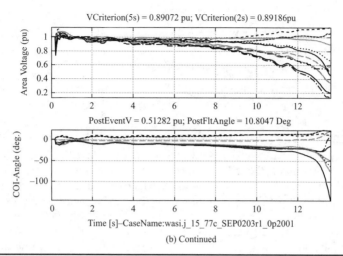

(b) Continued

Figure 4.14 (Continued).

Table 4.4 summarizes the candidate features shown in Figs. 4.12 to 4.14, assuming the following definitions:

1. SlowWASI(5s) = long window-based system-wide WASI, sampled value at 5 s after fault clearing [Log(deg.Hz)]

2. PostEventWASI = end value of long window-based system-wide WASI [Log(deg.Hz)]

3. FastWASI(1s) = short window-based system-wide WASI, sampled value at 1 s after fault clearing [Log(deg.Hz)]

4. FastWASI(2s) = short window-based system-wide WASI, sampled value at 2 s after fault clearing[Log(deg.Hz)]

5. VCriterion5s = system-wide minimum voltage over a 2-s sliding window, sampled value at 5 s after fault clearing (pu)

Features	C1	C2	C3	C4	C5	C6
1-SlowWASI(5s)	−0.45	−4.2	−1.47	0.52	0.054	−1.60
2-PostEventWASI	−0.42	−10.8	−1.17	−3.45	−2.30	−1.13
3-FastWASI(2s)	−2.71	−6.98	−3.42	−1.98	−2.10	−4.39
4-FastWASI(1s)	−2.25	−4.75	−3.31	−1.44	−1.63	−3.09
5-VCriterion5s	0.90	0.99	0.89	0.94	0.87	0.89
6-VCriterion2s	0.82	0.95	0.84	0.91	0.80	0.89
7-PostEventV	0.61	0.99	0.74	0.96	0.91	0.51
8-PostFltAngle	6.3	5.1	6.0	62.1	6.0	10.8
WASI Class	3	4	3	1	2	3
Voltage Class	2	4	2	1	1	2

Table 4.4 Values of the Features of Contingencies Illustrated in Figs. 12 to 14

6. VCriterion2s = system-wide minimum voltage over a 2-s sliding window, sampled value at 2-s after fault clearing (pu)

7. PostEventV = end value of system-wide minimum voltage over a 2-s sliding window (pu)

8. PostFaultAngle = system-wide maximum difference between the pre-fault and post-fault COI-angle (deg.)

Transient Energy versus Voltage-Based Classification

It was shown in the previous section that the fast WASI is a good attribute for splitting a batch of contingencies into two classes: stable and unstable. For the Hydro-Québec system, the threshold value is around FastWASI2s > −2.5, with WASI given in log(deg.Hz). However, some of the cases classified this way as "stable" are better seen as "severe yet stable" cases, while most cases classified as "unstable" actually lost synchronism in the 20-s observation time frame after fault clearing. In addition, some of the well-damped cases, looking very nice at the outset, may, in fact, be transient-voltage weak, and this important pitfall should be identified as well.

Since in vulnerability assessment we are mostly interested in "qualitative" statements about the contingency (such as "Very severe," "Don't care," "Not secure," "Going unstable," etc.), it seems that a clustering scheme based on unsupervised learning may suit the need. Given the 120 contingencies selected for this study and the heuristic observations in Figs. 4.18 through 4.20, four energy-based features were selected for clustering: SlowWASI(5s), FastWASI(2s), PostEventWASI, and PostFaultDTH. Using the *K-means* clustering function of a statistical software package, we discovered four classes, illustrated in Fig. 4.15, as functions of the last four features. Basically, the first class encompasses topologically stressed post-contingent grids. Even if it looks stable from the outset, it is very likely that such a post-contingent grid would fail the N-1 criterion and this information could be brought to the operator's attention without delay. By contrast, the fourth class, with 45 elements, represents the so-called "Don't care" contingencies, with no action required for many of them. The last two classes are equally split into about 30 cases each: Class No. 3 consisting essentially of severe transient cases or cases with slowly deteriorating voltages, while almost all cases in class No. 2 are stable but with transient-voltage violations.

Even if voltage and transient phenomena are always interrelated to some extent, there are cases that can be energy safe but voltage unsafe. Therefore, both analyses should ideally be performed in parallel on all contingencies. Applying the same clustering method as before on the following features: VCriterion(2s), VCriterion(5s), PostEventV, and PostEventWASI, yields the four-class map shown in Fig. 4.16. Basically, all second-class cases are either unstable or at risk of short-term voltage collapse. By contrast, the third and fourth

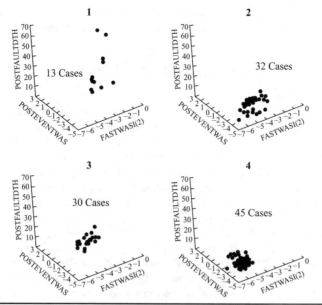

Figure 4.15 Automatic clustering of 120 contingencies into four classes based on four features: FastWASI(2s), SlowWASI(5s), PostEventWASI, and PostFaultDTH.

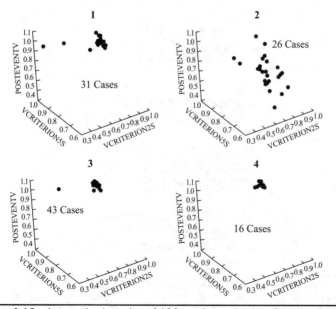

Figure 4.16 Automatic clustering of 120 contingencies into four classes based on four features: VCriterion(2s), VCriterion(5s), PostEventV, and PostEventWASI.

classes are voltage-safe. The first class has some overlap with others, but all its constituent contingencies have transient-voltage problems.

Data Mining–Based Catastrophe Prediction

Background

In this section, we present a comprehensive data-mining scheme for achieving catastrophe prediction, which is a key goal of any EWS. As shown in Fig. 4.17, the development of data mining-based predictive analytics starts invariably with the adoption of a process of data collection and preparation. Initially, we have to collect information about network configurations and contingencies. Next, a reduced-size database of stability results obtained through a high-performance computation center is used to partition the grid into coherent areas using fuzzy clustering of bus phasor data [9]. As a by-product, this partitioning also results in medoid busses, which are the most suitable locations for PMUs. In a second stage, an extended set of simulations is performed, and the bus phasor signals collected are aggregated by area and referred to the center-of-inertia reference frame. Finally, the

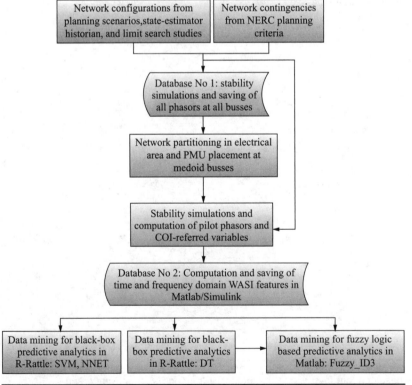

FIGURE 4.17 Model-building process for catastrophe anticipation in power systems.

data-mining features in the time and frequency domains are derived according to the algorithms in the section "Wide-Area Severity Indices" and saved in a large database for subsequent data-mining investigations.

However, developing predictive models from historical or simulated PMU records is no small task because power system responses are highly nonlinear and therefore very context-dependent. Although data mining has been shown to be very apt at this task of synthesizing such massive databases into predictive models, they require thousands of configurations and thousands of thousands of instances to minimally capture the intrinsic characteristics of the underlying power grid dynamics [21]. In this context, we are witnessing two competing trends: the fuzzy-logic rules-based approach [22], which has the advantage of transparency and interpretability [23], and the machine learning-based black-box approach, which relies on several complementary tools from statistical learning: neural networks (NNET) [11], support vector machines (SVM) [24], decision trees (DT) [25], and ensemble decision trees, referred to as "random forests" (RF) in this chapter [21]. It is not the intent of the authors to rank the usefulness of these approaches. They all have strengths and weaknesses [25] in terms of computational burden, training time, accuracy, robustness to noise, etc., and, indeed, they are often selected purely on the basis of their immediate availability to the user or a prior familiarity with them.

Data Organization for Training and Testing

In the proposed study, two different power system models are considered. The first is a 783-bus representation of the Hydro-Québec system used for operations planning. The same configurations and contingencies described in [21] are used. They represent winter and summer operations planning models with about 1000 load-flow patterns generated by the transfer limit search and critical clearing-time search processes based on 32 carefully chosen 735-kV contingencies. The second grid model is a 67-bus, nine-area test system used in [20] to demonstrate a PMU placement method. From five base load-flow configurations, many others are generated by stressing the system by means of a power transfer limit search under 32 contingencies. Figure 4.18 shows the actual system and the test system, and the complete setup for data generation to design the catastrophe predictors based on data-mining models. In this study, we will combine the two sets of data arising for the Hydro-Québec system and the test system in the hope of finding a single more general predictor instead of applying the methods developed to the two systems separately. The complete idea can be explained as follows.

Let us suppose that we have a set of examples $A = \{a_1, a_2 ..., a_k\}$ describing the general concept of system security, $C = \{S, I\}$, where S and I stand for secure (stable) and insecure (unstable) states, respectively. For each case, a set of attributes (A) is stored along with the status (C)

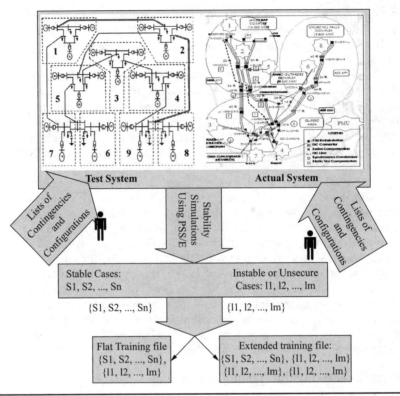

FIGURE 4.18   Database generation process applied to test [16] and actual systems [9]. The *m* insecure cases are replicated three times in the extended training file.

of the case derived by analyzing the simulation. Learning the concept of "security" consists in inferring its general definition from the set of given examples. However, it is well-known that the insecurity threat is very system-dependent. Years ago, the Hydro-Québec system was transient stability-limited while today it is voltage stability-limited in its most likely configurations. Similarly, the analysis in [16] showed that the nine-area system in Fig. 4.18 is very oscillatory but not overly sensitive to voltage instability. Therefore, learning the "security concept" on networks with widely different security attributes will result in a more general predictor encapsulating a broader security definition. Given that the original dataset includes only 22.5% of unstable cases, it is said to be skewed or unbalanced. A naïve classifier learned from these skewed datasets is always biased toward the majority classes, which constitute a major percentage of the samples in the dataset. As a result, the accuracy on the minority classes is hampered. We have, therefore, balanced the dataset artificially by replicating unstable cases three times ($m = 3$ in Fig. 4.18), which brings the dataset to 98,800 cases.

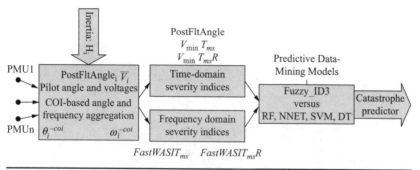

Figure 4.19   Principle of data mining model-based catastrophe prediction.

Selected PMU-Based WASI Features

From the algorithm in Fig. 4.19, the selected wide-area severity indices are defined in the time and frequency domains as follows:

1. $V \min T_{ms}$ — System-wide minimum voltage over the time span of T = 150 ms or T = 300 ms after fault clearing

2. $V \min T_{ms} R$ — Area-wide minimum voltage over the time span of T = 150 or T = 300 ms after fault clearing, considering only the busses in the faulted area

3. $FastWASIT_{ms}$ — System-wide frequency-domain severity index defined over the time span of T = 150 ms or T = 300 ms after fault clearing

4. $VLowPassT_{ms}$ — Filtered system-wide minimum voltage for T = 300 ms

5. $FastWASI\_T_{ms}$ — Filtered system-wide severity index defined for T = 300 ms

6. $TDEF$ — Fault duration

7. $PostFltAngle$ — System-wide maximum COI angle deviation from steady-state to fault-clearing time

In this work, the subset of features, which considers system-wide variables only, in contrast to [22], is used in the following for predictive catastrophic event analysis based on data-mining predictive models, as shown in Fig. 4.19. However, principal-component analysis is first applied to determine the number of most significant features among the candidate set [26]. Although we have 10 candidate features, we observe from Fig. 4.20 that five components alone could explain 93% of the variability in the data. Next, the second plot (a biplot) remaps the data points from their original coordinates to coordinates of the first two principal coordinates. The vectors drawn give an indication of how much of a role each variable plays in each of the two components,

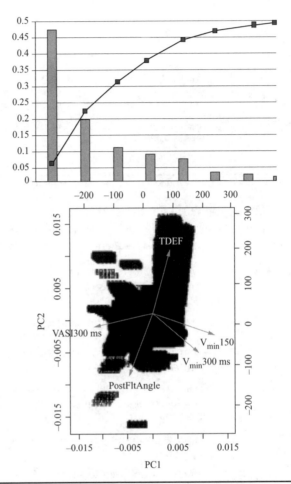

Figure 4.20 Principal-component analysis of the features space.

showing their correlation to the components. The axes are labeled with the correlation, to be interpreted for the variables, and the values of the principal components, to be interpreted for the data points. The six variables with the largest magnitude in the two-way plot are shown. According to their angle (or vector orientation), these variables are geometrically decoupled, and appear therefore as good candidates for predictive analysis of system stability.

Black-Box versus Comprehensible Predictors

Model Training

The black-box catastrophe predictors proposed in this study use open-source software R [26], which includes the implementation of conventional DTs, RFs, NNET, and SVM. The training is done using the extended data files separately for each scenario and the testing using the performance files, which are flat files. The aforementioned data-mining techniques were configured to randomly select only 70% of the assumed data file to build the model during training. The NNET-based predictor has a feed-forward structure with 10 hidden layers. Similarly, the SVM is set with a radial basis kernel. The decision tree is trained with the "R" package, which is called with the following settings: convergence error of 10^{-4}, with a minimum bucket, before and after a node split, equal to 600 and 300 samples, respectively. The DT obtained for classifying stable and unstable cases is shown in Fig. 4.21.

For all RF models, a limit of 210 trees was set with $n_{try} = 10$ essentially to avoid memory overflow. However, it was later observed that the

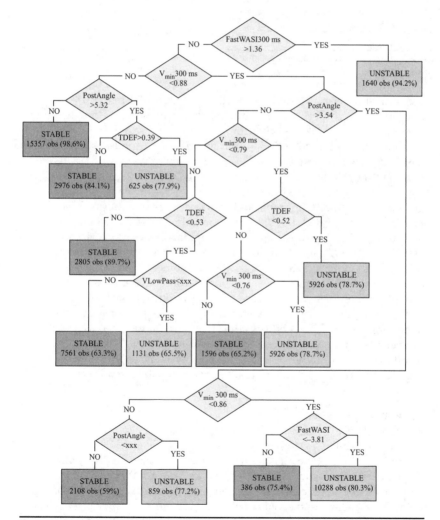

FIGURE 4.21   Decision tree generated for classifying stable (OK = −1) and unstable cases (OK = 1).

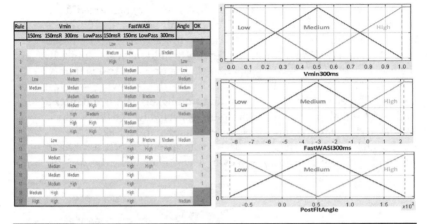

Figure 4.22   Fuzzy_ID3 catastrophe predictor based on combined Hydro-Québec and test systems.

generalization error started to stabilize at around 50 trees [10]. To develop a fuzzy decision tree model, the input features are first fuzzified and Quilan's ID3 algorithm is applied next to the resulting symbolic data representation in order to derive a fuzzy decision tree [26]. The corresponding fuzzy rule-based predictor was trained using the open-source software *fispro* described in [23]. The result for the combined HQ and test system is given in Fig. 4.22 where 19 rules generated for classifying stable and unstable cases are described. The Fuzzy_ID3-based decision tree is shown in Fig. 4.23, where the various membership functions have triangular shapes categorized as "HIGH," "LOW," and "MEDIUM." For implementing this fuzzy inference system, the Sugeno model, combined with a centroid defuzzification scheme, is chosen to obtain a smooth boundary of the stability conditions. In this case, the crisp output decision after defuzzification is a number varying from −1 to 1 in order of the increasing degree of instability. A threshold is, therefore, necessary to ascertain whether a given case is stable or not. We will present the results with a threshold of 0 and −0.2 respectively, to illustrate the impact of this choice on the trade-off between reliability and security.

Performance Assessment

The performance assessment is based on three important indices, namely accuracy, reliability, and security, as defined as follows [11, 21]. The complete statistics of the catastrophe predictors developed using NNET, SVM, RF, DT, and Fuzzy_ID3 are depicted in Table 4.5. While analyzing the statistics of accuracy, reliability, and security indices, the RF-based predictor emerges as the only tool with a more than 99% level of performance on all three indices. It is observed that, while switching from black-box solutions such as NNET, SVM, and RF to semitransparent solutions such as DT, and transparent solutions such as Fuzzy_ID3 with

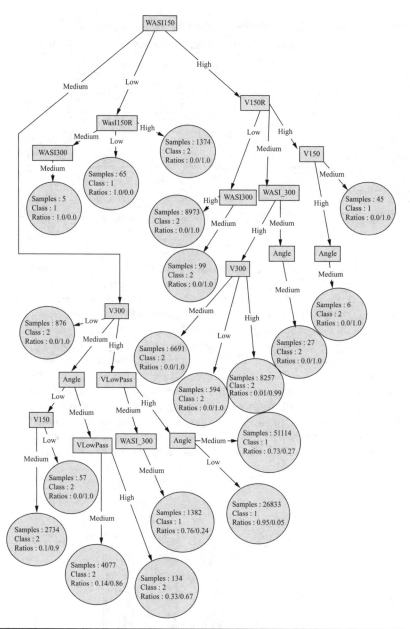

Figure 4.23 Fuzzy_ID3-based decision tree (Fuzzy_ID3) for classifying stable (OK = −1 = Class no. 1) and unstable cases (OK = 1 = Class no. 2).

various thresholds, there is a trade-off between accuracy, reliability, and security for the combined HQ and test, the HQ, and the test systems.

To provide a few clues to this trade-off, Fig. 4.24 compares the performance of the transparent (Fuzzy_ID3) and the best black-box

Data-mining tools	Accuracy (%)	Reliability (%)	Security (%)
	Wasiall_300ms	Wasiall_300ms	Wasiall_300ms
NNET	85.7	82.0	87.9
SVM	87.7	83.9	92.5
RF	99.2	99.9	99.0
DT	84.3	80.1	86.0
Fuzzy_ID3 (Threshold = 0)	83.7	86.4	82.9
Fuzzy_ID3 (Threshold = −0.2)	82.0	77.1	83.4

TABLE 4.5   Performance (Accuracy) Comparison between NNET, SVM, DT, Fuzzy–ID3, and RF

solutions (RF). The classification accuracy is 99.0% with RF compared to 82.0% and 85.5%, respectively, for Fuzzy_ID3 with a threshold of 0.2 and 0. Similarly, the reliability and security measurements are close to 99.0% with RF compared to 64.2% and 92.1%, respectively, with Fuzzy_ID3 having a threshold of 0.

Although the most accurate solution is the RF black-box solution, the Fuzzy_ID3 appears to be relatively flexible compared to other black-box models, given its adjustable decision threshold, which allows the classification to be offset toward more or less reliability or security. In fact, the fuzzy model output is a crisp number with a continuous variation between –1 and 1. It can, therefore, be interpreted as a fuzzy measure of stability, which is quite convenient for making practical trade-offs. Figure 4.25 illustrates the histogram of the fuzzy decision

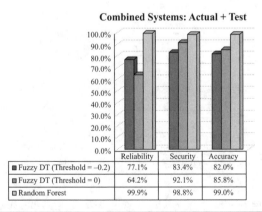

FIGURE 4.24   Results of a comparison between Fuzzy–ID3 and random forests for reliability, security, and accuracy measurements for combined actual and test systems.

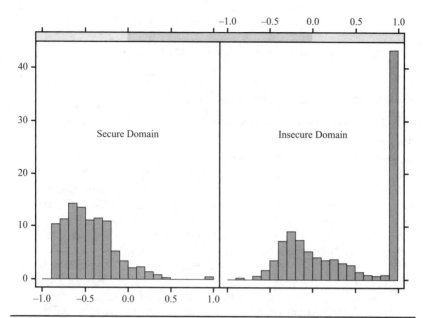

Figure 4.25   Histogram of the probability of secure and insecure contingencies based on the crisp decision output of the Fuzzy–ID3 based predictor.

based on Fuzzy_ID3. It is clear that high (positive) values of the decision match unstable cases, while negative values match stable cases. However, the two categories overlap in the tails. Furthermore, Fig. 4.26 provides a different view of the output decision in the form of two-dimensional

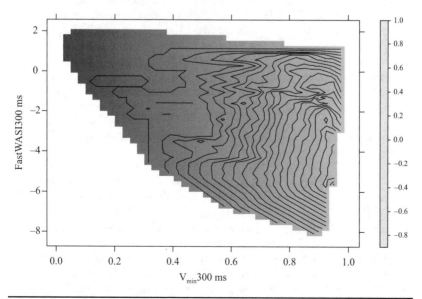

Figure 4.26   Fuzzy_ID3-based insecurity probability of using Sugeno defuzzification scheme (60,000 contingencies).

contour level plots. This confirms the intuitive view that low-voltage and low-energy disturbances with a voltage close to 1 are almost stable while high-energy and/or low-voltage disturbances are almost insecure.

Early Detection of System Oscillations and Damping Condition Assessment

Background

Oscillatory stability problems are one of the major threats to grid stability and reliability in many power systems worldwide. An unstable oscillatory mode can cause large-amplitude oscillations and may result in system breakup and large-scale blackouts. There have been several incidents of system-wide oscillations. Of these, the most notable is the August 10, 1996, western system breakup produced as a result of undamped system-wide oscillations [27]. More recently, unstable oscillations requiring early online detection schemes were reported in Australia [28] and Colombia [29]. In these cases, it was discovered that oscillation alarms could be issued when the power system is lightly damped to provide time for operators to take remedial action and reduce the probability of a system breakup as a result of a light damping condition.

To address this need, several measurement-based modal analysis algorithms have been developed. They include Prony analysis [30–31], the Eigen-Realisation Algorithm (ERA) [32–33], stochastic state space identification (SSSID), Yule-Walker, and the N4SID algorithm [27, 28, 34–35]. Each has been shown to be effective for certain situations but not as effective for others. For example, the traditional Prony analysis and ERA work well for disturbance data but not for ambient data, while Yule-Walker is designed for ambient data only. Even in an algorithm that works for both disturbance data and ambient data, such as SSSID, latency results from the time window used in the algorithm are an issue in timely estimation of oscillation modes. For ambient data, the time window needs to be longer to accumulate information for a reasonably accurate estimation, while for disturbance data, the time window can be significantly shorter, so the latency in estimation can be much less.

We can broadly state the online modal tracking problem in terms of identifying state-space models of signals with forced or random excitations, followed by eigenanalysis of the resulting model (Fig. 4.27). In the early-warning context, we can assume that the data are output-only measurements from a PMU-based wide-area measurement system. The data can be one-shot post-disturbance ringdowns or ambient noise. The only point to note is that, in the former case, we can use a standard ERA-based modal identification method, while in the latter case, stochastic identification methods will somehow be mandatory [32].

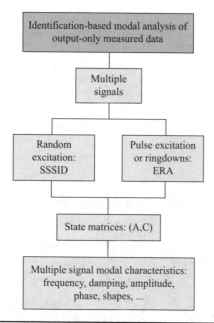

Figure 4.27   Online modal analysis for undamped oscillation detection.

Multiband Modal Analysis

As advocated in [36], the performance of the ERA and SSSID methods can be improved when band-pass pre-filtering is applied to the incoming signal before the identification stage. This is achieved by linear filter bank preprocessing, which can deal with closely spaced natural modes embedded in significant noise. In addition, filter bank processing makes it possible to determine the energy channel by channel, using the Teager-Kaiser energy operator (TKEO) concept to test the level of within-band energy or activity before even attempting any modal identification on that channel. This approach, which performs the modal identification separately for each frequency band provisional to the within-band energy exceeding a given threshold, has been shown to be effective in minimizing the ambiguity of modal parameters in the presence of significant noise, while offering a relatively fast response time to modal parameter changes (Fig. 4.28).

Principle

The power system response signals can generally be described as the sum of the combined amplitude (AM) and frequency modulated (FM) primitive signals:

$$x(t) = \sum_k A_k(t)\cos(\Theta_k(t)) = \sum_k a_k e^{\sigma_k}\cos(\omega_k(t)t + \phi_k) \qquad (4.9)$$

where a_k, ω_k, σ_k, and ϕ_k are, respectively, the amplitude, frequency, damping, and phase of the k-th modal component. It should be stressed that all parameters a_k, ω_k, σ_k, and ϕ_k can be slowly time-varying without posing any significant complication to the ongoing development.

Figure 4.29 illustrates the overall concept of first decomposing the multicomponent power system signal (Eq. 4.9) using a filter bank, and then performing a time-frequency analysis on each channel component using the energy separation algorithm (ESA) [36, 37] or, alternatively, the Hilbert transform [37, 38]. The ESA provides the amplitude (A_i) and frequency (ω_i) of the dominant modes with decent

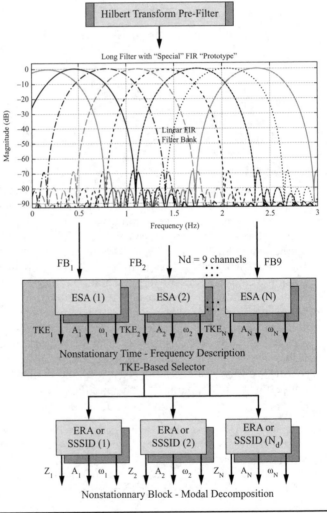

Figure 4.28 Overview of the MBMA scheme with a TKEO-based criterion to detect channels with the strongest oscillatory activity.

accuracy after filter-bank preprocessing, but, when accurate damping information is required or the filter bank output signal is not monochromatic enough, a more detailed modal analysis, using a parametric method, for instance, should be preferred to the ESA. In Fig. 4.29, ERA or SSSID is shown as the electromechanical-mode identification tool, even though any alternative identification method could fit just as well [27, 28, 32–35]. An optional discrete Hilbert transform (DHT) [30] is first applied to the incoming signal to reject quasi-steady-state components more efficiently. The filter bank then splits the possibly multicomponent signal, as in [36], into $N = 9$ essentially orthogonal components. After decomposing the incoming signal, each channel's energy (TKE_i, $I = 1...9$) is computed using TKEO [37], and N_d (typically $N_d <= 4$) energy-dominant signals are selected for parametric modal analysis based on an energy threshold test. Modal analysis is then performed on sliding nonoverlapping data blocks whose size can be adjusted for each channel, according to the channel center frequency, and the results are damping (Zi), amplitude (A_i), and frequency (ω_i), $I = 1,...N_d$.

Linear SIMO Signal Identification

Background on State-Space-Based Modal Analysis

Without any loss of generality, a time series given in the form $\{y_k, k = 1, 2, ..., N\}$, with y_k, a column vector of dimension n_y, can be considered the output of a linear filter with suitable input excitation [39]. If the signal is deterministic, such as a ringdown power system response, e.g., a convenient input is the discrete impulse. On the other hand, if the signal is stochastic, like the ambient power system response to random load switching, the input is naturally random noise excitation. Whatever the assumption, it is convenient to assume a single input for the multiple signals in order to derive a more compact transfer function representation with a single denominator. The signal will, therefore, share the same poles but, for each pole, the residue will vary with respect to the output.

For illustration, consider the following ringdown signal:

$$y(t) = Ae^{\sigma t} \cos(\omega t + \psi), \qquad t \geq 0 \tag{4.10a}$$

which is the impulse response of the following transfer function:

$$H(s) = \frac{(s + a)\cos(\psi) - \omega \sin(\psi)}{(s + a^2)^2 + \omega^2} \tag{4.10b}$$

with s, the Laplace transform symbol. Using the Tustin method of discretizing a continuous time transfer function [32], the relationship between H(s) and H(z) in Fig. 4.29 becomes obvious. Assuming that these transfer functions are appropriate, i.e., the numerator polynomial orders are strictly less than the order of the denominator polynomial,

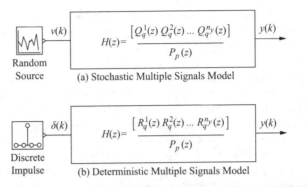

FIGURE 4.29 Application of ESA on multicomponent AM-FM signals.

the deterministic single input multiple output (SIMO) model of the system is equivalent to the following state-space model, $\{A, B, C\}$:

$$\begin{cases} X_{k+1} = Ax_k + Bu_k \\ y_k = Cx_k \end{cases} \tag{4.11}$$

with $u_k \in \mathbb{R}^{n_u \times 1}$ and $y_k \in \mathbb{R}^{n_y \times 1}$, the input and output vectors, respectively, (and $n_u = 1$ for a SIMO system).

To solve this problem, a simple idea very common in experimental modal analysis [32] is to build two shifted Hankel matrices encompassing, in a compressed form, the system-state matrices. This approach generally starts with organizing the input-output data in Hankel matrices. Hence, noting that the impulse response of Eq. 4.11 is given by $H_0 = D, H_k = CA^{k-1}B, k = 1, 2, \ldots, N$, the associated Hankel matrix is defined at the reference time $k > 0$ as follows:

$$H_{k|(i,j)} = \begin{bmatrix} H_k & H_{k+1} & \cdots & H_{k+j} \\ H_{1+k} & H_{1+k+1} & \cdots & H_{1+k+j} \\ \vdots & \ddots & \ddots & \vdots \\ H_{i+k} & H_{i+k+1} & \cdots & H_{i+k+j} \end{bmatrix} = \vartheta_i A^{k-1} \wp_j \tag{4.12a}$$

where i and j are fixed integers, and

$$\vartheta_i = \begin{bmatrix} C \\ CA \\ \vdots \\ CA^{i-1} \end{bmatrix}, \wp_j = [B \quad AB \quad \cdots \quad A^{j-1}B] \tag{4.12b}$$

are the extended observability and controllability matrices, respectively.

Deterministic SIMO-ERA

If $\{y_k, k = 1, 2, \ldots, N\}$ are assumed to be measurements of the impulse responses describing the system $\{A, B, C\}$, we first organize them in a sequence of impulse block matrices, $\{H_k\}, k = 0, 1, \ldots, N$. Then, $\{H_k\}$ is used to build the Hankel matrices $H_{1|(i,j)}$ and $H_{2|(i,j)}$. If $i >> n = dim(A)$, the singular value decomposition (SVD) of $H_{1|(i,j)}$ can be partitioned as follows:

$$H_{1|(i,j)} = \begin{bmatrix} U_n & U_o \end{bmatrix} \begin{bmatrix} \Sigma_n & O \\ O & \Sigma_o \end{bmatrix} \begin{bmatrix} V_n^T \\ V_o^T \end{bmatrix} = \vartheta_i \wp_j \qquad (4.13)$$

where Σ_n contains the n dominant singular values and $U_n^T U_n = V_n^T V_n = I_n$, with I_n, the n-dimensional unit matrix. Therefore, the minimal-dimension estimates of the observability and controllability matrices are given by:

$$\vartheta_i = U_n \Sigma_n^{1/2} \qquad \wp_j = \Sigma_n^{1/2} V_n^T \qquad (4.14)$$

Using the expression of $H_{2|(i,j)} = \vartheta_i A \wp_j$ and thanks to the orthogonality of U_n and V_n, the state matrix is obtained:

$$A = \vartheta_i^{\dagger} H_{2|(i,j)} \wp_j^{\dagger} = \Sigma_n^{1/2} U_n^T H_{2|(i,j)} V_n \Sigma_n^{1/2} \qquad (4.15a)$$

where the symbol † denotes the pseudo-inverse. Therefore, the input matrix is given by the first m columns of $\wp_j$ while the output matrix equals the first p rows of ϑ_i. Formally, we have:

$$B = U_n \Sigma_n^{1/2} E_{n_u}^T \qquad C = E_{n_y} \Sigma_n^{1/2} V_n^T \qquad (4.15b)$$

where $E_{n_u}^T = [I_{n_u} \quad O]$ and $E_{n_y}^T = [I_{n_y} \quad O]$ are two special matrices consisting of unit and null matrices of appropriate dimensions.

Stochastic SIMO-SSSID

The state-space representation of the SIMO stochastic model of the signals is obtained by setting the forced input $\{u_k, k = 1, 2, \dots, N\}$ equal to zero in Eq. 4.11:

$$\begin{cases} x_{k+1} = Ax_k + w_k \\ y_k = Cx_k + v_k \end{cases} \qquad (4.16)$$

The process noise w_k is the input that drives the system dynamics, whereas the measurement noise is the direct disturbance of the $n_y \times 1$ system response vector $\{y_k, k = 1, 2, \dots, N\}$, which is, therefore, a mixture of the observable part of the state and some noise modeled by the measurement noise v_k. The state matrices $\{A, C\}$ can be determined using SSSID [34–35]. Although several variants of this method have emerged in recent years, they all seem to be more or less related to the following fundamental algebraic relationship between Hankel matrices of input and noise data [34]:

$$Y_{i|(i-1,j)} = \vartheta_i X_{i|j} + \Gamma_i^s W_{i|(i-1,j)} + V_{i|(i-1,j)} \qquad (4.17a)$$

where $X_{i|j}$ is the extended state-matrix:

$$X_{i|j} = [x_i \quad x_{i+1} \quad \dots \quad x_{i+j-1}] \qquad (4.17b)$$

while:

$$\Gamma_i^s = \begin{bmatrix} O & O & \cdots & O \\ C & O & & O \\ \vdots & & \ddots & \vdots \\ CA^{i-2} & CA^{i-3} & \cdots & O \end{bmatrix} \qquad (4.17c)$$

is a lower block triangular Toeplitz matrix. Let us define the projection of the matrix $A \in \mathbb{R}^{p \times j}$ on the row space of matrix B as [34–35]:

$$A/B = AB^T(BB^T)^{\dagger}B \qquad (4.18)$$

Using projections of the Hankel matrix of future outputs $Y_{i|(i-1,j)}$ on the space of the Hankel matrix of past outputs, $Y_{0|(i-1,j)}$, Eq. 4.17a can be rewritten as follows [34–35]:

$$W_c(Y_{i|(i-1,j)}/Y_{0|(i-1,j)})W_r = \underbrace{W_c \vartheta_i (X_{i|j}/Y_{0|(i-1,j)}W_r)}_{1 \qquad\qquad 2}$$
$$+ \underbrace{W_c(\Gamma_i^s W_{i|(i-1,j)} + V_{i|(i-1,j)})W_r}_{3} \qquad (4.19a)$$

where the input-output data-related weighting matrices are selected essentially in order to cancel term 3 in Eq. 4.19a above:

$$W_c = I_{in_y} \qquad W_r = Y_{0|(i-1,j)}^T \Phi_{[Y_{0|(i-1,j)}, Y_{0|(i-1,j)}]}^{-\frac{1}{2}} Y_{0|(i-1,j)} \qquad (4.19b)$$

Therefore, Eq. 4.24 contains no reference to noise terms. An SVD of its right term, which is completely defined by the input-output measurements, then allows us to determine the extended observability (term 1 in Eq. 4.24) and state (term 2 in Eq. 4.24) matrices as follows:

$$W_c\left(Y_{i|(i-1,j)}/Y_{0|(i-1,j)}\right)W_r = \begin{bmatrix} U_n & U_o \end{bmatrix}\begin{bmatrix} \Sigma_n & O \\ O & \Sigma_o \end{bmatrix}\begin{bmatrix} V_n^T \\ V_n^o \end{bmatrix} \qquad (4.20a)$$

with:

$$\vartheta_i = U_n {\textstyle\sum_n}^{1/2} = W_c \vartheta_i$$
$$\tilde{X}_{i|j} = {\textstyle\sum_n}^{1/2} V_n^T = X_{i|j}/Y_{0|(i-1,j)}W_r \qquad (4.20b)$$

From Eq. 4.12b, the matrix C is the first n_y lines of ϑ_i, while A is determined using the shift invariance property of ϑ_i:

$$A = \underline{\vartheta}_i^{\dagger}\overline{\vartheta}_i \qquad C = \vartheta_i(1:n_y,:) \qquad (4.21a)$$

in Matlab notations, with $\overline{\vartheta}_i$ and $\underline{\vartheta}_i$ denoting ϑ_i without the first n_y and last n_y lines, respectively, and $\vartheta_{i-1} = \underline{\vartheta}_i$:

$$\vartheta_i = \begin{bmatrix} \vartheta_i \\ CA^{i-1} \end{bmatrix} = \begin{bmatrix} C \\ \overline{\vartheta}_i \end{bmatrix} = \begin{bmatrix} \vartheta_{i-1} \\ CA^{i-1} \end{bmatrix} \qquad (4.21b)$$

Modal Characteristics from State-Space Matrices

The first step is to convert the discrete-time state-space model of the previous sections into an equivalent continuous-time model using the inverse bilinear transform or any other suitable mapping function between the two domains. For instance, in the case of ERA-based state-space matrices (A,B,C) [32]:

$$\hat{A} = \log(A)/\Delta t \qquad \hat{B} = F_n \left(\hat{A}^{-1}(B - I) \right)^{-1}$$
$$\hat{C} = C * \Delta t \tag{4.22}$$

where Δt is the sampling rate. Naturally, Matlab contains all the necessary ingredients to achieve this process. Transforming the continuous-time system into modal space results in the following alternate representation:

$$\{\hat{A}, \hat{B}, \hat{C}\} \Leftrightarrow \Lambda, \Psi^{-1}\hat{B}, \hat{C}\Psi \tag{4.23}$$

where $\Lambda = diag\ (\lambda_1, \lambda_2 ..., \lambda_n)$ is a diagonal matrix of eigenvalues and Ψ, a matrix whose columns are the corresponding eigenvectors. $\Sigma = \Psi^{-1} B$ is the modal shapes matrix and $\Pi = C\Psi$, the initial modal amplitudes matrix [32–33]. From these definitions, the residues associated with the kth mode can be collected in a $n_y \times n_u$ dimensional matrix R_k with entries:

$$r_{ij}(k) = \Pi(i, k) \times \Sigma(k, j) = A_{ij}(k)e^{i\phi_{ij}(k)} \tag{4.24}$$

where $i = 1, 2, ..., n_y, j = 1, 2, ... n_u, k = 1, 2, ..., n$, and n is the number of distinct natural modes of Λ, with damping $\sigma_k (sec^{-1})$ and natural frequency $\omega_k (rad/sec)$. The corresponding transfer function matrix takes the form:

$$G(s) = \sum_{k=1}^{k=n} \frac{R_k}{s - \lambda_k} = \sum_{k=1}^{k=n} \frac{R_k}{s - \sigma_k - i\omega_k} = \left[g_{ij}(s) \right] \tag{4.25}$$

To facilitate modal recognition and interpretation, the following results are expressed mainly in terms of the zero-initial-condition impulse response components [inverse Laplace transform of $g_{ij}(s)$] according to the following formula:

$$h_{ij}(t) = \sum_{k=1}^{k=n} 2A_{ij}(k)e^{\sigma_k t} \cos \left(\omega_k t + \phi_{ij}(k) \right) \tag{4.26}$$

In the SSID configuration mode, $B = 0$ and the residue and impulse response matrices simplify to vectors defining the free responses of the output-only system:

$$r_i(k) = \Pi(i, k) = A_i(k)e^{i\phi}(k)$$

$$h_i(t) = \sum_{k=1}^{k=n} 2A_i(k)e^{\sigma_k t} \cos \left(\omega_k t + \phi_i(k) \right) \tag{4.27}$$

Illustration of Real-Time Modal Analysis of Output-Only Signals

Post-contingency Ringdowns

To demonstrate the ringdown analysis, we used the IEEE nine-bus test system [40] shown in Fig. 4.30. The base case consists of 1000 MW exported from area 2 and imports of 100 and 900 MW to areas 1 and 3, respectively. Since it is inherently unstable, we installed IEEE4B power system stabilizers at generators 1, 2, and 4. This was enough to stabilize all system modes. A three-phase fault was then applied on line C at bus 7 for 0.5 s, followed by outage of the faulted line. The system remained stable under such conditions, and the speed deviations of the four generators were recorded to determine the post-fault modal characteristics of the system. Applying the SIMO version of the ERA method to these ringdown signals resulted in an accurate single-input, four-output, state-space representation (A,B,C) with an order of $n = 14$.

A sample illustration of this SIMO model goodness-of-fit is shown in Fig. 4.31a for the responses of generators 2 and 4. The simulation is simply the impulse response of the ERA model (A,B,C) starting from zero initial conditions. On each plot, the modal parameters are shown in decreasing amplitude and the filtering is done according to the same rule as in the previous section: retain a mode if its damping is less than 0.3 or its amplitude is greater than 1/50 that of the top mode in the list. To further check the SIMO model accuracy, Fig. 4.31b superimposes the power spectrum density of the actual and reconstructed fault responses. This amounts to comparing the spectrum of the two signals in each plot of the previous figure. These results confirm that the SIMO model with 14 eigenvalues can simultaneously reproduce the four post-fault speed spectra in the 0–5 Hz frequency range.

The illustration (Fig. 4.32) pertaining to this application shows the observability mode shapes for the six dominant complex eigenvalues of the SIMO model. The common frequency mode (0.06 Hz) is observed

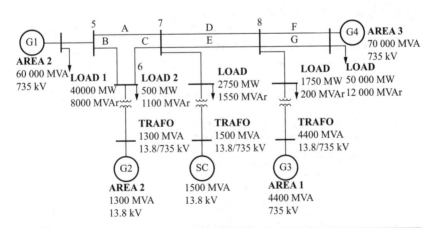

FIGURE 4.30 The nine-bus test network [40].

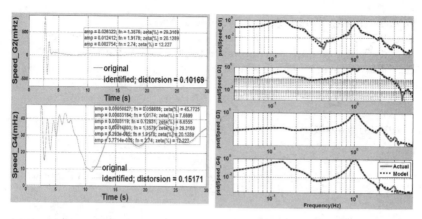

(a) Time domain validation   **(b) Frequency normalized PSD validation.**

FIGURE 4.31 ERA-based modal decomposition of multiple ringdown signals in response to a 0.5-s fault at bus 7 with IEEE PSS4B at generators 1, 2, and 4.

with the same phase and magnitude at all generators, while the interarea mode, which shifts from 0.22 to 0.12 Hz after the line outage, involves generators 1 and 2 against 3 and 4. The post-PSS local mode of generator 2 is 1.36 Hz, while the local mode of generator 3 (without PSS) remains unchanged at 1 Hz. The mode at 1.92 Hz involves generators 2 and 3 while the last mode, 2.74 Hz, sited at generator 2, seems too weak to have any special significance.

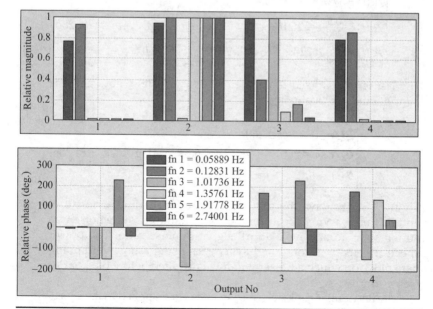

FIGURE 4.32 Multiple-signal observability mode shapes from IEEE nine-bus test system with IEEE PSS4B at three generators.

Time-Varying Ambient Signals Embedded in Noise

To better establish the performance of the stochastic state space-based modal analysis on time-varying ambient signals, we will first consider the following synthetic time series:

$$y(t) = \sum_{k=1}^{3} A_k(t) \sin(2\pi f_k(t) \times t + \theta_k(t) \times \pi/180) + v(t) \qquad (4.28a)$$

where $v(t)$ is a white noise process and:

$$
\begin{aligned}
A_1(t) &= 1/3 \ pu, \ \theta_1(t) = 0 \ \text{deg} \\
f_2(t) &= 1\text{Hz}, \ \theta_2(t) = 0 \ \text{deg} \qquad \forall t \geq 0 \qquad (4.28b) \\
A_3(t) &= 1/3 \ pu, f_3(t) = 1.5 \ \text{Hz}
\end{aligned}
$$

The first component is a frequency chirp, ramped from 0.15 to 0.4 Hz in 400 s:

$$
\begin{cases}
f_t(t) = 0.15, & t \leq 10 \\
f_t(t) = \frac{0.25}{400}(t-40) + 0.15, & 40 \leq t \leq 440 \qquad \text{Hz} \quad (4.28c) \\
f_t(t) = .40, & t \geq 100
\end{cases}
$$

The second component is a fixed undamped sine (at 1 Hz) whose amplitude ramp-modulated starting at 260 s:

$$
\begin{cases}
A_2(t) = 1/6, & t \leq 260 \\
A_2(t) = \frac{0.5}{200}(t-260) + 1/6, & 260 \leq t \leq 440 \qquad pu \quad (4.28d) \\
A_2(t) = \frac{0.5}{200} \times 220 + 1/6 = 0.71667, & t \leq 440
\end{cases}
$$

The last term is a fixed undamped sine (at 1.5 Hz) whose phase is ramp-modulated beginning at 260 s:

$$
\begin{cases}
\theta_3(t) = 0, & t \leq 260 \\
\theta_2(t) = t - 260, & 260 \leq t \leq 600 \qquad \text{deg} \quad (4.28e) \\
\theta_2(t) = 400, & t \geq 600
\end{cases}
$$

Figure 4.33 displays the previously mentioned time series (top) and its time-varying spectrum (bottom). Random white noise was added so as to achieve a 15 dB signal-to-noise ratio. The power spectral density is computed every 15 s using a nonoverlapping data block on which a Kaiser window is applied before the Fourier transform.

The linear filter bank described in [36] is first applied to the signal. The dominant channel output signals, which fulfill the TKEO energy threshold criterion, are shown in Fig. 4.34 along with their energy.

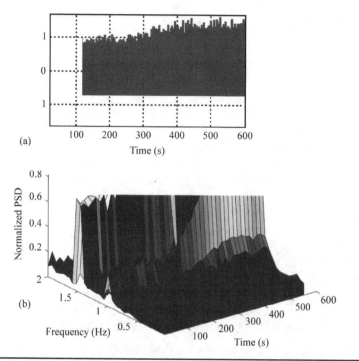

Figure 4.33   Synthetic time type-varying electromechanical oscillation embedded in noise (15 dB SNR). Top: Time series; Bottom: Spectrogram.

Figure 4.34 provides the time-varying frequency and amplitude according to the TKEO theory [36]. Even under a 15 dB signal-to-noise ratio, the results shown are clearly in accordance with the analytic signal definitions (Eq. 4.28). However, the results for the low-frequency channels, FB1 and FB2, need some discussion. Initially, the lowest-frequency component was 0.15 Hz, which is the center frequency of FB1, while at t = 300 s, the same frequency has been ramped up to 0.3 Hz, which is now the center frequency of FB2. Therefore, the TKEO amplitude of FB1 makes a smooth transition to zero while the amplitude of FB2 increases to its maximum value of 1/3. The multibank analysis relies on TKEO energy to detect the time for channel change, and this is reflected in the frequency plot at the top of Fig. 4.24: the frequency estimate is from channel 1 (black) up to about t = 200 s and thereafter from channel 2 (magent).

The SSSID in section IV-B was applied to 15-s consecutive nonoverlapping blocks of noisy data filtered through a nine-channel filter bank. A threshold was applied to the TKEO energy [20] in order to select just the four dominant channels in each given 15-s data block under analysis for further modeling. The SSSID algorithm was performed on a SISO basis, for each selected channel, using the parameters I = 25 and n = 3. After identifying matrices A and C, extraction

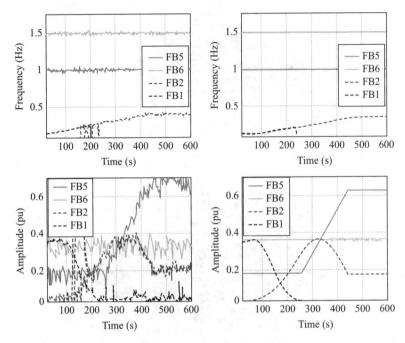

(a) ESA decomposition with/without noise (left/right)

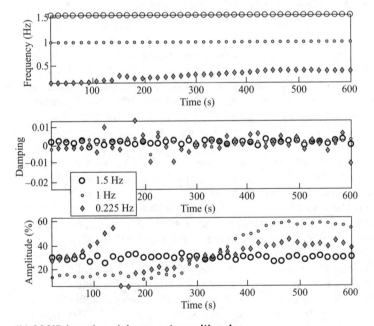

(b) SSSID based modal parameters with noise

FIGURE 4.34 Multiband ESA-based tracking modal analysis of a synthetic signal embedded in a 15-dB SNR noise.

of modal information yields the results illustrated in Fig. 4.34. It is obvious that, despite the low 15-dB SNR and the multiple time-varying parameters of the signal, the estimates of mode frequency are very accurate. The amplitude and damping are more uncertain during periods of rapid frequency change, especially when the dominant channel changed from FB1 to FB2. Interestingly, the damping and amplitude are not significantly affected by the phase modulation within the ramping rate considered (1 degree/s). It should be noted that, in Fig. 4.34, amplitude attenuation due to the pass-band response of the filter bank is not compensated (in contrast to [36]).

Conclusion

In the context of smart grids, development of EWS will enable a paradigm shift in the engineering of power grid monitoring and controls, which will undergo a transition from a human-centric to a data-centric decision-making process. In the new framework, an abundance of historical and real-time quality data, generally collected through the PMU infrastructure, is being deployed worldwide; thanks to massive smart-grid investments. In a data-rich power grid, converting the data to models and then to information and controls, becomes a fundamental enabling step, requiring the development of new predictive analytical tools that can help predict the onset of catastrophes following disturbances, or quantify how far the system is from a safe operating boundary (Fig. 4.35).

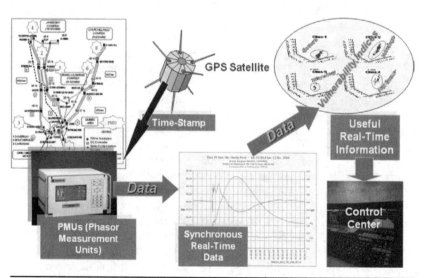

FIGURE 4.35 Converting wide-area PMU-based data into actionable information for human-made online decision and automation systems.

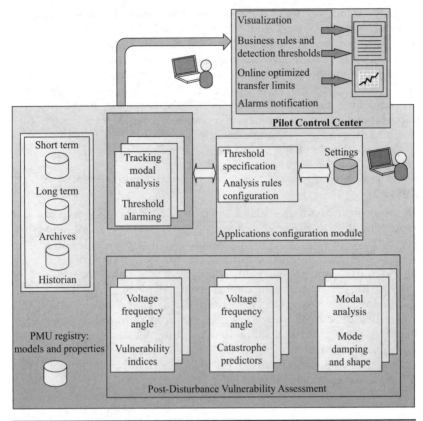

Figure 4.36   Integrating EWS in the smart grid context.

However, developing predictive models from historical or simulated PMU records is no small task because power system responses are highly nonlinear and, therefore, very context-dependent. Although data mining has been shown to be very apt at this task of synthesizing such massive databases into predictive models, they require thousands of configurations, and thousands of thousands of instances to minimally capture the intrinsic characteristics of the underlying power grid dynamics. On the other hand, tracking modal analysis, which is also a key part of any EWS for modern heavily stressed grids, is difficult to tackle using a data-mining approach.

In this chapter, we have presented a comprehensive methodology that addresses all facets of this new tool, as summarized in Fig. 4.36. Within a unified framework based on network partitioning and center-of-inertia referencing concepts, the mathematical underpinning of PMU-based early warning systems was developed in an integrated top-down approach. We have highlighted the need for essential two modules: the first dedicated to tracking modal analysis of ambient signals, and the second focusing on post-disturbance vulnerability assessment with respect to disturbance energy, transient voltage or frequency dips, and oscillation damping. Given the data rate and extensive geographical coverage achieved thanks to widespread use of PMUs, an integrated EWS is likely to become a key application of smart and more agile control centers, enabled with new alert and preemptive action capabilities. For this reason, the FERC smart-grid policy of 2009 identified wide-area situational awareness as a high-priority functionality and, as a matter of fact, EWS is at the core of it.

References

[1] Steps to Establish a Real-Time Transmission Monitoring System for Transmission Owners and Operators Within the Eastern and Western Interconnection, *DOE-FERC Report to Congress Pursuant to Section 1839 of the Energy Policy Act 2005,* 3 February 2006 [available on-line]: http://energy.gov/oe/downloads/steps-establish-real-time-transmission-monitoring-system-transmission-owners-and (accessed on 5 June 2012)

[2] North American SynchroPhasor Initiative (NASPI), [on-line] http://www.naspi.org

[3] J. Warichet, T. Sezi, and J.C. Maun, "Considerations about synchrophasors measurements in dynamic system conditions," *Int. J. of Electrical Power and Energy Systems,* pp. 1–13, 2009.

[4] A.G. Phadke, B. Kasztenny, "Synchronized phasor and frequency measurement under transient conditions," *IEEE Trans. on Power Del.,* vol. 24, no. 1, pp. 89–95, Jan. 2009.

[5] I. Kamwa, K. Pradhan, G. Joos, "Adaptive Phasor and Frequency-Tracking Schemes for Wide-Area Protection and Control," *IEEE Trans. Power Del., vol.* 26, 2011 (in print).

[6] I. Kamwa, J. Béland, G. Trudel, R. Grondin, C. Lafond, D. McNabb, "Wide-Area Monitoring and Control at Hydro-Québec: Past, Present and Future", Panel Session on PMU Prospective Applications, *2006 IEEE/PES General Meeting,* Montreal, QC, Canada, June 18–22, 2006 (paper 06GM0401, 12 pages)

[7] U.S.-Canada Power System Outage Task Force, "Final Report on the August 14, 2003 Blackout in the United States and Canada: Causes and Recommendations," April 2004. https://reports.energy.gov/BlackoutFinal-Web.pdf

[8] FERC Smart Grid Policy, 19 March 2009, [on-line]: http://www.ferc.gov/whats-new/comm-meet/2009/071609/E-3.pdf

[9] I. Kamwa, A.K. Pradhan, G. Joos, S.R. Samantaray, "Fuzzy Partitioning of a Real Power System for Dynamic Vulnerability Assessment," *IEEE Trans. on Power Systems,* **24**(3), Aug. 2009, pp. 1–10.

[10] I. Kamwa, R. Grondin, "PMU Configuration for System Dynamic Performance Measurement in Large Multi-area Power Systems," *IEEE Trans. on Power Systems,* **17**(2), pp. 385–394, May 2002.

[11] I. Kamwa, R. Grondin, L. Loud, "Time-varying Contingency Screening for Dynamic Security Assessment Using Intelligent-Systems Techniques," *IEEE Trans. on Power Systems,* **PWRS-16**(3), pp. 526–536, Aug. 2001.

[12] I. Kamwa, J. Beland, D. McNabb, "PMU-Based Vulnerability Assessment Using Wide-Area Severity Indices and Tracking Modal Analysis," in presented at the Panel Session on advanced PMU applications in system dynamics, in *proc. 2006 IEEE PES Power Systems Conference and Exposition,* PSCE '06, Atlanta, GA, pp. 139–149, Oct. 29 2006.

[13] J. Lambert, D. McNabb, and A.G. Phadke, "Accurate voltage phasor measurement in a series-compensated network," *IEEE Trans. on Power Del.,* vol. 9, no. 1, pp. 501–509, Jan. 1994.

[14] I. Kamwa, M. Leclerc, and D. McNabb, "Performance of demodulation-based frequency measurement algorithms used in typical PMUs," *IEEE Trans. on Power Del.*, vol.19, no. 2, pp. 505–514, Apr. 2004.

[15] G. Trudel, J.-P. Gingras, J.-R. Pierre, "Designing a reliable power system: The Hydro-Québec's integrated approach," *IEEE Proc. Special Issue on Energy Infrastructure Defense Systems*, **93**(5), pp. 907–917, May 2005.

[16] I. Kamwa, A.K. Pradhan, and G. Joos, "Automatic segmentation of large power systems into fuzzy coherent areas for dynamic vulnerability assessment," *IEEE Trans. on Power Systems*, vol. 22, no. 4, pp. 1974–1985, 2007.

[17] R. Krishnapuram, A. Joshi and O. Nasraoui "Low-complexity fuzzy relational clustering algorithms for web mining," *IEEE Trans. on Fuzzy Systems*, vol. 9, no. 4, pp. 595–607, 2001.

[18] G. Pison, A. Struyf, P.J. Rousseeuw, "Displaying a clustering with CLUSPLOT," *Computational Statistics & Data Analysis*, vol. 30, pp. 381–392, 1999.

[19] T. Lie, R.A. Schlueter, P.A. Rusche, R. Rhoades, "Method of Identifying Weak Transmission Network Stability Boundaries," *IEEE Trans. on Power Systems*, **PWRS-8**(1), pp. 293–301, Feb. 1993.

[20] D.R. Ostojic, "Spectral Monitoring of Power System Dynamic Performances," *IEEE Trans. on Power Systems*, **PWRS-8**(2), pp. 445–451, May 1993.

[21] I. Kamwa, S.R. Samantaray, G. Joos, "Catastrophe Predictors from Ensemble Decision-Tree Learning of Wide-Area Severity Indices" *IEEE Trans. on Smart Grid*, **1**(2), pp. 144–158, Sep. 2010.

[22] I. Kamwa, S.R. Samantaray, G. Joos, "Development of Rule-Based Classifiers for Rapid Stability Assessment of Wide-Area Post-Disturbance Records," *IEEE Trans. on Power Systems*, **24**(1), pp. 258–270, Feb. 2009.

[23] S. Guillaume, "Designing fuzzy inference systems from data: An interpretability-oriented review," *IEEE Trans. Fuzzy Syst.*, vol. 9, pp. 426–443, June 2001. [on-line]: http://www.inra.fr/internet/Departements/MIA/M/fispro/FisPro_EN_doc.html

[24] F. R. Gomez, A. D. Rajapakse, U. D. Annakkage and I. T. Fernando, "Support Vector Machine Based Algorithm for Post-Fault Transient Stability Status Prediction Using Synchronized Measurements", *IEEE Transactions on Power Systems*, vol. 26, Feb. 2011.

[25] T. Hastie, R. Tibshirani, J. Friedman, *The Elements of Statistical Learning*, 2nd Ed., Springer-Verlag: New York, 2009, p. 745.

[26] Rattle (the R Analytical Tool to Learn Easily), by D. Williams, ver. 2.3, May 2008: http://rattle.togaware.com/

[27] N. Zhou, F. Tuffner, Z. Huang, S. Jin, "Oscillation Detection Algorithm Development Summary Report and Test Plan," Pacific Northwest national Laboratory Report, October 2009, PNNL-18945, [on-line]: http://www.pnl.gov/main/publications/external/technical_reports/PNNL-18945.pdf

[28] R.A. Wiltshire, Analysis of Disturbance in Large Interconnected Power Systems, PhD Thesis, Queensland University of Technology, Brisbane, Australia, 2007, [on-line]: http://eprints.qut.edu.au/35773/1/Richard_Wiltshire_Thesis.pdf

[29] O.J. Arango, H.M. Sanchez, D.H. Wilson, "Low Frequency Oscillations in the Colombian Power System – Identification and Remedial Actions," paper C2-105_2010, *Cigré 2010*.

[30] S. Zhang, X. Xie, J. Wu, "WAMS-based detection and early-warning of low-frequency oscillations in large-scale power systems," *Electric Power Systems Research*, vol. 78, pp. 897–906, 2008.

[31] T.J. Browne, V. Vittal, G.T. Heydt, A.R. Messina, "A comparative assessment of two techniques for modal identification from power system measurements," *IEEE Trans Power Syst.*, vol. 23, no. 3, pp. 1408–1415, Aug. 2008.

[32] Jer-Nan Juang, *Applied System Identification*. New Jersey: PTR Prentice-Hall, 1994, p. 394.

[33] I. Kamwa, R. Grondin, E.J. Dickinson, S. Fortin, "A minimal realization approach to reduced-order modelling and modal analysis for power system response signals," *IEEE Trans. on Power Systems*, **PWRS-8**(3), pp. 1020–1029, Aug. 1993.

[34] P. Van Overschee, B. De Moor, "N4SID: Subspace Algorithms for the Identification of Combined Deterministic-Stochastic Systems," *Automatica*, **30**(1), 1994, pp. 75–93.

[35] T. Katayama, *Subspace Methods for System Identification*, New-York: Springer, 2005, p. 390.

[36] I. Kamwa, A.K. Pradhan, G. Joos, "Robust Detection and Analysis of Power System Oscillations Using the Teager-Kaiser Energy Operator," *IEEE Trans. on Power Systems*, vol. 26, no.1, pp. 323–333, Feb. 2010.

[37] A. Potamianos and P. Maragos, "A Comparison of the Energy Operator and Hilbert Transform Approaches for Signal and Speech Demodulation," *Signal Processing*, vol. 37, pp. 95–120, May 1994.

[38] D.S. Laila, A.R. Messina, B.C. Pal, "A refined Hilbert–Huang transform with applications to interarea oscillation monitoring," *IEEE Trans Power Syst.*, vol. 24, no. 2, pp. 610–620, May 2009.

[39] M.H. Hayes, *Statistical Digital Signal Processing and Modeling*, New York: John Wiley & Sons, 1998.

[40] P.M. Anderson, R.G. Farmer, *Series Compensation of Power Systems*, PBLSH! Inc., Encinitas, CA, USA, 1996, Appendix B: pp. 519–530.

[41] I. Kamwa, R. Grondin, Y. Hebert, "Wide-Area Measurement Based Stabilizing Control of Large Power Systems – A Decentralized/ Hierarchical Approach," *IEEE Trans. on Power Systems*, **PWRS-16**(1), pp. 136–153, Feb. 2001.

[42] Nawab S.H., T.E. Quatieri, "Short-Time Fourier Transform," in: *Advanced Topics in Signal Processing* (J.S. Lim, A.V. Oppenheim, eds.), Prentice Hall, Englewoods Cliff, NJ, 1988.

[43] Y.N. Zhou, L.L. Zhu, K.K.Y. Poon, D. Gan, H. Zhu, Z. Cai, "Area Center of Inertia—A potential unified signal for synchronous and frequency stability control of interconnected power systems under short and long time spans," *The International Conference on Electrical Engineering (ICEE)*, 8–12 July 2007, Hong Kong, paper no. 390, pp. 1–6 [on-line]: http://www.icee-con.org/papers/2007/Oral_Poster%20Papers/08/ICEE-390.PDF

CHAPTER 5

The Integration of Renewable Energy Sources into Smart Grids

Mietek Glinkowski, Jonathan Hou, Dennis McKinley, Gary Rackliffe, Bill Rose
ABB, Inc.

Human beings have always been intrigued by natural elements and the weather that nature creates. It is no surprise that human nature would inevitably find a way to utilize these natural elements, such as wind, solar, and water, to create alternative forms of energy. Wind power, solar power, hydropower, biomass, and other "renewable" energy sources are quickly becoming less alternative and more accepted, mainstream power sources in society today.

These renewable energy sources are also becoming just as important as the concept of the smart networks themselves. Renewable energy is already, in fact, an integral part of the emerging smart grid revolution that is just beginning to take shape. There are unique challenges and opportunities related to the integration of these renewable resources into power systems. Smaller installations, on the order of 10 to 50 MVA, are often tied directly to medium-voltage distribution grids, whereas the larger projects—100 MVA and up—require an interface with the transmission grid. From voltage regulations to ride-through, from predictability and scheduling to management of reactive power, these renewable sources, if integrated properly, can greatly support and improve network performance.

There are several key market drivers for renewable energy today. One key driver is the push for a renewable portfolio standard (RPS). For example, at least 33 out of 50 states in the United States today require that some sort of RPS be in place within the next decade. There have also been concerted efforts to create a national RPS that would take advantage of vast homegrown renewable energy resources, like wind or solar, and interconnect them to the power grid through new and enhanced transmission routes. The wind industry continues to push for a permanent renewable energy production tax credit (PTC) that would enhance national job growth and increased percentages of wind as part of the American energy mix. Renewable energy companies and utilities continue to work to promote better standards, interoperability, partnerships, and policies.

Another key market factor driving renewables among many utilities and power organizations includes creating sources that have a positive environmental impact, which includes carbon-footprint reduction and greenhouse gas emissions. This also affects the uncertainty and the risks that come from oil and other fossil fuels of oil, with the goal of reducing one nation's dependence on foreign countries for oil and gas exports.

Wind power is much more cost-effective now than in the latter part of the 20th century. The cost of solar power equipment and components remain fairly high, but as pricing continues to fall it is becoming more feasible and cost-effective to build solar farms. Hydropower and its power systems, centered around large rivers, dams, and lakes, have always been extremely cost-effective, and is perhaps the most widespread of any form of alternative energy, at least where water bodies are present nearby.

Exponential growth of renewable energy has been enabled in recent years, thanks to multiple technological breakthroughs and advances, making these energy sources more practical than ever. Some examples of how smart technologies—and the practices they enable— can impact the operation and overall health of the grid include:

- Real-time situational awareness and analysis of the distribution system that can drive improved system operational practices that will, in turn, improve reliability

- Fault location and isolation that can speed recovery when outages do occur by allowing work crews to drastically narrow the search for a downed line

- Substation automation (SA) that enables utilities to plan, monitor, and control equipment in a decentralized way, which makes better use of maintenance budgets and boosts reliability

- Smart meters that allow utility customers to participate in time-of-use pricing programs and have greater control over their energy usage and costs

- SCADA (supervisory control and data acquisition)/DMS (distribution management systems) that put more analysis and control functions in the hands of grid operators

- Voltage control, through reactive power compensation and the broader application of power electronics that increases transmission capacity of existing lines and improves the resiliency of the power system as a whole

More specifically, there have been significant technological advances in:

- Power electronics (PE) and their ability to control power flow, both active and reactive

- High-voltage direct current (HVDC) and flexible AC transmission systems (FACTS) technologies for power systems

- Photovoltaics and other solar energy technologies

- Automation technologies that enable the renewable energy resources to be connected to other smart grid initiatives and developments

The Smart Grid Connection

These technology breakthroughs are increasingly intersecting with overall advancements with the smart grid. This is a "natural" connection that the smart grid has with renewables. There is still debate in the power industry about what a smart grid is or what makes a grid truly intelligent. The answer given is often dependent upon the person or utility who is asked the question. One definition is based on a broader consensus of many widely varying opinions.

The consensus answer among many opinions seems to present an expansive view of the smart grid and what it can do, as defined by its capabilities and operational characteristics rather than by the use of any particular technology. The smart grid's broadest characteristics make it

- *Adaptive*, with less reliance on operators, particularly in responding rapidly to changing conditions

- *Predictive*, in terms of applying operational data to equipment maintenance practices and even identifying potential outages before they occur

- *Integrated*, in terms of real-time communications and control functions

- *Interactive* between customers and markets

- *Optimized* to maximize reliability, availability, efficiency, and economic performance

- *Secure* from attack and naturally occurring disruptions

Renewable energy technologies and systems in wind, solar, and biomass in particular have continued to grow in terms of scalability and large, utility-scale commercialization. Renewables offer many positive, attractive, unique characteristics that other energy sources do not. For example, these natural energy sources offer "free" fuel, as well as a little to no carbon footprint, and a virtually endless supply seemingly recycled over and over (thus, the term "renewable").

Of course, there are new and long-standing challenges to renewable energy adaption and acceptance. As mentioned earlier, the cost (and cost-effectiveness) of many renewables is still too high and is not competitive with fossil or nuclear fuel sources. Renewables are often observed by many as intermittent and therefore inconsistent as well as having a low predictability of supply, particularly wind and solar. Hydropower is seen as more consistent as long as the river continues to flow. These renewable energy sources often require large land requirements. For example, wind = 3.7–4.0 W/m^2 and solar = ~1 W/m^2, and both have low-to-moderate "fuel" efficiency (wind = ~35%, solar = ~13%). Variability, scheduling, the need for energy storage, and the ability to ride through system disturbances in large-scale networks in particular are also key challenges.

This chapter will briefly examine the history, status, and ongoing efforts of various renewable energy sources, especially the two fastest-growing of these energy sources worldwide, wind and solar power (in photovoltaic plants in particular).

Advances in Smart Grid–Enabled Wind Power Technologies

Technological advancements throughout the entire power system have led to the integration of wind power into today's emerging smart grid in a way that's increasingly efficient and reliable.

This integration works both ways. Advanced wind turbines with power electronic controls and other devices can support a grid with reactive power and protect the equipment during severe grid disturbances, while a smart grid allows connectivity of the wind turbines as intermittent sources of energy. In recent years, large wind parks have become increasingly connected with the grid using technologies such as HVDC systems, FACTS, and static var compensators (SVC) with energy storage. These technologies have added new degrees of controls to the power system, mitigating the intermittency of the wind energy production and the impact of electrical disturbances.

This section discusses advanced wind energy solutions from an electrical perspective, both present and future, from the smart controls and automation of the wind turbines to the novel designs of large-scale wind farms and interconnections. The new technology of double-fed induction generators and permanent magnet generators allows for a wide range of controlling both real and reactive power outputs due to changing power system conditions and wind resources. More efficient power electronic drives based on the pulse-width modulation (PWM) technology have become almost standard components of wind project designs.

The challenge of remote locations of wind farms and the trend to go offshore with large wind parks is being met with the application of thyristor-based HVDC and transistor-based HVDC interconnections, allowing for better controllability and utilization of wind energy. These technologies improve stability within the grid and provide the ride-through capability during external faults and severe voltage sags. There is more work to be done in the area of wind power energy system delivery. New technology solutions are being investigated and implemented, particularly in the area of power interconnection and wind farm protection and control. This paper illustrates some examples of what the future might bring and how these technological advances will contribute to a power system that enables more wind power energy to be delivered in a more efficient, reliable, and smarter way.

Utilities appear to be making smart grid investments in phases. Transmission investments are typically based on projects following demonstration of need to meet system reliability and capacity requirements. Smart grid transmission investments can incorporate power electronics for HVDC and FACTS applications. Advanced monitoring and control technologies, such as phasor measurements, and the integration of this information to SCADA/EMS and operations systems are a current focus of the U.S. Department of Energy (DOE) and utilities.

For distribution investments, the initial phase is to get core information technology (IT) systems in place, and many utilities are implementing advanced metering infrastructure (AMI) to engage customers and to provide timely data on power consumption and distributed energy resources. In parallel, utilities are making investments in distribution grid management to deliver reliability through fault detection, isolation, and restoration (FDIR) using DMS, distribution SCADA, communications to substations and feeders, and substation and distribution automation (SA and DA) with controllable devices. Distribution grid management also delivers efficiency with volt/var optimization (VVO) to minimize feeder losses and to reduce feeder peak demand.

Another focus of the smart grid is to leverage the distribution grid management and transmission SCADA/EMS systems to manage the interconnection of wind generation and other renewable generation. As shown in Table 5.1, a shift in the nation's generation mix from centralized to a combination of centralized and distributed with substantial renewable resources is being implemented. Supported by tax credits, the development of wind energy is expected to rapidly grow. This growth will create operating challenges from the variability of wind generation and will require a smarter grid.

	Current Grid	Smart Grid
Communications	None or one-way	Two-way, real-time (fast)
Customer Interaction	None or limited	Extensive
Operation and Maintenance	Manual, time-based	Remote monitoring and diagnostics, predictive
Generation	Centralized	Centralized and distributed, *substantial renewable resources*, energy storage
Power Flow Control	Limited	More extensive

TABLE 5.1   Moving Down the Road to the Smart Grid of the Future

Wind Energy in the Context of Smart Grids

The wind energy in the context of smart grids becomes important due to primarily three factors: economic, environmental, and maturity.

The economy of wind energy is still an ongoing debate. As the wind industry is still in its early stage, the economics are not as optimized as they should be. Although the "fuel" is free, the wind projects come at a cost that is still on the border line with conventional energy sources. Tax credits (PTC and ITC, or investment tax credits) are still needed to move many projects forward. The scale of wind penetration and the cost associated with the need of grid reinforcement due to the added wind generation are all influencing the economics. In several national studies [1] the estimated cost for grid reinforcement due to wind energy is analyzed. Figure 5.1 illustrates the findings for several European countries. Clearly, higher degrees of wind energy penetration drive higher costs of grid reinforcement to support this amount of new energy. The cost ranges from approximately 50 to 100 Euros/kW of wind power

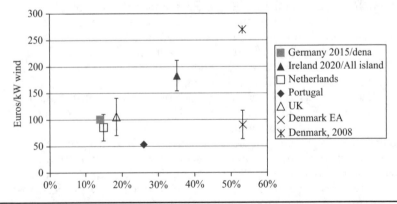

FIGURE 5.1   Estimated cost for grid reinforcement due to wind power for different degrees of wind energy penetration [1].

for 10% to 25% penetration to almost 270 Euros/kW for penetrations of more than 50%. The latter case (from Denmark studies) requires an additional comment. The high number comes with the assumption of allocating approximately 40% of total grid reinforcement cost to wind power. A second Denmark study, however, indicates only approximately 80 Euros/kW at approximately 55% of penetration when only the additional 2250 MW of offshore wind is to be connected to the grid.

The U.S. studies appear to suggest that the cost of grid reinforcement due to the addition of wind is not dependent on the degree of penetration. Dollars per kilowatt ($/kW) stays flat with increasing wind penetration.

The coming years will clearly see more commercialization, higher volume productions, and therefore lower prices of wind energy projects, especially from larger projects (economy of scale). The other important factor of economics is improved security of the power supply. A more secure system will be lower in cost.

The environmental aspect of wind energy is clear. No pollution and minimal environmental impact (birds, noise, and aesthetics) make wind one of the most suitable choices for a renewable portfolio.

The maturity of the wind technology is also important. It is clearly the most mature and the most utility-scale-ready alternative among all the renewable energy solutions. Wind turbines of 1.5 to 3.0 MW are almost standard, and new products on the market target the range of up to 10 to 20 MW [2, 3].

One of the criticisms of wind energy is its variability and therefore poor predictability over time. Indeed, the wind is variable, and often predictions of wind are dependent on the prediction of weather patterns and therefore are good only for about 48 to 72 hours. But the loads are also very variable (Fig. 5.2). So, if a smart grid was able to optimize the

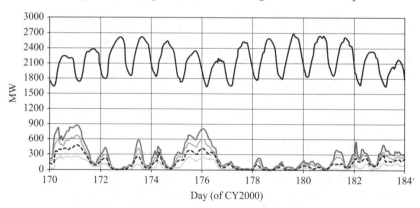

FIGURE 5.2   An example of typical 2 weeks of summer Idaho power load from calendar year 2000 and wind generation. Top curve: Load profile. Bottom curves: Wind generation (Idaho Power Wind Integration Study, February 2007, Idaho Power Company for the Idaho Public Utilities Commission, http://www.idahopower.com/AboutUs/PlanningForFuture/WindStudy).

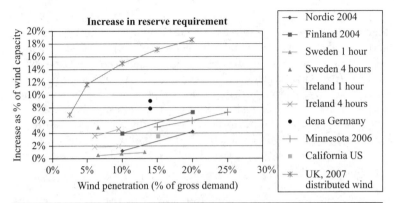

FIGURE 5.3   Increase in reserve requirement for increasing wind penetration [1].

fluctuations in the wind and match them with the load, constantly adjusting the distribution of the load to the wind farms able to produce power, the penetration of the wind in the system could increase.

There is also a challenge with backup generation. Again, the wind variability requires that backup generation be provided when wind energy is not available. In the extreme, 1:1 backup reserve would be required (1 MW of backup for every 1 MW of wind energy). However, smart utilization of different wind farms in different locations and matching them to the load could reduce the need for backup to the much lower level. Reference [1] provides an excellent overview of the required reserves due to wind power.

In Fig. 5.3, all case studies confirm that higher wind penetration will need a higher degree of backup reserves. The numbers vary from 1% to 2% to around 18% reserve. One should also notice that the degree of reserve is strongly influenced by the variability of wind. For 1 hour of variability, the degree of reserve could be around 1%, i.e., almost one-fifth of the required reserves of 5% if 4-hour variability is considered. For the Germany, Minnesota, and California cases, 24 hour wind uncertainty has been taken into account. This rather wide range of reserve estimates comes from several factors, like the start-up and ramp-up times of different reserves (fossil, hydro, nuclear, gas turbine, etc.) as well as the flexibility of the grid itself to handle the power shifts and changing load flows. This second aspect relates to the need for a smart grid.

The other aspect of variability is energy storage. Presently commercial, utility-scale energy storage is available only in the form of the pumped-storage hydro, where water is discharged during the peak load (day hours) and pumped up to the upper reservoir at light load (at night).

This technology is limited to areas where both land and water are abundantly available. Battery energy storage has been making good inroads into commercialization, and the first systems are beginning to show up in utility trials [4].

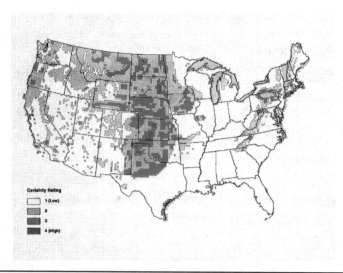

FIGURE **5.4**   Wind certainty rating for the WPC of 3 and higher [2].

As to the predictability of wind energy, two factors are the most noticeable. First, the predictability itself is much better now, thanks to the advances of the meteorological sciences. The certainty ratings for Wind Power Class of 3 and higher, as shown in Fig. 5.4, indicate that in areas where wind is the most attractive solution (central states of the United States), the certainty rating is high, mostly between 3 and 4 [5]. Second, a smart grid will be more adaptive and interactive, so it can better adapt to the changing wind conditions at different wind farms in real time with a much shorter response time than the existing systems.

In addition to the mid-central states of the United States, where wind resources are abundant, the offshore coastal areas are receiving growing attention. These wind resources are often close to the load centers on the east and west coasts, and the surface turbulences are much lower due to no obstacles like buildings, mountains, trees, etc. The technological challenge to be solved is underwater connectivity and the harsh, corrosive environment of the sea.

All these factors demonstrate that a smart grid can and should easily and efficiently embrace growing wind energy production sites [9]. Moreover, a smart grid can have a significant, although indirect, impact (of up to 5%) on reducing the CO_2 emissions by supporting penetration of renewable wind and solar generation (25% RPS) [6].

Energy efficiency and renewable energy such as wind are also intertwined in a different relation within the smart grid vision. Since energy efficiency is sometimes called the fifth fuel of electricity, it, as well as the wind energy, has to be accommodated by the smart grid. Therefore, both the energy efficiency and renewable resources demand a grid that has much more sophisticated and flexible functionality, the ability to send and receive power in all directions, aggregation of the renewables (wind) to lower the impact of their variability, transfer of loads and generation sources to better optimize the transmission and distribution systems, better measure, monitor, curtail load (if necessary), and control of the flow of energy [7].

The advanced technologies available today and in the near future will play a major role in wind energy penetration. These new technologies are described in the following sections from the component, equipment, and product levels to the advanced solutions for grid connectivity and at the power system level.

Turbine Level Solutions: Intelligent Wind Converters

The wind turbine is at the heart of every wind energy project. To maximize efficiency, all modern turbines are variable speed. Therefore, their generators, be it double-fed induction generators (DFIG) or variable speed permanent magnet (PM) synchronous machines, turn at asynchronous speed with respect to the grid frequency. In order to produce electrical energy at 50/60 Hz at an optimum operating point of the machine, the DFIG uses a partial converter to supply the rotor with the variable frequency [8]. The PM machines need full-scale converters. For growing offshore applications where low maintenance is of critical importance, PM machines are becoming the preferred choice. Large power ratings of 5 MW or higher are best realized by higher voltages to limit the rated current. Full-load converters tend to be very efficient at partial load operations, characteristics that will be dominating the wind turbines for most of their life (average turbine capacity factor is 30%–35%). All of these factors combined, offshore application, partial load efficiency, high rated power, and low maintenance requirements, call for medium voltage (MV) full-scale converters that can tie directly to the grid (see Fig. 5.5). This approach has several advantages, including lower continuous current, easier usage of MV switchgear, and fewer transformation steps to connect to the transmission network [9].

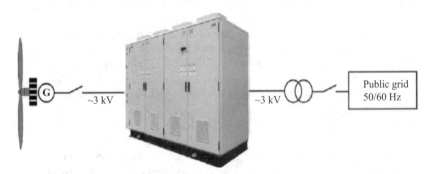

FIGURE 5.5   Concept of PM generator with full-scale converter at medium voltage.

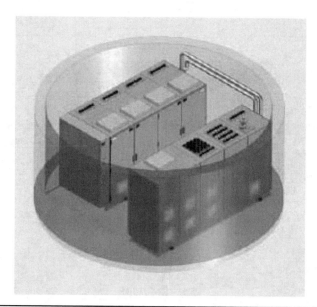

FIGURE 5.6   Full-scale PE converter rated 5 MW installed inside the wind tower [10].

As with any PE converters, the selection and design are guided by three objectives: reliability, efficiency, and cost. The result is often a PE system-based on proven industrial solutions for large-scale industrial machines.

One such example is based on a simple, modular structure based on the PEBB (power electronics building block). These PEBBs result in a compact design and are based on high-power semiconductors, IGCT (integrated gate commutated thyristors). The complete converter of 5 MW can fit inside the turbine tower with the water cooling, grid harmonic filters, and generator harmonics filters, as shown in Fig. 5.6.

The converters consist of three subsystems: PEBBs, control system, and mechanics. The four-quadrant, three-level converter topology combines two neutral point connected (NPC) phases for high-power density. A basic circuit diagram is shown in Fig. 5.7.

The control subsystem is connected with fiber-optic (FO) cables to reduce the electromagnetic compatibility (EMC) interference, and the entire system is condensation- and vibration-proof for harsh wind turbine offshore environments.

In normal operation of the four-quadrant, three-level converter, two IGCTs in each phase are always in the blocking state. This allows the operation of the DC link in the middle at twice the DC voltage of a two-level converter with the same components. Also, the current ripples are much lower (four times), which reduces the torque ripples to the generator and its gearbox and shaft.

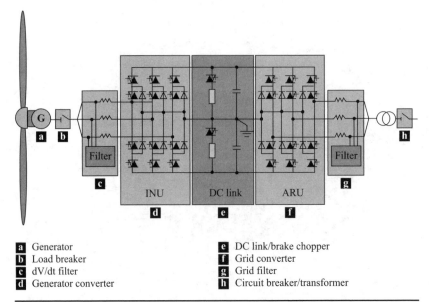

a Generator	**e**	DC link/brake chopper
b Load breaker	**f**	Grid converter
c dV/dt filter	**g**	Grid filter
d Generator converter	**h**	Circuit breaker/transformer

FIGURE 5.7 Basic diagram of the full-scale, four-quadrant, three-level wind-power converter [10].

Figure 5.8 and Table 5.2 summarize the different functions and technical parameters that a modern MV power electronic converter (as in Fig. 5.7) can exhibit. Many of the functionalities are very relevant to the smart grid and complementary to the smart grid objectives: no in-rush, ability to ride through, harmonic cancellation, remote monitoring, advanced diagnostics, and others.

In recent years more transmission system operators require the wind-generating plants to ride through the severe network faults and to support the system with reactive power (var). In the United States, for example, the Federal Energy Regulatory Commission (FERC) has

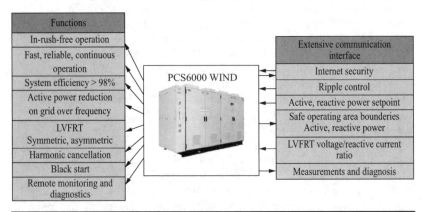

FIGURE 5.8 Functions of the full-scale converter as in Fig. 5.7 for wind turbine applications relevant to the smart grid. LVFRT (low-voltage fault).

Input Values (Generator input)	
Input voltage	0 … 4kv
Input power	Up to 8.5 MVA
Input Power factor	−1.0 … + 1.0
Input frequency	0 … 100 Hz
Generator type	Asynchronous/synchronous
Output Values (to grid)	
Output transformer	Standard MV transformer for 6 pulses
Output voltage	0 … 3.5kV
Output power	Up to 8.7 MVA
Output frequency	50/60 Hz +/−5%
Output Power factor	1.0 (controlled, reactive power setpoint or power factor setpoint can be specified)
Output filter	Series filter and tuned shunt filter
Output power response time	< 20 ms
Output power setting time	< 80 ms
Losses and Efficiency	
Frequency converter losses	< 1.2%
Filter losses	< 0.6%
Other components	< 0.2%
Overall converter efficiency	> 98% at rated output

TABLE 5.2 Technical Data of the MV Converter Shown in Fig. 5.7

implemented Order 661 and 661-A. Regional transmission organizations (RTOs) and independent system operators (ISOs)(for example, MISO, CAISO, PJM, ISO-NE, SPP, and NYISO) are required to create wind interconnection criteria for the low-voltage ride-through requirement [11]. Although different operators have different specific requirements, the modern turbines and their converters have to be able in principle to operate successfully during and after the system faults and supply up to full (rated) reactive power.

In the example, this is accomplished by the voltage limiter unit (VLU—brake chopper) that can dissipate active power during faults and still provide the safe and continuing operation of the wind turbine generator.

The current waveforms from the generator (bottom curves) and the generator output voltage (second curve from the bottom) are steady during the collapse of voltage from the grid side (top curve). No disturbances or transients are experienced, leaving the generator free of electrical and mechanical (torque) stresses.

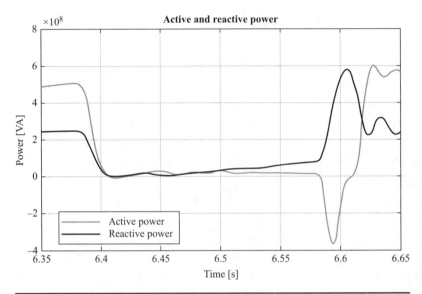

Figure 5.9   Active (top ———) and reactive (bottom ———) power from the PE converter during the voltage dip to 18%. Positive reactive power corresponds to overexcitation.

Reactive power compensation, power factor improvement, and dynamic var support become more important than ever due to the continuing efforts to improve the voltage profile and power quality, and to reduce power losses. Smart grids will require the energy sources, like wind turbines, to provide them with full var management, from leading to lagging power factors, dynamically, and with fast response times. Figure 5.9 illustrates just one example of the dynamic response of the PE converter to a system voltage disturbance resulting in voltage dip to 18% lasting 0.2 s.

The converter is capable of delivering 5 MW and 2.5 Mvar of power during normal condition; it rides through the voltage disturbance for approximately 0.2 s, and then recovers quickly to normal conditions generating P and Q as desired.

Inside the Tower: Medium Voltage Switchgear

Wind farms, or wind parks as they are sometimes called, require many different systems and components for power to be produced and delivered. Many of these components and products are often taken for granted, as they have existed elsewhere in the power systems for many years. However, the unique characteristics of wind energy and wind turbines demand some different approaches. For example, MV switchgear, which is used to protect MV systems from faults and interruptions, in wind turbine applications has to meet additional constraints, namely, fitting through the narrow door of the wind tower. Wind towers have narrow door openings so as not to

Figure 5.10   MV (36/38 kV) switchgear under installation through the wind tower door.

compromise the mechanical strength and rigidity of the entire tower construction. The switchgear, which would be more cumbersome and more expensive to be placed outside, has to go inside each tower, typically at its bottom platform. In addition, in case of servicing or future replacement, the same equipment has to be able to be taken out of the tower. Considering that the typical wind farm MV rating is 36/38 kV (IEC/ANSI), the design challenge is to reduce the width (and often the height) of the MV switchgear cubicle without compromising the dielectric clearances necessary for this voltage rating. The equipment must also provide the highest level of safety for the operating personnel by providing arc-resistant construction against arc flash and internal arcing [12].

One example of such a switchgear is an SF6 insulated, vacuum-breaker–based panel with 420 mm width (see Fig. 5.10).

Grid Interconnection Solutions

There are fundamentally two ways of interconnecting wind farms to the power grid: AC and DC. Each has its special characteristics, unique advantages, and limitations. From the smart grid point of view, both have advanced technologies that are utilized to maximize the wind energy production and to maintain or even improve the grid reliability and stability.

AC: AC Wind Interconnection Solution

The interconnection of wind generation for land installations is typically based on an AC substation located near the wind generators to step up the voltage from the wind farm collector system to a transmission-level voltage. (For smaller plants and where utility systems operate at 36/38 kV, a direction connection to a distribution grid is possible.) Under optimal conditions, a transmission line is located near the wind generation and the substation is connected directly to the transmission line, or a short overhead transmission line extension is used to connect the wind generation to the grid. Because of the cost of transmission lines, the difficulty in obtaining transmission rights-of-way, and the time required for siting and permitting, close access to the transmission grid can be as important as the availability of wind when developing wind generation.

In Europe, a massive amount of wind generation is planned with a significant portion located offshore. Locating wind generation offshore does increase the cost of the generation equipment, installation, and maintenance costs because of the harsh seawater environment. Offshore wind generation also changes the transmission grid interconnection and requires high-voltage (HV) cables to transmit the power from the wind generation collector system to the onshore grid. HV cables may also be required for onshore applications if the right-of-way for an overhead line is not available.

The system to interconnect the offshore wind generation to the grid also requires the local collector grid at sea, but the interconnection system is more complex. The local collector system consists of MV subsea cables that feed the wind generation to a substation mounted on an offshore platform. This substation raises the voltage for the subsea transmission cable that feeds an onshore substation (see Fig. 5.11).

The onshore substation is a transmission-switching station and may include transformers if the local transmission grid voltage differs

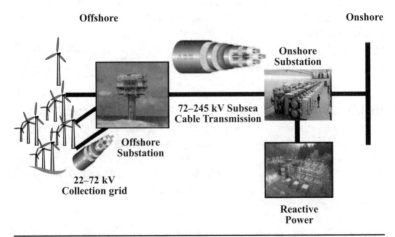

Offshore   **Onshore**

Onshore
Substation

72–245 kV Subsea
Cable Transmission

Offshore
Substation

22–72 kV
Collection grid

Reactive
Power

Figure 5.11 Subsea cable transmission connecting offshore wind farm with onshore power grid.

from the cable. Finally, reactive compensation for var support is typically located within the onshore substation to provide voltage support, which may be required for transmission grid stabilization because of the variability of the wind output.

In many wind farm installations, solid dielectric cables have emerged as the preferred technology choice for HV transmission, and they offer the following advantages:

- Absence of insulating fluids eliminates the risk of accidental release of hazardous materials and substances into the environment.

- Lower maintenance costs (solid dielectric cables are virtually maintenance free).

- Cable capacitance per mile and phase is less than 60% of the capacitance of fluid-filled cables.

- Ability to splice cables in discontinuous shifts.

- Cables are suitable for direct burial in open trenches at a lower overall installation cost than using concrete-encased duct bank systems.

- Fiber-optic control and communication cables can be embedded in the cable.

An example of a solid dielectric cable technology is shown in Fig. 5.12. Use of fiber-optic control and communication cable is worth noticing, as it is a technology trend that simplifies the system architecture and reduces the number of individual components by introducing new multifunctional products.

Figure 5.12   Solid dielectric submarine cable design up to 230 kV AC. Conductor material: Copper; Conductor screen material: Conductive PE; Insulation type/material: Dry cured triple extruded XLPE; Insulation screen: Conductive PE; Longitudinal water seal: Swelling tapes; Metallic sheath material: Lead alloy; Inner sheath material: Conductive PE; Assembling: Polymeric profiles; Cable core binder: Polymeric tape; Bedding: Impregnated tape; Armor material: Galvanized steel; Outer serving material: Polypropylene yarn.

However, AC cables have technical limitations for HV transmission applications over long distances. For example:

- Charging currents in AC cables consume capacity cumulatively with distance. For example, 40 km of 345-kV XLPE cable requires approximately 600 A of charging current. This also generates so many vars (of the order of 360 Mvars) that large power transfer becomes impractical.

- Capacity rating for the HV cable diminishes with distance, which limits the maximum practical distance of AC underground and submarine transmission circuits.

- Technical feasibility of HV AC cables, and this capacity limitation as a function of cable length, will impact the evaluation and consideration of underground and submarine HV AC cable systems.

DC: HVDC Interconnection for Offshore Installations

As explained earlier, more and more emphasis in the wind industry is being placed on offshore farms due to their proximity to load centers, more steady winds, and less aesthetic impact.

However, typically the offshore installations will require more (or all) new electrical infrastructure per MW than onshore. The equipment has to be more robust due to the harsh sea environment, and its availability has to be higher due to the higher cost of servicing and weather-dependent access to the turbines. In addition, the spacing of large offshore turbines could reach 0.5 km or more. These requirements demand higher technology solutions. The individual turbines are connected at the 36 to 38 kV voltage level via underwater cables, but the connection to the shore might require HVDC rather than an AC solution. For longer distances to the wind farm (50–100 km or more), HVDC presents an attractive alternative [14]. DC cables can be used with voltage source converters (VSC) that are built with power transistors, rather than thyristors, connected in series to provide the voltage rating required. Figure 5.13 illustrates a typical converter station with its DC valves, AC and DC switchyards, filters, and cooling systems.

The heart of the system is the two-level bridge with mid-point grounded capacitors (Fig. 5.14). This design provides low ground currents at both substations (converters) and assures good dynamic operation of the interconnection.

This low-ground current is even more important for the sea-level environment of offshore wind farms. The insulated-gate bipolar transistor (IGBT)–based valves are controlled by PWM controls that can synthesize the power frequency sinusoidal waveform with a fast, almost instantaneous, response.

This technology combination enables full control of power inputs and outputs, both in active and reactive domains. The result is a fully adjustable amplitude, phase angle, and frequency of the voltage. The

a AC power area
b Converter reactors
c HVDC light valves
d DC power area
e Cooling system
f Chopper resistor

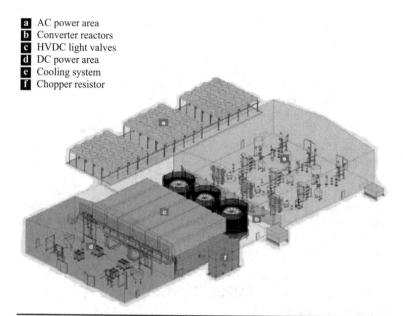

FIGURE 5.13   HVDC converter station.

use of IGBTs is also advantageous in the network restoration or at startup of the offshore wind farm. An offshore converter station can be started in the frequency mode, and when the AC side of the voltage is ramped up and stabilized, the individual wind turbines can be automatically connected to it.

Full, independent control of active (P) and reactive (Q) powers in each converter station plays an important role in optimal operations and therefore smart utilization of the offshore wind farm. Active power can be dialed and changed from full-rated import to full-rated export instantly as long as both converters have the same command. If the converters did not have the same active power throughput, it would result in accumulating the active (real) power in the voltage of the DC link, quickly charging them beyond their rated limits. Reactive power, on the other hand, can be dialed at will independently at each station.

FIGURE 5.14
HVDC VSC bridge
with mid-point
grounded
capacitors.

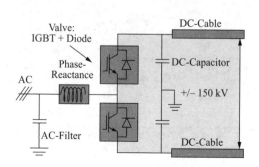

It is worth mentioning the characteristics of the HVDC VSC-based system at abnormal system conditions. Fault ride-through (during voltage collapse) has already been mentioned. The HVDC converters can also respond to the frequency changes on both sides of the DC link. If a system-side frequency is dropping, the power output of the converters can be increased to counteract it.

Future possibilities of the technology go even further. One can envision that with full control of the HVDC VSC-based converters, the wind farm can operate at a variable frequency—corresponding to the variable rotor speeds and variable winds—while the grid-side converter is fully synchronized with the constant system frequency of the power system. This would enhance the efficiency of the wind power conversion even further. Another technologically smart proposition may be to design the wind farm entirely for DC operation, simplify the conversion equipment within the individual turbines, and transport it as DC to the converter station on the grid side.

Support of the Grid Operation: Wind SVC with Energy Storage

The variable nature of the solar and wind renewable energy is unquestionable. Figure 5.15 shows an example of the hourly breakdown of renewable energy resources in California [15] with clearly variable contributions of wind and solar.

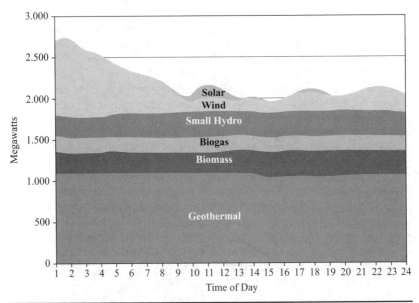

Figure 5.15 Example of hourly breakdown of renewable energy resources illustrating the variable nature of wind and solar [15].

With such challenges, a smart grid should be able not only to accommodate these resources for strategic environmental reasons, but also take advantage of these energies to the maximum extent possible. Conventionally, the variable energies require reserves that would be available when the variable energy resource is not. Wind experts estimate that, at a minimum, 5% to 18% [1] and, at an extreme, 80% of reserves are required to back up the wind farms. This is where energy storage could substantially reduce the need for the reserve power. Electrical energy storage has been a challenge for decades, although the recent focus on energy storage using thermal and mechanical storage and new battery chemistries shows promise.

In the previous sections, the need for control of active and reactive power has also been outlined with the advanced technologies of wind turbine converters and HVDC. A combination of dynamic control of P and Q, as well as the energy storage for variable generation of power, is yet another technological advance that would be utilized more in the context of a smart grid.

Connection to the power grid is accomplished by a power transformer and a series reactor. The VSC converts AC to DC. In a conventional SVC the DC side is supplied by a capacitor—in this case, two capacitors with a grounded midpoint. In the combination (called SVC-VSC with energy storage) batteries are in parallel to the capacitors. There is a technological synergy of these two solutions. Both battery storage and SVC operate on the same power electronics converter [16].

Batteries provide the active power support, and the capacitors provide the reactive power support (either consuming or injecting vars into the system). The VSC, as in the case of the HVDC before, is based on IGBTs, and the energy storage uses Li-ion battery technology.

Clearly, the battery energy storage is not limitless, and every injection of energy into the network has to be followed by the recharging period. But here again lies the advantage of the combined solution: The converter acts as a battery charger so there is no need for additional devices.

Another advantage of the SVC with energy storage is active harmonic filtering. SVC can inject a harmonic current into the grid with such a magnitude as to cancel the harmonic component of the system voltage.

Wind Power Summary

The increasing demand for wind energy will continue to grow, as long as it is fueled by the production tax credit and investment tax credit in the United States. The demand of renewable energy is further supported by the RPS required by regulators in many countries and in most of the United States. Additional legislation in climate changes and energy bills support increasing the production of wind energy sources in the United States and worldwide.

Paired with commercially available grid management systems, power electronics, enhanced medium voltage switchgears for wind farm application, grid interconnection solutions, cable technologies, and SVCs with energy storage, advanced wind energy technology can enable a reliable, environmentally friendly, efficient smart grid operation. A cost-effective wind energy production and delivery system can be implemented with a systematic approach and end-to-end design. These enhancements result in a smart grid that is an adaptive, integrated, and optimized electricity generation and delivery system, which benefits both customers and the environment.

Solar Power in the Context of Smart Grids

The sun is a virtually unlimited renewable energy source with enormous potential. It is sufficient to think that instant by instant the surface of our terrestrial hemisphere exposed to the sun receives power exceeding 50,000 terawatts (TW). The quantity of solar energy is about 10,000 times the total electric energy used throughout the world.

Among the different systems using renewable energy sources, photovoltaics is a promising technology due to the intrinsic qualities: it has no fuel costs (the fuel is free of charge) and limited maintenance requirements (no moving parts); it is noiseless and quite easy to install. Moreover, photovoltaics (PV), in some stand-alone applications, are definitely convenient in comparison with other energy sources, especially in those places that are difficult and uneconomic to reach with traditional electric lines and yet abundantly exposed to solar radiation, such as Africa.

Of course, there are challenges associated with PV that must be overcome. For example, in Nevada, a U.S. Department of Energy–backed study shows that adding hundreds of megawatts of solar power could cost the utility NV Energy millions of dollars in fossil fuel backup. (See the similar discussion in the preceding section on wind power.) About 150 to 1000 MW of intermittent PV penetration requires a lot of fossil fuel-fired backup generators to help smooth it out on cloudy days, adding up to approximately \$3 to \$8 per MW-h in the hot Nevada desert [17].

In Italy, for example, photovoltaics is strongly increasing thanks to a feed-in tariff policy, a mechanism to finance the PV sector, providing the remuneration, through incentives granted by the Electrical Utilities Administrator (EUA), of the electric power produced by plants connected to the grid.

Solar irradiation refers to the integral of the solar irradiance over a specified period of time [kWh/m^2]. The solar radiation on a horizontal surface consists of three components: a direct radiation, indirect diffused radiation from the sky, and radiation reflected by the ground and the surrounding environment.

For example, in winter, the sky is often overcast and the diffuse radiation is greater than the other two.

The average annual irradiance varies from the 3.6 kWhm² a day of the Po Valley to the 4.7 kWh/m²/d in the south-central region and the 5.4 kWh/m²/d of Sicily. Therefore, in the favorable regions it is possible to draw about 2 MWh (5.4 * 365) per year from each square meter—that is, the energy equivalent of 1.5 petroleum barrels for each square meter.

General Considerations of Photovoltaic Plants

Starting with a general description and basic concepts of PV plants, it is important to analyze the methods of connecting to the grid, along with protection against overcurrents, overvoltages, and indirect contact.

Because of their relative small size and lower voltage outputs, PV panels offer an opportunity to be a distributed generation (DG), generating close to the loads and therefore reducing transmission and distribution losses. Solar radiation power, when mapped out for most countries, can vary from 1000 to 2000 kWh/m²/y. Seasonal, daily, and hourly changes occur.

The advantages of photovoltaic plants are many, and can be summarized as follows:

- Distributed generation where needed
- Saving of fossil fuels
- No emission of polluting materials
- Reliability of the plants since they do not have moving parts (useful life usually over 20 years)
- Reduced operating and maintenance costs
- System modularity (to increase the plant's power, it is sufficient to raise the number of panels) according to the real requirements of users

However, the initial cost for the development of a PV plant is quite high due to a market that has not reached its full maturity from a technical and economical point of view. Moreover, the generation of power is fluctuating (dynamic) due to the variability of the solar energy source. See Fig. 5.16 as an example.

The annual electrical power output of a PV plant depends on different factors. Among them

- Solar radiation incident on the installation site
- Inclination and orientation of the panels
- Presence or lack of shading

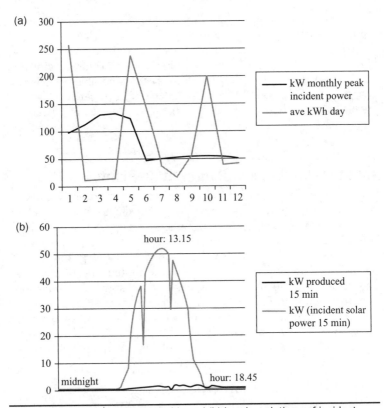

Figure 5.16 Example of (a) monthly and (b) hourly variations of incident and generated solar energy in North Carolina. Data from October 2010. Graph (a) month 1-January, month 12- December. Graph (b) hourly variations were measured at 15-min intervals. Two down spikes on the hourly chart are likely due to measurement error. Vertical scale as indicated in graph legends. (*Courtesy of North Carolina State University Solar House*).

- Technical performances of the plant components (mainly modules and inverters)

The main applications of PV plants are:

- Installations, with storage systems, for users isolated from the grid
- Large installations for users connected to the low-voltage (LV) grid
- Larger solar PV power plants, usually connected to the MV grid

This section takes a brief look at the larger installations.

A PV plant consists of a generator (PV panels), with a supporting frame to mount the panels on the ground, on a building or on any building structure, a system for power control and conditioning, a possible energy storage system, electrical switchboards and switchgear assemblies housing the switching and protection equipment, and the connection cables.

Photovoltaic Generation

One of the main components of a photovoltaic plant is the PV generator. The elementary component of a PV generator is the photovoltaic cell where the conversion of the solar radiation into electric current takes place. The cell is composed of a thin layer of semiconductor material, generally silicon properly treated, with a thickness of about 0.3 mm and a surface from 100 to 225 cm^2.

The energy balance for a typical PV generator shows the considerable percentage of incident solar energy, which is not converted into electric energy.

100% of the incident solar energy

- **3%** reflection losses and shading of the front contacts
- **23%** photons with high wavelength, with insufficient energy to free electrons; heat is generated
- **32%** photons with short wavelength, with excess energy (transmission)
- **8.5%** recombination of the free charge carriers
- **20%** electric gradient in the cell, above all in the transition regions
- **0.5%** resistance in series, representing the conduction losses

= **13% usable electric energy**

Under standard operating conditions (1 W/m^2 irradiance at a temperature of 25°C), a PV cell generates a current of about 3 A with a voltage of 0.5 V and a peak power equal to 1.5–1.7 W.

Grid-Connected Plants

Permanent grid-connected plants draw power from the grid during the hours when the PV generator cannot produce the energy necessary to satisfy the needs of the consumer. When the PV system produces positive net electric power, the surplus is injected into the grid, which operates as a big accumulator. As a consequence, grid-connected systems don't need accumulator banks.

These plants offer the advantage of distributed—instead of centralized—generation: In fact, the energy produced near the consumption area has a value higher than that produced in traditional large power plants, because the transmission losses are reduced and

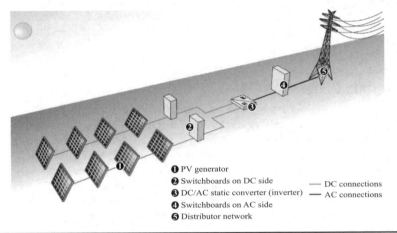

① PV generator
② Switchboards on DC side   —— DC connections
③ DC/AC static converter (inverter)   —— AC connections
④ Switchboards on AC side
⑤ Distributor network

FIGURE **5.17**   Principle diagram of a grid-connected photovoltaic plant.

the cost of the transmission and distribution are reduced. In addition, the energy production when the sun is shining allows the requirements for the grid to be reduced during the day, and that is when the demand is high.

Figure 5.17 shows the principle diagram of a grid-connected photovoltaic plant.

Intermittence of Generation and Storage of the Produced Power

On a large scale, the PV utilization is affected by the intermittency of production. The distribution network can accept a limited quantity of intermittent input power, after which problems with the stability of the network can rise. The acceptance limit depends on the network configuration and on the degree of interconnection with the adjacent grids.

On average it is considered dangerous when the total intermittent power introduced into the network exceeds a value from 10% to 20% of the total power of the traditional power generation plants. See also the corresponding info in the wind energy section of this chapter. As a consequence, the intermittency of solar power generation restricts somewhat the extension of a PV contribution to the national energy balance. This remark is valid to all intermittent renewable sources.

As a first remedy to get around this negative aspect, it would be necessary to store energy for a sufficiently long time so as to be able to inject the solar power into the network in a more continuous and stable form. Electric power can be stored in a number of ways: big superconducting coils, converting it into other forms of energy; kinetic energy stored in flywheels or compressed gases; gravitational energy in water basins; chemical energy in synthetic fuels; and electrochemical energy in batteries. To maintain energy efficiently for days and/or months, two storage systems emerge: batteries and hydrogen. At the present state of the art of these two technologies, the electrochemical storage seems feasible in the short to medium term, to store the energy for some hours to some days.

Therefore, in relation to photovoltaics applied to small grid-connected plants, the insertion of a small-to-medium battery storage subsystem may improve the situation, thus allowing a partial overcoming of the acceptance limit of the network. With regard to the seasonal storage of the huge quantity of power, hydrogen seems to be the most suitable technology. The excess energy stored in summer could be used to optimize the annual capacity factor of renewable power plants, increasing it to the average one of the conventional power plants (about 6000 hours).

As an added alternative one can approach the problem of intermittency by mixing different sources of renewable energy—solar, wind, biomass, hydro, ocean-wave, etc.—and utilize the differences, i.e., the degree and timing of their intermittency. This could reduce, although not completely eliminate, the need for backup base generation.

Voltage-Current Characteristic of the PV Module

A photovoltaic cell can be considered as a current generator. The voltage-current characteristic of a PV module is shown in Fig. 5.18. Under short-circuit conditions the generated current is the highest (I_{sc}), whereas with the open circuit the voltage (V_{oc}) is the highest. Under the two conditions the electric power produced is null. Under all other conditions, the produced power rises: It reaches the maximum power point (P_m) and then it falls near to the no-load voltage value.

Therefore, the characteristic data of a solar module can be summarized as follows:

- I_{sc}—short-circuit current
- V_{oc}—no-load voltage
- P_m—maximum power produced under standard conditions (STC)
- I_m—current produced at the maximum power point
- V_m—voltage at the maximum power point
- FF—filling factor: It is a parameter that determines the form of the characteristic curve V-I, and it is the ratio between the maximum power and the product ($V_{oc} * I_{sc}$) of the no-load voltage multiplied by the short-circuit current

If a voltage is applied from the outside to a PV cell in reverse polarity with respect to standard operation, the current remains

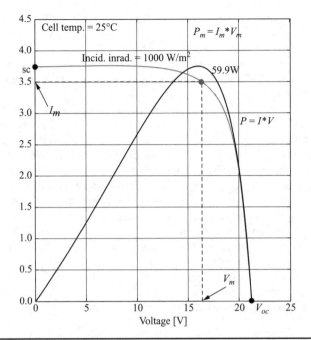

FIGURE 5.18   The voltage-current characteristic curve of a PV module.

constant and the power is now absorbed by the cell. When a certain value of inverse voltage ("breakdown" voltage) is exceeded, the junction P-N is punctured, as in a diode, and the current damages the cell. In the absence of light, the generated current is null at reverse voltages up to the "breakdown" voltage.

A PV plant connected to the grid and supplying a local load can be represented in a simplified way by the scheme shown in Fig. 5.19. The supply network (assumed to be at infinite bus) is represented as an ideal voltage source. The PV generator is represented by an ideal current generator (with constant current and equal incident solar power), whereas the local load is shown as a resistance R_u.

The currents I_g and I_r, which come from the PV generator and the network, respectively, converge in the node N of Fig. 5.19 and the current I_u absorbed by the consumer plant comes out from the node:

$$I_u = I_g + I_r \tag{5.1}$$

Since the current to the load is

$$I_u = U / R_u \tag{5.2}$$

the relation among the currents becomes:

$$I_r = U / R_u - I_g \tag{5.3}$$

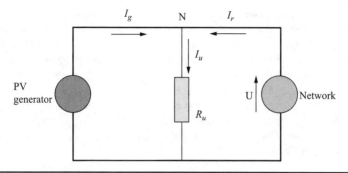

FIGURE 5.19   A PV plant connected to the grid and supplying a local load.

If in the Eq. [5.3] $I_g = 0$, as it occurs during the night hours, the current absorbed from the grid results in:

$$I_r = U/R_u \qquad (5.4)$$

On the contrary, if all the current generated by the PV plant is absorbed by the local load, the current supplied by the grid shall be null:

$$I_g = U/R_u \qquad (5.5)$$

When the incident solar energy increases, the generated current I_g exceeds that required by the load I_u, and the current I_r becomes negative, that is flowing into the grid.

The previous considerations can be made also for the powers:

- $P_u = U * I_u = U2/R_u$ the power absorbed by the user plant
- $P_g = U * I_g$ the power generated by the PV plant
- $P_r = U * I_r$ the power delivered by the grid

Expected Energy Production per Year

From an energy point of view, the design objective is usually to maximize the collection of the available annual solar radiation. In some cases (e.g., stand-alone PV plants) the design criterion could be optimizing the energy production over some definite periods of the year.

The electric power that a PV installation can produce in a year depends on:

- Availability of the solar radiation
- Orientation and inclination of the modules
- Efficiency of the PV installation

Since solar radiation is variable, to determine the electric energy that the plant can produce in a fixed time interval, the solar radiation relevant to that interval is taken into consideration, assuming that the performance of the modules is proportional to the incident energy.

The annual solar radiation for a given site may vary from one source to the other by as much as 10%, since it is statistical and subject to the variation of the weather conditions from one year to the next.

Starting from the mean annual radiation E_{ma}, the following formula is applied to obtain the expected produced energy per year E_p for each kWp:

$$E_p = E_{ma} \cdot hBOS \ [kWh/kWp] \qquad (5.6)$$

where:

hBOS (*Balance Of System*) represents the overall efficiency of all the components of the PV plants on the load side of the panels (inverter, connections, losses due to the temperature effect, losses due to dissymetries in the performances, losses due to shading and low solar radiation, losses due to reflection . . .). Such efficiency, in a plant properly designed and installed, may range from 0.75 to 0.85.

Instead, taking into consideration the average daily irradiation E_{mg}, to calculate the expected produced energy per year for each kWp:

$$E_p = E_{ma} \cdot 365 \ hBOS \ [kWh/kWp] \qquad (5.7)$$

Inclination and Orientation of the Panels

The maximum efficiency of a solar panel would be reached if the angle of incidence of solar rays were always 90°.

Outside the tropics, the sun cannot reach the zenith above the earth's surface, but it is at its highest point (depending on the latitude) with reference to the summer solstice in the northern hemisphere and in the winter solstice in the southern hemisphere. Therefore, if we wish to incline the panels so that they can be struck perpendicularly by the solar rays at noon of the longest day of the year, it is necessary to know the maximum height (in degrees) in which the sun reaches above the horizon in that instant.

Finding the complementary angle of α ($90° - \alpha$), it is possible to obtain the tilt angle β of the panels with respect to the horizontal plane (IEC/TS 61836) so that the panels are oriented perpendicularly to the solar rays.

However, it is not sufficient to know the angle α to determine the optimum orientation of the panels. It is also necessary to take into consideration the sun's path through the sky over the different

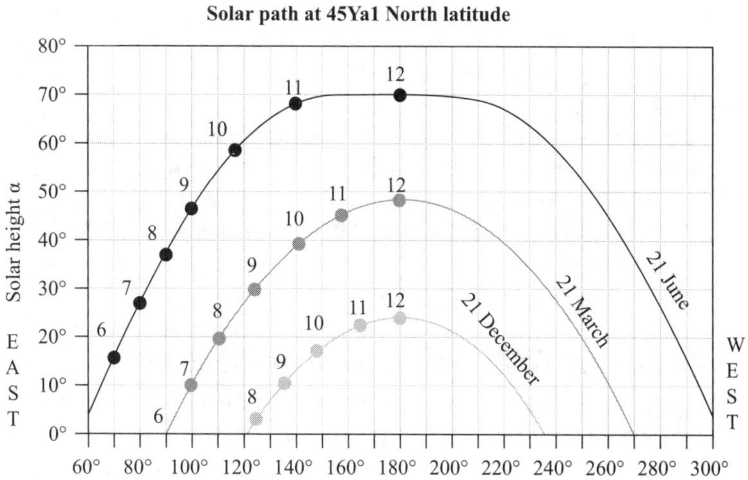

Solar path at 45Ya1 North latitude

FIGURE 5.20   The tilt angle should be calculated taking into consideration all the days of the year.

periods of the year, and therefore the tilt angle should be calculated taking into consideration all the days of the year (Fig. 5.20). This enables us to obtain an annual total radiation captured by the panels (and therefore the annual energy production) higher than that obtained under the previous irradiance condition perpendicular to the panels during the solstice.

The fixed panels should be oriented to the south as much as possible in the northern hemisphere, so as to get a better exposure of the panel surface at noon local hour and a better global daily exposure.

The orientation of the panels may be indicated with the *Azimuth5 angle* (γ) of deviation with respect to the optimum direction to the south (for the locations in the northern hemisphere) or to the north (for the locations in the southern hemisphere).

Voltages and Currents in a PV Plant

Typically, PV modules generate a current from 4 to 10 A at a voltage from 30 to 40 V.

In order to get the projected peak power, the panels are electrically connected in series to form the strings, which, in turn, are connected in parallel. The current trend is to build strings with as many panels as possible, given the complexity and cost of wiring, in particular of the paralleling switchboards between the strings. This also means that the string voltage increases in modern PV installations.

The maximum number of panels that can be connected in series (and therefore the highest reachable voltage) to form a string is determined by the operation range of the PE inverter and by the availability of the disconnecting and protection devices suitable for the voltage. It should be reminded that the strings produce DC voltage, which the PE inverter translates to the power frequency of the AC system.

The voltage of the inverter is related, due to reasons of efficiency, to its power: Generally, when using an inverter with power lower than 10 kW, the voltage range is from 250 to 750 V, whereas if the power of the inverter is greater than 10 kW, the voltage range is usually from 500 to 900 V.

Connection to the Grid and Measure of the Energy

A PV plant is allowed to connect to the public distribution network if the following conditions are complied with:

- The parallel connection shall not cause disruptions to the continuity and quality of the public network service and shall preserve the level of service for the other users connected.

- The PV plant must not be connected or the connection must be immediately and automatically interrupted in the absence of the supply voltage from the distribution network, or if the voltage magnitude and frequency of the network are not in the range of the allowed values.

- The production plant must not be connected or the connection to the system must be immediately and automatically interrupted if the phase unbalance of the power generated by three-phase plants consisting of single-phase generators exceeds the maximum value allowed by the system for single-phase connections.

This is in order to avoid that:

- In case of lack of voltage in the grid, the connected active (generating) users supply the grid.

- In case of fault on the MV line, the grid itself will be supplied by the PV plant connected to it (and feed the fault).

- In case of automatic or manual reclosing of the circuit breakers of the distribution network, the PV generator may be out of phase with the network voltage, which would likely damage the generator.

The PV plant can be connected to the LV, MV, or HV grid, depending on the generated peak power:

- Connection to the LV grid for plants up to 100 kW
- Connection to the MV grid for plants up to 6 MW

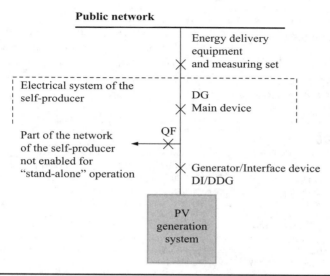

FIGURE 5.21   The generation system in parallel with the public network. (Guide CEI 82-25, II ed. (= second edition) CEI 82-25: Guide to the construction of photovoltaic generation systems connected to medium- and low-voltage electrical grids.)

In particular, the connection of the PV plant to the LV network:

- Can be single-phase for powers up to 6 kW
- Must be three-phase for powers higher than 6 kW and, if the inverters are single-phase, the maximum unbalance between the phases must not exceed 6 kW

The principle diagram of the layout of the generation system in parallel with the public network is shown in Fig. 5.21.

Conclusion

Renewable energy sources are, by nature, intermittent in their generation. However, many technological advancements in recent years, particularly within the realm of power electronics, have enabled these power sources to be connected to almost any power grid, whether at low-voltage, medium-voltage, or high-voltage levels. A mix of various renewable energy sources, as well as various energy storage initiatives, can help to offset the intermittency of solar and wind in particular. On a large scale, these renewables require power grid upgrades, particularly with regard to interconnectivity to transmission networks. Smart switching, bidirectional power flow, advanced metering, and dynamic reactive power compensation (Var), characteristics of a smart grid, are all enablers of renewable energy. Solar power will continue to grow, particularly closer to the loads. Onshore and offshore wind power will continue to be scaled up, utilizing large areas of land. With more and better transmission technologies in place to interconnect with this wind, this power will be much more easily accessible to load centers in metropolitan areas far away. With these technological breakthroughs and with the true emergence of the smart grid, the future is certainly bright for all aspects of renewable energy.

References

[1] H. Holttinen et al., "Impacts of large amounts of wind power on design and operation of power systems, results of IEA collaboration," in *Proc. 8th Int. Workshop on Large-Scale Integr. Wind Power Into Power Syst. Well as on Transm. Networks of Offshore Wind Farms*, Bremen, Oct. 14–15, 2009.

[2] *Annual Wind Industry Report 2009*, American Wind Energy Association. [Online]. Available: http://www.awea.org/publications/reports/AWEA-Annual-Wind-Report-2009.pdf.

[3] *EU's Sixth Framework Programme (FP6). Upscaling*, 2008. [Online]. Available: http://www.upwind.eu/Shared%20Documents/WP11%20-%20Publications/leaflets/080311_WP1B4%20final.pdf.

[4] Pacific Northwest National Laboratory, *Wide-Area Energy Storage and Management System to Balance Intermittent Resources in the Bonneville Power Administration and California ISO Control Areas*, Jun. 2008. [Online]. Available: http://www.pnl.gov/main/publications/external/technical_reports/PNNL-17574.pdf.

[5] U.S. Department of Energy, 20% Wind Energy by 2030, Jul. 2008.

[6] "Smart Grid: An Estimation of the Energy and CO2 Benefits," U.S. Dept. Energy, Pacific Northwest National Laboratory, PNNL Rep. 19112, Jan. 2010, Rev. 1.

[7] R. Wiser and M. Bolinger, B2009 Wind Technologies Market Report, U.S. Dept. Energy, Energy Efficiency and Renewable Energy, Aug. 2010.

[8] N. Janssens, G. Lambin, and N. Bragard, Active power control strategies of DFIG wind turbines,[in Proc. IEEE Power Tech 2007. [Online]. Available: http://www.labplan.ufsc.br/congressos/powertech07/papers/167.pdf.

[9] P. Maibach, A. Faulstich, M. Eichler, and S. Dewar, Full-Scale Medium-Voltage Converters for Wind Power Generators Up to 7 MVA. [Online]. Available: http://www05.abb.com/global/scot/scot232.nsf/veritydisplay/9847712acf892432c125740f003c3d65/$File/Full-Scale_MediumVoltage_Converters%20for_%20Wind_%20Power%20Generators_%20up_%20to_%207_%20MVA.pdf.

[10] M. Eichler, Offshore but online,[ABB Review 3/2008, pp. 56–61. [Online]. Available: www.abb.com/abbreview.

[11] National Renewable Energy Laboratory, Generation Interconnection Policies Ad Wind Power: A Discussion of Issues, Problems, and Potential Solutions, Jan. 2009.

[12] R. McDermott and G. Hassan, Investigation of Use of Higher AC Voltages on Offshore Wind Farms, Mar. 2009. [Online]. Available: http://www.gl-garrad-hassan.com/assets/technical/283_EWEC2009presentation.pdf.

[13] A. Sannino, P. Sandeberg, L. Stendius, and R. Gorner. (2008). Enabling the power of wind. ABB Rev. 3/2008, pp. 62–66. [Online]. Available: www.abb.com/abbreview.

[14] National Renewable Energy Laboratory, *Electrical Collection and Transmission Systems for Offshore Wind Power*, Mar. 2007.

[15] *California ISO (CAISO) Daily Renewables Watch*. [Online]. Available: http://www.caiso.com/green/renewrpt/20100422_DailyRenewablesWatch.pdf.

[16] E. Muljadi, C.P. Butterfield, R. Yinger, and H. Romanowitz, *Energy Storage and Reactive Power Compensator in a Large Wind Farm*, Oct. 2003. [Online]. Available: http://www.nrel.gov/docs/fy04osti/34701.pdf.

[17] Jeff St. John, "Cost of Big PV to the Grid: $3 to $8 per Megawatt Hour," Greentech Media, Jan. 5, 2012. Available: http://www.greentechmedia.com/articles/read/cost-of-big-pv-to-the-grid-3-to-8-per-megawatt-hour.

CHAPTER 6

The Microgrid in the Electric System Transformation

Steven Pullins
President, Horizon Energy Group

Internal and external factors have changed the U.S. electric system's future trajectory from a resource-centric model to a technology-centric model in order to survive. The business-as-usual approach is no longer sustainable; thus, a more intelligent, more distributed, more integrated system is required. With this comes complexity.

The issue for industrial firms, commercial parks, and universities is that the cost of the electric service is increasing and the reliability of the grid is decreasing. In addition, the transformation for reducing power plant emissions is progressing very slowly, which does not meet the "green" objectives of many commercial and industrial consumers and university campuses. A microgrid is one of the smart grid solutions to these problems. A microgrid is a small version of the main grid that has been customized to the industry complex or university campus with intelligent controls, optimization schemes, and generation resources to offset purchasing power from the utility.

Horizon Energy Group has found that a microgrid with its diverse resources can deliver a 10% to 15% reduction in the annual energy costs, improve reliability of the electric service, and reduce the emissions footprint of the campus [1]. Industry research firm Pike Research has estimated that 2000 microgrids will be built in the United States over the next several years, mainly at business and university sites [2].

141

What Is a Microgrid?

"A *microgrid* is a group of interconnected loads and distributed energy resources within clearly defined electrical boundaries that acts as a single controllable entity with respect to the grid. A microgrid can connect and disconnect from the grid to enable it to operate in both grid-connected or island mode."

DOE Microgrid Exchange Group, October 2010

Not all microgrids are islands or remote villages. Most microgrids (e.g., Fig. 6.1) will emerge to help consumers deliver a common set of community- or campus-based resources to optimize economic, reliability, and environmental objectives while connected to the main grid. Sometimes the microgrid will operate as an island if the objectives are challenged and the microgrid controller control functions determine that the microgrid will perform better as an island. In an island mode of operation, a microgrid is capable of providing the core electric needs of customers or facilities within the microgrid in the absence of electricity being supplied by the local utility. In the case of an unanticipated outage from the local utility, a microgrid will have the capability of performing "black start" operations to restore power supply within the microgrid.

As a smart grid solution, microgrids most closely address all the smart grid characteristics laid out by the U.S. Department of Energy

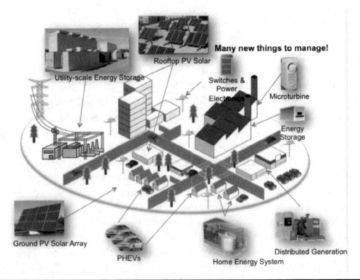

Figure 6.1 Community-based microgrid.
Source: Horizon Energy Group.

Characteristic	Retail Market	BAU Central Station Supply	DR, EE, Conserve	Clean Energy	VPP (DG Integration)	Microgrid
Enable active participation by consumers	X		X		X	X
Accommodate all generation and storage options				X	X	X
Enable new products, services, and markets	X		X		X	X
Provide power quality for the digital economy	X	X	X			X
Optimize asset utilization and operate efficiently	X		X		X	X
Anticipate & respond to system disturbances (self-heal)						X
Operate resiliently against attack and natural disaster						X

FIGURE 6.2   Mapping key grid strategies to smart grid characteristics.

(DOE) Modern Grid Strategy team [3]. See Fig. 6.2. The smart grid will

- Enable active participation by consumers
- Accommodate all generation and storage options
- Enable new products, services, and markets
- Provide power quality for the digital economy
- Optimize asset utilization and operate efficiently
- Anticipate and respond to system disturbances (self-heal)
- Operate resiliently against attack and natural disaster

When compared to other key strategies in tomorrow's grid, microgrids provide alternatives and flexibility to accommodate a changing marketplace and changing consumer.

Advantages of a Microgrid

Microgrids provide advantages to both the local utility and the microgrid consumers. Microgrid advantages can be realized by commercial and industrial (C&I) businesses, as well as university campuses that are mainly served by conventional electric and gas utilities but have a strong desire to improve their economic, reliability, and/or emissions base.

Microgrid advantages also accrue for utilities, where there is a need to meet high renewable energy obligations established by the state. While not always obvious to utilities, microgrids also improve economics and reliability for their electric service in general. The microgrid can optimize a "troublesome" business complex or university campus, making it more economic and reliable to serve, essentially transforming it from a reliability challenge to a strong point in the distribution network. However, utilities are not prepared for a complex shift in distribution design unless pushed by the high renewable energy obligations. Therefore, utilities will benefit from the economic and reliability advantages as a secondary driver. In general, the utility benefits include reduced congestion, improved reliability, improved power quality under conditions of high levels of renewable generation sources, and support of demand response.

The microgrid is proving to be the best way to increase the use of zero-cost fuel resources, namely wind and solar, at the distribution network and consumer level. The key benefits of the microgrid are as follows:

- **Savings**   The microgrid portfolio of resources is tuned to the university campus or business complex to provide economic savings through active participation in the electric market and the addition of renewable generation resources.

- **Sustainability**   The microgrid portfolio enables a hedge against traditional fuel cost increases.

- **Stewardship**   The microgrid enables a high penetration of renewable energy resources that reduce emissions and provide an opportunity for green marketing (corporate branding).

- **Reliability**   The microgrid converts the campus electrical network from a passively controlled system at the utility distribution level to an actively controlled and optimized system at the resource and load level, leading to better reliability.

These benefits are maximized with consumers who already have some energy resources (typically underutilized) and combined objectives. For example, a university campus may have a strong sustainability culture and the need to provide exceptional reliability of the electric service to sophisticated laboratories where experiments run for months or years. Power interruptions of such an experiment can render months of work meaningless. An industrial firm may have the need to reduce its cost of goods sold where energy costs play a major role, while at the same time reduce the total emissions footprint to meet local or state environmental standards.

Architecture and Design of a Microgrid

There are several architecturally significant elements of the microgrid that deliver the multiobjective needs of the consumer or utility. Multiobjective control and optimization is no small challenge. The normal operating environment can be complex as the microgrid optimizes the use of many resources and loads to meet the multiobjective environment. The upset, transient, and emergency operations can be even more complex [4].

Resource integration   The ease of integration and use of multiple resources with differing operational characteristics is a primary factor in the success of a microgrid. The optimal use of diesel generators, wind turbines, photovoltaic (PV) arrays, fuel cells, energy storage devices, etc., each with a unique operating characteristic, is a control system challenge. In addition, over time different resources will be added and subtracted from the microgrid creating the need for the microgrid to be scalable and adaptable.

Volt, var, variability (VVV) management   Distribution management systems today are mainly built to manage voltage and reactive power, but not manage the variability associated with renewable energy sources, especially when there is high penetration of these variable resources. The microgrid architecture recognizes the significance of integrating variability management into the voltage and reactive power management for the system. The common misconception is that renewable energy resources are not predictable and not dispatchable. The industry's traditional paradigm is that predictable means steady state and only steady-state resources are dispatchable. The microgrid is designed to maximize the penetration of renewable-generation resources and effectively manage those resources.

Control modes and signal design   Whether in a main grid-connected overlay mode or an island mode, the operations and risk profiles must be understood by the microgrid control system. Economic, reliability, and environmental operating objectives must be properly structured to make sense to the microgrid controls, resources, and loads, and communicated to these effectively. For example, is the driving signal to a load a price or a different voltage signal? Is the driving signal to the environmental optimization a CO_2 price, target CO_2 amount, or resource setpoint previously correlated to an emissions level?

Distributed control   The microgrid leverages the use of existing diesel generator controls, PV inverters, energy storage power conversion systems (PCS), etc., to leverage resource integration and to provide local protection of equipment. Some resources have sophisticated controls and some do not. The architecture should take advantage of the sophisticated resource controls and provide a default level of microgrid controls where the resource controls are not present.

Sharing information in real time (two-way)   The amount of information available in a microgrid is significantly more than currently available to consumers and utilities at the distribution level. The information available can range from overview types (instantaneous or average cost of operations, emissions to date, day-ahead resource dispatch plan, etc.) to operational types (hour-ahead estimated load and resource mix, grid device commands, etc.). This information needs to be shared with the appropriate stakeholders (consumer, utility, and marketplace) in an understandable and usable manner.

Market influences   The market provides more than just a price at a point at a time. The market information influences the purchase of power, the production of power, and the selling of power by the microgrid control space. The influence on the optimization will focus on the various operational objectives of the microgrid, such as energy arbitrage or providing regulation services to the wholesale or retail market. The dispatch decision is very dynamic (optimization of economic, reliability, and environmental dispatch signals) and is highly influenced by the real-time market and day-ahead market trends.

Grid overlay   It is important for the microgrid design to recognize that the microgrid mostly operates in parallel with the main grid somewhat like an overlay, but it also has the ability to seamlessly island (separate from the grid) and reconnect with the grid when the microgrid objectives dictate the need to support operations. This is more than a controls design requirement, but also a microgrid resource portfolio design requirement.

The design of the microgrid has two main considerations:

- Designing the microgrid resources and devices portfolio based on capital costs, operating costs, emissions, and grid stability
- Designing the microgrid controls for optimal performance of the stated objectives of the consumer or utility

Each main consideration has significant complexities as discussed next; however, there are some common threads in the design of the portfolio and the design of the controls.

First, the underlying theme is that the portfolio and the controls must support a real-time, constantly optimizing manner of performing. The upfront design and build-out of the resources portfolio must work in concert with the optimization schemes in the microgrid controls to deliver the multiobjective basis manifested in the economic, reliability, and environmental (ERE) dispatch:

- Reducing and flattening costs of energy over time
- Improving and holding electric reliability to a very high standard
- Minimizing the total emissions footprint of the campus or business

Second, the complexity of the microgrid is a departure from the existing manner of design and control of the distribution network. From a traditional distribution utility operations perspective, the variables for performance are fundamentally voltage, reactive power, and load. This can be simply managed as long as the balancing scheme is like a seesaw with load on one side and power sources and grid devices on the other side. The balancing scheme is to vary the power sources to accommodate the varying load while maintaining additional power resources running in standby to absorb other small fluctuations. Now, from a smart grid future perspective where consumers are more active and engaged, the variables for performance are not only voltage, reactive power, and load, but also variable cost, emissions footprint, use of embedded resources (sunk cost items), and quality of service (potentially variable if there is a cost advantage).

Finally, the value proposition for businesses and campuses moves from being homogenous to locally optimized. Today, the traditional distribution utility operations design the system, resources, and controls for homogenous reliability and cost of operations applied across all consumers, but that does not provide the best economics, reliability, and emissions footprint for a specific business or campus. The costs are spread across all the consumers in a homogenous way (sometimes for many years) through the state regulatory process. The result is a perceived fair and just approach to sharing the burden, but the business and campus are left with mismatches on economic, reliability, and environmental objectives. With a microgrid, a more direct relationship between these consumer objectives and the costs required to deliver them can be achieved through a specific resource portfolio design and microgrid controls design.

The general design of the microgrid starts with the specific consumer objectives, while the general design of the distribution network starts with the homogenized objectives of many varied consumers. This is a fundamental difference.

In general, the design of the microgrid takes into account the best portfolio of existing and new resources, grid devices, markets, and load management tools that can deliver the specific consumer objectives. In this case, those objectives are established by the university campus or C&I business. The best portfolio for a specific campus or business will be an optimum marriage of the following considerations:

- Desired investment period of payback or internal rate of return on the investment
- Resources for reducing emissions (wind, solar, biomass conversions, geothermal, etc.)
- Capital costs and operating costs of resources for reducing emissions (installed costs of each resource, fuel costs of each, typical operating and maintenance costs of each, utility and state incentives, tax benefits, etc.)
- Limitations on resources for reducing emissions (wind, sun, weather limitations, biomass availability, capacity factors, maintenance outages, etc.)
- Resources for baseload energy (coal, natural gas, fuel cells, geothermal, biomass conversions, etc.)
- Capital costs and operating costs of resources for baseload energy (installed costs of each resource, fuel costs of each, typical operating and maintenance costs of each, etc.)
- Limitations on resources for baseload energy (emissions, maintenance outages, capacity factors, etc.)
- Market transactions that can reduce annual cost of energy (buying wholesale energy at low market prices, selling excess energy production at high wholesale or retail market prices, providing regulating services, etc.)
- Capital costs and operating costs of enabling market transactions (certified market participation, physical access to the market, communications and controls, etc.)
- Limitations on market transactions (minimum dispatch size, access to wholesale or retail markets, etc.)
- Resources for improved reliability (storage, capacitors, regulators, remote enabled switches, relays, etc.)
- Capital cost and operating costs of resources for improved reliability (installed cost of each resource, cost of operation of each, typical operating and maintenance costs of each, etc.)

A design tool is typically used to build a financial model of the microgrid in accordance with the objectives established by the consumer and taking into account the operational characteristics of the generation resources and facility energy usage patterns. The financial model enables various scenarios to be explored in order to understand the relationship between the economics, reliability, emissions footprint, and the cost (capital and operations expense) as the relative value of the objectives are considered.

FIGURE 6.3
Microgrid master
controller design.
Source: Horizon
Energy Group.

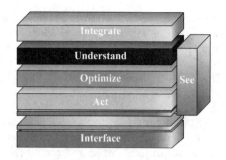

FIGURE 6.3
Microgrid master
controller design.
Source: Horizon
Energy Group.

The microgrid master controls design, recognizes the consumer objectives, portfolio of resources, devices, load management tools, and market influences, and then optimizes all these elements into a real-time solution that delivers the objectives in the best way possible. In addition, the microgrid master controller (MMC) will try to present the microgrid to the distribution network as a good citizen of the electrical network by becoming a firm, reliable node in the network and offering grid support services when needed by the network.

The MMC design has several layers (see Fig. 6.3) where algorithms reside to perform the various duties, computations, and comparisons necessary to deliver a reliable, optimal state of performance of the microgrid at any given moment in time. This organized structure is necessary to address the numerous computational efforts that are taking place in different time regimes. That is, interfacing with a running diesel generation resource requires a different time cycle than integrating with an off-board financial transaction software solution for monthly billing. The MMC's high-level architecture has the control layers as presented in Fig. 6.3:

Interface This layer provides the interface to all resources and devices in the microgrid network.

Act This layer provides the actions or decision-making steps that are needed in real time.

Optimize This layer provides the primary means of continuously optimizing the microgrid network for all inputs and variables.

Understand This layer provides the more computationally intensive decision services that support optimization but are not needed in real time.

Integrate This layer provides access to complex off-board engines that support the overall microgrid operations or network design but are not needed in real time. This includes market transactions to the local ISO or regional transmission operator (RTO).

See This layer provides visualization and presentment of key information to operators, management, maintenance, and customers. Some information is in real time while most is in near real time.

This design was derived from use cases and functional requirements developed by the Borrego Springs Microgrid team (San Diego Gas & Electric, Horizon Energy Group, Lockheed Martin, and Gridpoint) in consultation with multiple utility groups and others.

There are still emerging new design elements to build into the MMC in order to maximize system benefits, such as

- State estimation (day ahead). This is different from an RTO or ISO state estimator; here the focus is in a day-ahead time frame and also takes into account an unbalanced three-phase load flow environment.

- State measurement (real time). This is similar to an RTO/ISO state estimator, but with fewer nodes and specific well-defined resources. This must take place in a real-time control environment.

- ERE dispatch tools. This is a tool with complex requirements as outlined earlier.

- Objective functions and algorithms for optimization.

- Anticipatory and response/corrective algorithms.

- Energy arbitrage algorithms.

- MMC structural flexibility. Enterprise vs. distributed; brain vs. agent community.

Barriers

There are several key barriers to the introduction of the microgrid, but the work proceeds. Currently, there are at least 20 microgrids being built today in the United States that meet the DOE Microgrid Exchange Group definition and most of the design features discussed earlier. Plus, Pike Research reported recently that they are tracking more than 140 microgrid projects worldwide totaling over 1100 MW of capacity [5].

The key barriers include: Knowledge and understanding   Even with this upward trend in microgrid projects, utilities and consumers don't really know what is possible yet. There are simply too few operating microgrid examples with full performance metrics presented in the public literature. This means utilities and consumers will continue to proceed with caution or simply wait until the risks are better understood before proceeding with a microgrid.

Complexity   As discussed earlier, complexity in building and operating a microgrid is a natural barrier until sufficient operating microgrid examples exist to help utilities and consumers understand the risks and quantify the benefits. The complexity is not an inherent fault of the microgrid per se, but a limitation of the existing distribution network. If the U.S. distribution network was on par with those of leading countries, then the leap to a microgrid would not be as great. The intricacies of active control of the distribution network vice the existing passive control would be better understood and ease the transition. The incorporation of multiple resources of different technologies would be a well-experienced method of achieving electric system goals.

Funding Since microgrids are new applications, traditional equity and debt financiers are not ready to invest. There are too few projects operating for the equity and debt community to understand the risks and business model. This is similar to the early days of wind, solar, and geothermal project investment where institutional investors did not invest or fund debt until other equity investors had driven the risk out of the projects. Now, major institutional investors make many routine project investments in wind, solar, and geothermal projects.

Regulation A consumer-owned or third-party-owned microgrid challenges the utility/regulator hold on consumers, especially commercial, industrial, and university consumers. This is not a technical challenge because nearly all of the design and operating issues have been resolved by the RTOs and ISOs at the transmission level. Only scaling the technical solutions to the distribution and consumer levels remains as an integration challenge. Even the RTOs/ISOs recognize the need to engage the consumer-level load and resources. There have been several changes to the market participant rules at most RTOs/ISOs in the United States to encourage small consumers with small resources (down to 1 MW) to participate in the wholesale market. Thus, the key regulatory barrier is a matter of challenging the monopoly franchise grip on the consumer.

Over time the process of authorizing consumer electricity rates and cost recovery for utility projects has become a utility-regulator transaction with little influence from the actual consumers. These processes have been codified in legislative action at the state level, which breeds a sense that the regulator is the customer and the utility works for the regulator.

U.S. utilities are resisting the use of distributed resources on the grid, citing safety issues. The claim is that putting energy on the distribution network from consumer locations will cause a safety issue if there is a fault on the network. The analogy to this is a trucking company asking the state police to keep other drivers off the highway because most of the trucking company drivers are blind, and the other drivers might get hurt if they drive. Instead of correcting the blindness, keep the other drivers off the highway. Instead of upgrading the distribution network with sensors and controls to actively manage the network, the utility's argument is to prevent others from using the network.

Design tools   Current distribution design tools do not support incorporation of local distributed resources, load management programs, or market influences into the utility's design of the distribution network. Thus, the utility engineers are not enabled to create newer, more actively managed networks. Today, utility engineers design the electrical system to serve whatever peak load may exist, which leads to overcapacity designs. This trend has resulted in lower utilization of expensive generation, transmission, and distribution assets, which is ultimately reflected in higher costs to consumers.

Are Microgrids Significant to the Electric System?

As previously stated, Pike Research has described a rapidly growing microgrid market over the next several years. Horizon Energy Group research [1] shows that there are roughly 26,000 potential sites in the United States where a microgrid would be an advantage (technically and economically) over the existing method of energy supply to a community, campus, or business.

Clearly, if a significant number of small and mid-size cities, commercial complexes, industrial parks, and university campuses were to rapidly transform to a microgrid, the effects on the existing distribution network would not be known. This is an area where study needs to take place to understand the local and national implications. The industry needs to recognize that stopping the growth of consumer-initiated microgrid sites is not a solution; rather, the industry should embrace ways to incorporate microgrids for greater benefit.

As a way of comparison (see Fig. 6.4), consider a business as usual (BAU) approach to serving a 10 MW campus in the Midwest and a microgrid approach serving that same 10 MW campus.

10 MW Campus*	Business As Usual	Microgrid
Annual Electric Bill	$8.6 M	$6.9 M
Total Generation Capacity Supplying Campus	~22 MW	~14 MW
Percentage Renewables	1%	39%
Demand Management	4%	30%
Emissions Reduction (T/yr)	–	23,824
Reliability (SAIDI/SAIFI)	120/1.2	12/0.5

*Using Horizon Energy Group Microgrid Model

FIGURE 6.4   Ten MW campus microgrid comparisons.

The modeling shows that the microgrid can be tuned to flatten and reduce annual energy costs, reduce the emissions footprint, and improve reliability by an order of magnitude.

Conclusion

Microgrids are proving to be a powerful tool for operating and managing the electrical grid. However, it will take the next few years to see just how popular and effective microgrids will be for consumers and utilities.

It appears that microgrids will be more beneficial to consumers with certain characteristics, such as those associated with midsize industrial firms, larger commercial entities, and university campuses, especially those with significant renewable resource objectives.

There are several barriers to widespread use of microgrids stemming from microgrids being a nontraditional solution for the industry. However, there are a sufficient number of sizable microgrid projects underway that should provide a basis for clarification of risks and benefits such that widespread use of microgrids should happen where the business case can be made by consumers.

Another important lesson from the introduction of microgrids will be expanding the utility and RTO/ISO view of available resource options at the distribution and consumer level.

References

[1]   "The Role of Microgrids in the Electric System Transformation," S. Pullins, keynote at District Energy Systems & Microgrids Conference, Pace Energy and Climate Center, Nov. 2010.

[2]   "More than 2,000 Microgrids to Be Deployed by 2015," report press release, Pike Research, Jan. 2010.

[3]   "A Vision for the Smart Grid," DOE Modern Grid Strategy team, DOE National Energy Technology Laboratory, June 2009.

[4]   "Autonomous Microgrids: Revitalizing Distribution Company Relevance," Pullins, Bialek, Mayfield, and Weller, DistribuTech, 2008.

[5]   "Microgrids," Asmus, Davis, and Wheelock, Pike Research, 2010 http://www.pikeresearch.com/research/microgrids

Enhancing the Integration of Renewables in Radial Distribution Networks Through Smart Links

A. Gómez-Expósito, J. M. Maza-Ortega, E. Romero-Ramos, A. Marano-Marcolini
Department of Electrical Engineering, Universidad de Sevilla, Seville, Spain

Introduction

Distribution systems, and more specifically medium voltage (MV) networks, have not reached the automation level and technological sophistication of transmission systems. In fact, the introduction of the smart grid concept simply recognizes the urgent need and opportunity to improve the performance of distribution networks, in view of the new challenges they are facing nowadays [1].

MV systems, being structurally meshed, are however radially operated. Until recently, the absence of any generation at this level has invariably led energy to flow from the substation to distribution transformers. This significantly simplifies both the operation of these systems and the design of the protection equipment. Owing to the need of assuring back-up supply, radial feeders are customarily oversized so that they can serve, at least partially, the load of neighboring feeders in case of fault. In this context, the growing presence of dispersed generation (DG) is, on the one hand, reducing the spare feeder capacity, leading in many cases to congestions, and, on the other, complicating the operation and protection schemes. The use of power electronic devices is seen as a promising way of enhancing both asset utilization and massive DG penetration, much in the same way high voltage direct current (HVDC) systems were introduced decades ago at the transmission level. The continuous improvement in the areas of medium-voltage power conversion systems ([2]–[4]) encourages this trend even more.

The introduction of power electronic devices to control power flows in real time affects both the operation and planning activities of distribution utilities. On the one hand, regarding the distribution system operation, it is important to note that most distribution systems are designed to provide power radially (in one direction) from the substation to the loads via arborescent feeders. As recently shown, power electronic devices could interconnect some of these feeders to improve the operation by power flow control. For example, a cascaded-type back-to-back (BTB) converter is used in [5] to obtain a flexible bridge between feeders. An optimization problem is proposed in order to simultaneously minimize ohmic losses and regulate voltages. The same objectives are pursued in [6], where use is made of the so-called unified power flow controller (UPFC). Two different schemes are proposed to control the UPFC series and shunt converters, tests being performed on a rather simple laboratory model of the distribution system. In [7], [8], cascaded BTB converters are used to mesh the MV network, and the series-type BTB converters installed between two feeders are considered to control the voltage profile, minimize losses, and balance power flows in substation transformers. Simulations are performed on small and simple distribution systems, focusing on the control strategies of the power electronic device. References [9]–[11] are mainly aimed at integrating dispersed resources to also improve the system operation. On a distribution network comprising DC links, the authors of [9] determine the optimal operating state from the point of view of the exploitation cost. The objective is to obtain the best proportion of DG to conventional power generation for specific scenarios. In [10], the series controller is used to allow for higher DG penetration levels and increased loading. A quite simple model is adopted for the power electronic converter. The same authors propose in [11] a more general control scheme of the power electronic converter in order to regulate voltages, a gradient method being used to solve the optimization problem. Only one control method for the power electronic converter is tested on a rather ideal system.

On the other hand, a key issue in distribution system planning is to minimize the required investment in order to match all the growing demand in a reliable manner, while integrating as much renewable DG as possible. This last consideration is a new constraint that distribution utilities must face nowadays, as renewable generation, such as photovoltaic (PV) farms, is being promoted by most governments all over the world owing to environmental and strategic reasons. Indeed, the introduction of DG in MV networks can bring about some well-known benefits, but a massive indiscriminate integration, which is not preceded by a careful technical analysis, could negatively affect the distribution business. In this context, the use of well-established power converters in distribution networks, such as DC links, capable of regulating power flows and voltage magnitudes, provides new means for planning engineers to solve the congestion problems they are facing right now in radially operated networks. In many cases, the cost of the required power devices will be a small fraction of the total investment associated with PV farms, typically ranging from 2 to 5 MW at these voltage levels. But the introduction of a new technology in a traditional business, such as the electrical distribution, is far from being quickly and widely accepted. Therefore, any tool intended to clearly show the payback and related technical benefits associated with DC links should be most welcome.

This work is organized as follows: The next section first introduces the applications and benefits that the use of DC links can report to distribution utilities. The third section describes the BTB voltage source converter (VSC) topology, by far the most frequently found in this type of applications. The fourth section details a general optimization framework, which can be useful for planning and operational issues. This is tested on the benchmark distribution system proposed by the CIGRE Task Force C6.04.02. Finally, the main conclusions of this work are summarized.

DC Links in Radial Distribution Networks

Figure 7.1 shows the typical structure of radial distribution systems in urban areas. Only three feeders of two neighbor substations have been represented in detail, while the remaining ones are shown as concentrated spot loads for the sake of simplicity (note that more than two substations could be involved in the general case). Each feeder reaches a certain number of secondary distribution transformers where dispersed generators and consumers are connected to. In this case, the three feeders can be interconnected to each other at the remote end through suitable mechanical switches placed at a switching substation. Those tie-switches are normally open to force the radial system topology but could be closed in case of failure of a feeder section located upstream. In its simplest form, the switching

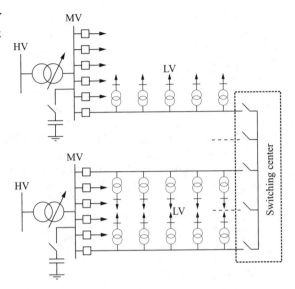

FIGURE **7.1** Typical distribution network topology.

substation reduces to a single normally open tie-switch between two neighbor feeders.

Much more flexibility could be achieved by installing DC links at the switching substation, simply replacing or in parallel with existing mechanical switches. DC links can be built around different configurations and the one considered here is based on the so-called VSC. Figure 7.2 shows a BTB device consisting of two VSCs sharing a common DC bus. In a more general case, n feeders can be connected through a compound DC link device composed of n VSCs sharing the same DC bus. For example, Fig. 7.3 represents a schematic three-terminal DC link that could replace the three mechanical switches in the switching substation of Fig. 7.1.

Bridges based on DC links provide new supply points among adjacent feeders and, consequently, a flexible meshed topology is established *de facto*. A set of two- or multi-VSC links properly located introduces additional degrees of freedom to the network operation. The simplest DC link based on two BTB VSCs is capable of controlling its active power flow along with the reactive power injected at both terminals (the remaining degree of freedom is used by the control system to keep the DC capacitor voltage within acceptable limits [4]). In the general case, a multi-VSC link connecting n feeders should be

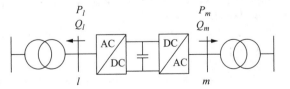

FIGURE 7.2 Back-to-back VSC.

FIGURE 7.3
Schematic diagram
of a three-leg
VSC-based DC link.

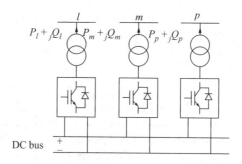

capable of controlling $n - 1$ active power flows and n reactive power injections. Note that the active power flows through all interconnected feeders are not independent variables, as the sum of all active powers should equal zero (ignoring power losses for simplicity) in order to keep the voltage of the DC link close to its reference value. The most significant advantages provided by these flexible "bridges," in opposition to mechanical switches, are

- The DC link is able to regulate in a continuous manner the power flow among the respective feeders, no matter the load and/or the DG penetration levels. Moreover, it is possible to reverse the "natural" power flows, i.e., the resulting power flows in case the loops were formed using null-impedance switches.

- The use of DC links based on VSC technology allows reactive power to be independently injected at both AC terminal nodes. This particular feature leads to a really interesting voltage control capability, taking into account that tie-switches are typically located at the remote end of feeders, where the voltage drop is larger for conventional passive feeders and the risk of voltage rise is higher in the presence of DG.

- The VSC technology can be used to mitigate voltage imbalances and, when typical commutation frequencies (1 to 2 kHz) are adopted, also low-order harmonics, improving in this way the power quality of the distribution network.

- The DC link can be used to connect any group of feeders regardless of the angular difference among them (this may be a crucial factor when the feeders are supplied from different substations). Moreover, it should be feasible to connect feeders with different rated voltages, which is not the case with mechanical switches.

- Existing short-circuit currents are not modified when adding multi-VSC links, due to their almost instantaneous current control capacity. This is a significant advantage to consider as the use of this technology does not involve any change of existing protecting devices.

Furthermore, it is possible for energy sources that are "naturally" DC generators, such as PV systems, fuel cells, or even storage devices, to be directly connected to the VSC DC bus. This implies the following are additional advantages:

- The power injected to the DC bus is delivered by the multi-VSC link to the set of connected feeders. The way such a power is split among all feeders can be dynamically controlled as required by the VSCs, depending on the system state.

- If one feeder is out of service, the remaining ones can be used to evacuate the generated power.

- This arrangement avoids the installation of additional transformers, inverters, and switchgears that would be needed in case the DC source had to be connected to the AC network. Note that not only is the investment reduced but also the total power losses.

However, this qualitative characterization is not enough to fully assess the technical and economic benefits derived from the use of this technology. The rest of this chapter provides a tool intended to quantify in a systematic manner the advantages that utilities and DG owners can get by using the additional network resources related to the presence of VSC-based DC links.

Back-to-Back VSCs Topology

The BTB VSC shown in Fig. 7.2 is the common topology employed to link distribution feeders due to its inherent simplicity and the fact that the short-circuit power is not increased, as explained before. The use of transformers is convenient in order to adjust the voltage of the distribution network to the rating of semiconductor devices. However, transformerless topologies have been proposed in the specialized literature [12]. A qualitative comparison between both alternatives can be done by attending to the following issues:

- *Space requirements* are lower in case of transformerless topologies. This point has to be taken into account for urban networks where usually switching centers are of reduced dimensions.

- *Galvanic isolation* is only achieved when a connection through a transformer is used. The main advantages are that a ground fault on the DC bus does not affect the distribution system and that any DC term of the VSC current, caused by a misoperation of the converter, is isolated from the AC network.

- *The VSC topology*, when connected through transformers, can be conventional and matured two-level or three-level structures,

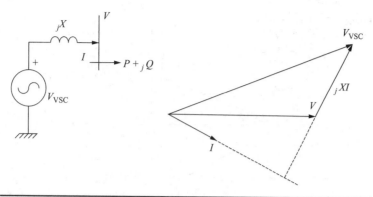

Figure 7.4 VSC single-phase steady-state model and its related phasor diagram.

depending mainly on the rated power. However, the use of innovative multilevel topologies is mandatory when using the transformerless option.

- *Standarization and feeder coupling flexibility* can be achieved when using the connection through a transformer. A standard design of the rated voltage of the BTB VSCs can be done regardless of the network voltage because the coupling transformer adapts the voltage levels. Finally, the use of coupling transformers is the only option to link distribution networks of rather different voltages.

The analysis of the operating characteristic of the BTB topology is required to perform a correct management of this new network asset. For this purpose the single-phase steady-state model of one VSC side and its related phasor diagram are shown in Fig. 7.4. The model comprises an AC voltage source, which is a function of the DC bus voltage and the modulation signals, and a coupling reactance needed to control the injected current to the distribution system.

Note that for a given network voltage it is possible to adjust the power flow depending on the value of the VSC voltage as:

$$P = \frac{V V_{VSC}}{X} \sin \delta \qquad (7.1)$$

$$Q = \frac{V V_{VSC}}{X} \cos \delta - \frac{V^2}{X} \qquad (7.2)$$

The active and reactive power locus can be easily obtained if the phasor diagram shown in Fig. 7.4 is multiplied by V/X as shown in Fig. 7.5. The operational limits of the VSC, given by the rated voltage and current values, are represented by dotted circles in this diagram.

FIGURE 7.5 VSC
PQ phasor diagram.

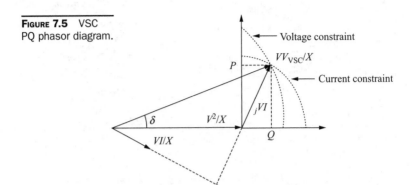

The active and reactive power limits for a given network voltage can be easily obtained using (1)-(4) and this diagram:

$$P_{max} = VI_{VSC}^{rat} \tag{7.3}$$

$$Q_{max} = \frac{VV_{VSC}^{rat}}{X} - \frac{V^2}{X} \tag{7.4}$$

The maximum active power is limited by the rated current of the VSC, I_{VSC}^{rat}, while the reactive power is limited by its rated voltage, V_{VSC}^{rat}. Therefore, if each VSC of the asynchronous link is dimensioned to provide the apparent power S_{max}, irrespective of the power factor at the rated network voltage, V^{rat}, the resulting operational limits are as shown in Fig. 7.6.

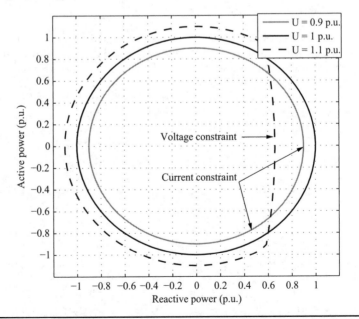

FIGURE 7.6 Operational limits of the VSC depending on the network voltage.

FIGURE 7.7
Reactive power
limit when the
voltage is higher
than V^{rat}.

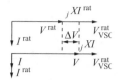

In this figure, the circumferences with center in the origin of the PQ diagram correspond to the rated current constraint. For the rated network voltage it is possible to establish any power flow below S_{max}. However, if the AC voltage is lower than the rated value, the maximum PQ locus is reduced because of the rated current limit. On the contrary, if the AC voltage is higher than the rated voltage, the maximum active power increases but the reactive power injected to the distribution network gets reduced due to the voltage constraint, which can be explained with the help of the phasor diagram in Fig. 7.7. Note that the VSC rated voltage, V_{VSC}^{rat}, is defined in terms of the rated network voltage, V^{rat}, and rated current, I^{rat}, for the case of pure reactive power injection, as this is the worst case (any other power factor leads to a lower VSC voltage). However, if the network voltage increases ΔV, the reactive current injection has to be reduced as the voltage limit of the VSC is V_{VSC}^{rat}.

Once the operational limits for one of the BTB VSCs have been studied it is straightforward to include both sides into the analysis. Assuming a lossless model, the active power of both VSC sides must be equal in steady state to maintain the DC bus voltage. Therefore, given the active power, the reactive power range of each VSC depends on the network voltage as shown in Fig. 7.8 for $V_m = 1$ p.u. and $V_n = 1.1$ p.u.

As the active power is a common variable for both VSCs, a three-dimensional representation of the operating region of the BTB VSCs comprising P, Q_l, and Q_m is possible, as shown in Fig. 7.9. The volume

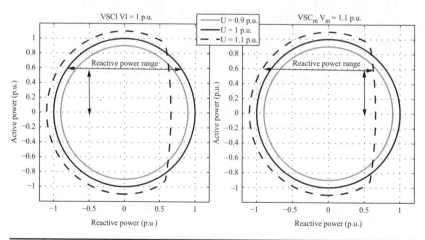

FIGURE 7.8   BTB VSCs reactive operating range for $V_l = 1$ p.u. and $V_m = 1.1$ p.u.

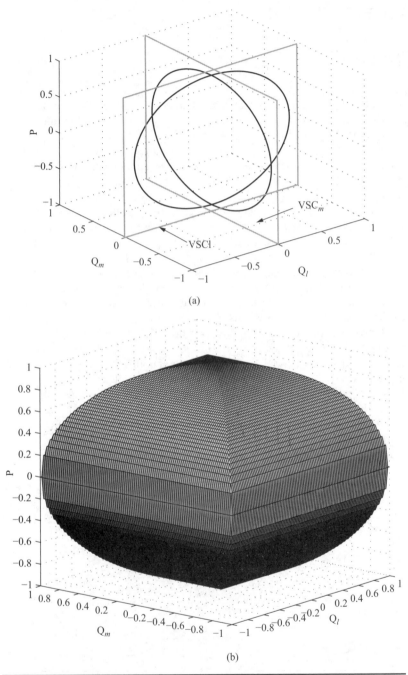

FIGURE 7.9 BTB VSCs three-dimensional operating region. (a) VSC PQ planes with common active power axis. (b) Feasible operating points.

within the represented surface contains the feasible operating points of the BTB VSCs for a given network voltage.

General Optimization Framework

Determining the optimal number, placement, sizing, and topology (including number of legs) of DC links, constitutes an ambitious planning issue whose rigorous formulation would lead to a hardly tractable mixed-integer, nonconvex problem. A simpler heuristic procedure is rather suggested in [14], which based on initial assumptions for a given set of DC links (rating, locations, topologies), repeatedly solves a conventional nonlinear optimization problem. This is intended to assess if the assumed set of links is economically justified to deal with load growth, DG penetration, or power loss reduction, when compared to the base case.

The formulation of any optimization problem consists of a scalar objective function f to minimize or maximize, along with equality and/or inequality constraints that must be enforced during the solution process. The mathematical problem can be formulated in a general way as,

$$
\begin{aligned}
\min \quad & f(x, u) \\
\text{s.t.} \quad & g(x, u) = 0 \\
& h(x, u) \leq 0
\end{aligned}
\tag{7.5}
$$

where vector x comprises the dependent or state variables and vector u the control variables. In this work, the complex bus voltages are the dependent variables, that is, the voltage magnitude and angle at each node, V_i and θ_i. The control vector u comprises the independent variables associated with the VSC-based DC links and, depending on the planning scenario, the remaining variables described later.

The following subsections describe the different issues related to the optimization problem, namely the objective function, the control variables, the detailed model of the DC link including losses, and the constraints imposed by the distribution network.

Objective Function

The following objective function constitutes the key common component of the optimization framework proposed in [14] to assess the benefits of DC links in active distribution networks:

$$
f(x) = \sum_{k=1}^{k=n_s} P_k^s
\tag{7.6}
$$

where n_s is the total number of distribution substations interconnected at the MV level by DC links and P_k^s is the total active power supplied through the distribution substation k to the set of radial feeders.

Note that the sum of the active power injected through the distribution substations to the respective feeders can be decomposed into three terms: load demand, DG injection, and system losses. As a consequence, Eq. 7.6 can be also rewritten as

$$f(x) = P_{\text{loss}} + \sum_{i=1}^{i=n} \left(\lambda P_i^l - P_i^g \right) \tag{7.7}$$

where P_i^l and P_i^g are respectively the active powers of consumers and distributed generators connected to bus i, n is the total number of buses for the entire set of feeders connected to the n_s substations, P_{loss} represents the total active power losses of the system, and λ is a variable (or constant parameter, depending on the scenario) intended to model the rate of load growth.

Taking into account Eq. 7.7, three figures of merit can be considered to quantify the benefits arising from the use of DC links:

- *Maximizing the penetration of DG.* For a given level of demand (constant λ), the goal is to increase as much as possible the amount of DG connected to the entire set of feeders. In those candidate buses (usually very few) where DG is to be connected, P_i^g is a variable (in the remaining buses $P_i^g = 0$). It can be easily shown that this problem is equivalent to minimizing $f(x)$ or maximizing $-f(x)$. Indeed, when the DG is maximized, the amount of power received through the substation to feed the existing load is minimized.

- *Maximizing the load growth.* Another interesting planning objective lies in the maximization of the load that can be served by the entire set of feeders for a given level of DG penetration (constant P_i^g). Although a single λ variable has been assumed in this work (homogeneous load growth), several λ's (for instance one per substation or even feeder) could be easily accommodated. In this case maximizing $f(x)$, or minimizing $-f(x)$, matches the planning objective.

- *Minimizing power losses.* Finally, an objective worth considering, perhaps more interesting in real-time operation than in planning environments, lies in reducing the active power losses for the entire set of feeders. This is achieved by simply minimizing $f(x)$ when both λ and P_i^g remain constant, the only control variables being those of the DC links. Note that this problem is closely related to the one traditionally faced by planning engineers, in which the best radial configuration is sought that minimizes power losses. The main difference lies in the binary (on/off) switching variables of the conventional problem being replaced in this case by the continuous power flow variables of DC links.

Once the case of interest is defined, the user can run the resulting optimization problem twice, with and without DC links. Then, the difference in DG penetration, load growth, etc., between both solutions will be due exclusively to the control capabilities of DC links. The cost of DC links could be eventually compared with the investment associated with conventional network reinforcement measures, for the same network scenario, in order to determine the real value of power converters.

DC Link Model

In order to simplify the presentation, a three-leg DC link connecting three feeders is considered (Fig. 7.3), the extension to the general DC link connecting n feeders being trivial. As mentioned previously, the configuration of the DC link is based on the pulse width modulation (PWM) VSC device. Furthermore, the most economical topology for the VSCs has been considered, namely the six-pulse two-level bridge.

When the AC sides of this converter are connected to three feeders via transformers, the power flow through the link can be adjusted to the operating requirements in all directions. When no storage device or active source is connected to the DC bus, the sum of the active powers extended to the three-branch cut-set must equal the converter power losses. Following the notation shown in Fig. 7.3, the mathematical formulation for this condition is

$$\sum_{j=p,l,m} \left(P_j + P_j^{\text{loss}} \right) = 0 \qquad (7.8)$$

where P_j^{loss} is the power losses associated with the j-th VSC.

Internal losses of the switching elements in the VSC, insulated gate bipolar transistors (IGBTs) and diodes, are quantified in P_j^{loss}. Both switching and conduction losses are considered. The model used in this work is the one proposed in [13], where a quadratic polynomial equation is used to approximate power losses. The loss equation formula for each IGBT-diode pair is

$$P_{\text{loss}} = aI^2 + bI + c \qquad (7.9)$$

where constants a, b, and c depend on the IGBT's manufacturer, and I is the root mean square (RMS) current through each IGBT. A three-MVA VSC, based on one of the most powerful existing IGBTs, is considered. Notice that Eq. 7.9 represents the loss of each IGBT in the VSC, six in total for the adopted two-level bridge.

Additionally, the self-commutating technology allows the reactive power absorbed/generated by each VSC to be independently controlled, providing an extra resource for enhancing the voltage profile. Consequently, the three-leg VSC-based DC link contributes with five new control variables, namely: reactive power injected at each AC bus of the converter, Q_l, Q_p, and Q_m, and two out of the three active power injections, P_l, P_p, and P_m (the third active power is determined by the power balance equation).

Network Constraints

Equality constraints included in the optimization problem are as follows:

1. Power balance equations at each bus:

$$P_i^g - P_i^l = V_i \sum_j (V_j G_{ij} \cos\theta_{ij} + V_j B_{ij} \sin\theta_{ij}) \qquad (7.10)$$

$$Q_i^g - Q_i^l = V_i \sum_j (V_j G_{ij} \sin\theta_{ij} - V_j B_{ij} \cos\theta_{ij}) \qquad (7.11)$$

where $P_i^g + jQ_i^g$ is the total complex power injected by dispersed generators connected to bus i, $P_i^l + jQ_i^l$ is the total complex load demanded at bus i, and $G_{ij} + jB_{ij}$ is the ij-element of the bus admittance matrix.

2. Voltage constraints at the main substations: HV/MV transformers at distribution substations are equipped with automatic tap changers to keep the voltage magnitudes at the head of feeders to a constant value. In consequence, Eqs. 7.10 and 7.11 are replaced by $V_i = V_i^{sp}$ for each node i, $i = 1, \ldots, n_s$. In this work it is assumed that the phase angle differences among buses of distribution substations are mainly determined by the HV grid, the effect of the load and/or DG increments downstream being neglected (if necessary, a more accurate model could be devised by adding an external equivalent of the HV grid linking all involved substations).

3. The DC link active power balance in Eq. 7.8 should be added if the three active powers are included as explicit variables.

4. Throughout this work it has been assumed that DG plants operate at unity power factor, as PV farms usually do.

Inequality constraints are defined as follows:

1. Conductor ampacity: thermal limit of overhead or underground lines:

$$0 \leq I_{ij} \leq I_{ij}^{max} \qquad (7.12)$$

2. Voltage magnitude limits:

$$V_i^{min} \leq V_i \leq V_i^{max} \qquad (7.13)$$

3. Converter capacity: the apparent power through each VSC cannot exceed its rated power, S_{max},

$$0 \leq S_j \leq S_{max} \qquad j = l, m, p \qquad (7.14)$$

The total apparent power associated with each converter must include the power losses defined by Eq. 7.9.

4. DG physical limits: Depending on the technology, land availability, etc., a maximum power could be eventually imposed to any DG power resource:

$$0 \le P_i^g \le P_i^{max} \qquad (7.15)$$

Results

This section is devoted to demonstrating the benefits of using DC links in distribution networks. It is structured in three subsections, one for each objective function (ohmic losses, loadability, and DG penetration).

The distribution system that will be used in the sequel is shown in Fig. 7.10. It is a simplified version of an actual MV rural network of Germany proposed by the CIGRE Task Force C6.04.02. The complete parameters of the line, as well as the daily load and DG profiles, are fully given in [15]. The 20-kV distribution network is supplied from a 110-kV substation, providing power to a nearby rural area and a small town with industrial and residential loads. The substation feeds two subnetworks, each through a 110/20 kV, 20-MVA transformer, with a total length of 15 and 8 km, respectively. This network includes 14 nodes where industrial and household loads are connected to. In addition to the loads, several DGs are also scattered over the feeders.

In the initial or base-case configuration the subnetworks are operated radially (the tie-switches between nodes 4–11 and 6–7 are open). The optimization problem is run by assuming that a three-leg BTB converter is located among nodes 6, 8, and 14. VSCs share a single DC bus and are connected through 20 kV/1100 V transformers of appropriate rated power.

The original network [15] is modified as follows: A new branch is added between node 6 and the third VSC terminal node, which becomes bus 15. In addition, the loads at buses 8 and 6 are modified: A large induction motor absorbing a constant 500-kW load is connected to bus 8, while the load of bus 6 is increased by 3 MVA keeping constant the original power factor. Finally a DC load is added on the DC side of the VSC, representing the future integration of electric vehicles in the distribution system. In this case it is assumed that the charger is directly connected to the DC bus through a DC/DC converter. The maximum power available for the electric car is 500 kW, allowing two cars to be charged in fast mode (250 kW each) or 10 cars in slow mode (50 kW).

Use of DC Links for Loss Reduction

The benefits of using a multiterminal DC link in steady-state operation are assessed for a whole day. The main objective is to compare the system losses and the voltages at each bus with and without the DC

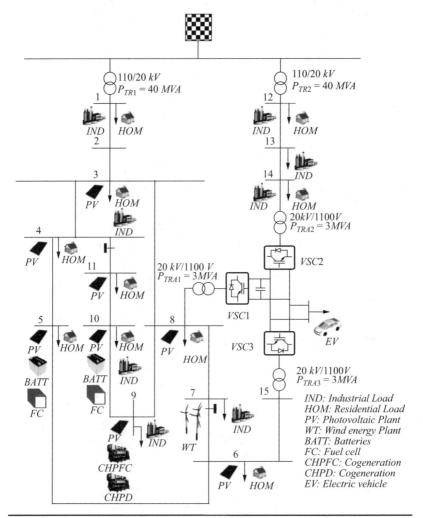

Figure 7.10 Layout of the distribution system used for the simulations.

link device. The losses and bus voltages are obtained at intervals of 5 minutes using a power flow algorithm. When using the DC link the previous optimization model provides the optimal set-points for the active power and voltages of each VSC. Then a power flow provides again the system state in the presence of these extra constraints.

Figure 7.11 shows a significant reduction of power losses when the three-leg VSC is active. The total losses for a typical day are 6680 kWh in the base case and 3465 kWh when the DC link is in operation (note that the latter figure takes also into account the three VSCs' losses). The energy saved represents almost half the total losses in the base case.

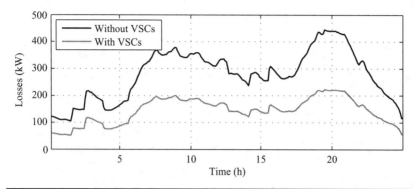

FIGURE 7.11 Comparison of network losses with and without VSCs.

To achieve these results, both the active and reactive power capabilities of the VSCs are used. Usually, the active power through the DC link device flows from the saturated feeders to the less loaded ones, thus balancing the currents throughout the network. This statement can be verified in Fig. 7.12. The leg VSC2 absorbs active power throughout the day, which is injected to the network by VSC1 and VSC3, both connected to less stressed feeders.

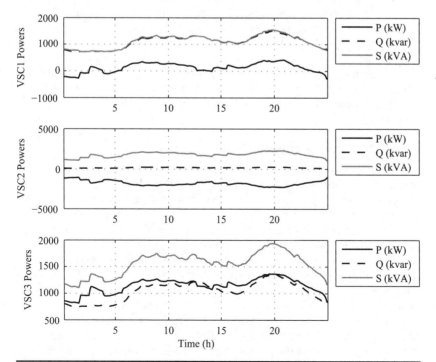

FIGURE 7.12 Active and reactive power injected by each VSC when the objective is loss minimization.

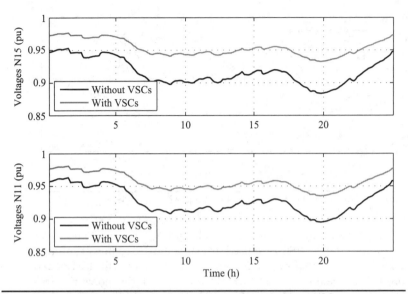

Figure 7.13   Evolution of voltages in two buses with and without VSCs.

It can be noticed in Fig. 7.13 that the DC link has an evident positive effect on the system voltages, keeping their values inside the feasible operating margin established by the grid code. Buses 11 and 15 are shown in the figure because they are most prone to suffer lower voltages according to the topology of the network and the loading conditions. Increasing the voltage at buses with lower values requires that the VSCs inject reactive power at their terminal buses, as shown in Fig. 7.12, which is directly related to the power loss reduction reported earlier.

Use of DC Links for Increasing the Load Level

Given the base-case load and DG daily patterns, the maximum network loadability is obtained for each scenario by finding the scaling factor λ, as described in the previous section. The loading level that the system can withstand can be improved using the VSCs. In this case, the network is constrained mainly by the line ampacities, and to a lesser extent by the low-voltage values reached at the end of the feeders. The DC link has a positive impact regarding both issues: on the one hand, it can shift active power from saturated feeders to the adjacent ones and, on the other, it can inject reactive power to increase the voltages of nearby buses (generally at the remote end of feeders).

Figure 7.14 compares the resulting λ values every 5 minutes for a whole day, with and without the DC link. Obviously, the maximum load increment depends on the initial loading condition, which explains why λ is much larger during valley hours in both scenarios. The minimum λ value in this system arises at about 8 P.M., which is

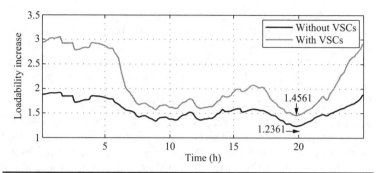

FIGURE 7.14 Comparison of the maximum loadability reached with and without VSCs.

also the moment when the VSCs are less influential from this point of view. Without VSCs the system load can be increased by 23.6% (beyond this load level a certain line of the system becomes congested). When the DC link is in operation, the maximum loadability increment at the same time is 45.6%. The loadability enhancement is even better for the rest of the day.

In this mode of operation the VSCs' active and reactive powers evolve as illustrated in Fig. 7.15. It can be noticed that the amount of active power that is transferred among the "bridged" feeders plays

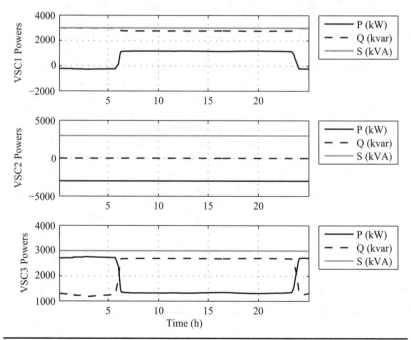

FIGURE 7.15 Active and reactive power through the VSCs when the objective is to increase the loadability.

an important role, the saturation of lines being the most important limiting factor in this case. The VSC2 always draws 3 MW, which are then injected according to the loading conditions by both VSC1 and VSC3 to their respective feeders. All VSCs inject reactive power to their connection buses to the extent allowed by their rated power.

Use of DC Links for Increasing the DG Penetration

The third possible goal of the DC link analyzed in this chapter is to increase the penetration of the DG. The test system used has several points of injection of active power from different generation technologies. It has been assumed that only renewable DG (photovoltaic and wind) may change, since the electrical power of the CHP units is related to the thermal load demanded by the industrial process and, therefore, cannot be modified. As in the previous case, two operational limits constraint this objective, namely the line ampacities and the voltage bounds. Figure 7.16 depicts the possible increase in DG with and without the DC link, along with the difference between these two quantities. It is remarkable that the increase is almost insignificant most of the day, with the exception of the interval from 10:00 to 15:00. This result is a consequence of the time evolution of the power injected by the photovoltaic and wind DG plants. The maximum power injected by the photovoltaic plants is produced mainly at that interval of time, being almost null in other hours. On the other hand, in spite of the more constant active power profile corresponding to the wind generator, the increment of DG penetration in this case is limited by the saturation of branch 7-8 (see Fig. 7.10). Figure 7.17 shows the active and reactive powers for each VSC. Once again, the three devices are working at their rated power almost the whole day, the difference with respect to the previous case being that the relative impact of active and reactive power injections varies along the day. Note that the VSC1 changes the sign of its active power based on the production of the photovoltaic DG. In absence of photovoltaic generation the VSC1 feeds active power to its terminal node. Otherwise, when

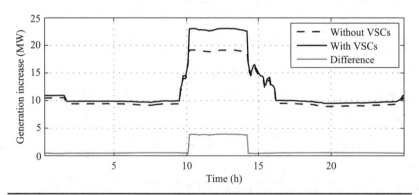

Figure 7.16 Comparison of DG penetration with and without VSCs.

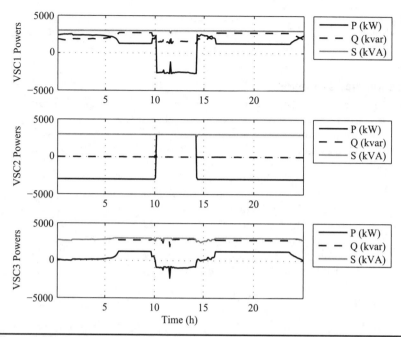

FIGURE 7.17 Active and reactive power through the VSCs when the objective is to increase the DG penetration.

the photovoltaic DG unit is injecting active power (from 10:00 to 15:00), the VSC1 partially draws this active power, which is injected back to bus 14 through the VSC2.

A careful analysis of this scenario shows that determining the optimal location of the DC link is not an easy task, as the resulting benefits are clearly affected by this choice. As stated earlier, the DG penetration is mainly limited in this case by the congestion of branch 7-8, which cannot be alleviated by the VSC1. As shown in Fig. 7.18,

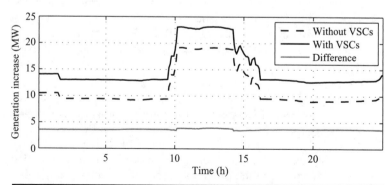

FIGURE 7.18 Comparison of the DG penetration with and without VSCs when the VSC1 is connected to bus 8.

better results can be obtained if this converter was rather connected to bus 8, enhancing in this way the capability to evacuate the active power produced by the wind-based DG.

Economic Assessment

Once the technical benefits that can be achieved using DC links in distribution systems have been reported, it is necessary to evaluate the economic viability of this proposal. The achieved benefits of using DC links depend on the specific aforementioned application as follows:

- Loss reduction. This application clearly represents a benefit for the utility. However, the use of a DC link involves an important investment that has to be justified. For this purpose, a rough economic study based on a simplified payback analysis is performed. The payback related to a candidate investment can be defined as follows:

$$PB = \frac{IC}{AI} \tag{7.16}$$

where PB is the payback ratio, in years, IC is the investment cost, and AI are the additional yearly incomes due to the investment. The additional income for the utility in this case is the cost related to the saved losses and the investment cost depends on the rated power of the installed DC link.

- Loadability increase. In this case the DC-link investment has to be compared with the traditional methods employed to reinforce the distribution network, such as increasing the ampacity of the feeder or adding new lines and/or substations. These strategies depend on the utility planning criteria and other factors such as the type of the distribution network (urban or rural), the expected load growth, power quality, and other regulatory issues that may affect the adopted decision for the network expansion. A fair economical assessment of the use of DC links for this purpose would have to consider all these factors, which are out of the scope of this work.

- Increase of DG penetration. As far as there are no economic incentives in the respective regulation to promote the massive penetration of DGs in distribution networks, the benefits of using DC links for this purpose fall back on the DG owners, which can increase the rated power of their power plants. Note that the payback analysis has to be slightly modified with respect to the previous ones. On the one hand, the DG promoter has to invest in both the DC link and the additional DG installed power. On the other hand, the benefits are due to the increment of generated energy. Therefore, the simplifed payback can be computed as follows:

$$PB = \frac{IC}{AI} = \frac{IC_0 + \Delta IC}{AI_0 + \Delta AI} \text{ (years)} \tag{7.17}$$

where IC_0 and AI_0 refer to the investment and annual incomes without DC links, and ΔIC and ΔAI represent the increment of those values when the DC link is installed. The previous expression can be expanded as:

$$PB = \frac{IC}{AI} = \frac{(S_{DG} + \Delta S_{DG})PC_{DG} + S_{DC}PC_{DC}}{(P_{DG} + \Delta P_{DG})EC_{DG}T} \tag{7.18}$$

where S_{DG} corresponds to the initial DG installed power, ΔS_{DG} is the increment of installed power, S_{DC} is the rated power of the DC link, PC_{DG} is the DG cost (euro/kVA), PC_{DC} is the DC link cost (euro/kVA), EC_{DG} is the DG energy price (euro/kWh), and T refers to the equivalent hours of operation at rated power of the DG. After some mathematical manipulations Eq. 7.18 can be simplified as:

$$PB = \frac{PC_{DG}}{EC_{DG}T} + \frac{S_{DC}PC_{DC}}{(P_{DG} + \Delta P_{DG})EC_{DG}T} = PB_0 + \Delta PB \tag{7.19}$$

where PB_0 is the payback of the DG promoter investment without installing a DC link and ΔPB is the increment of the payback due to the DG additional installed power and DC link cost.

Taking these considerations into account and the results of the simulations performed in the previous subsection it is possible to give some rough numbers about the profitability of the DC links The following data have been considered:

- Cost of each VSC of the DC link has been estimated as 250 euros/kVA including the VSCs, switchgears, measurement, control, and coupling transformers.

- Cost of energy for the utility has been estimated as 0.1 euros/kWh. This is the cost to be used in case of analyzing the reduction of losses of the distribution system.

- Energy prices of the photovoltaic and wind plants are 0.32 euros/kWh and 0.073228 euros/kWh (Spanish case).

When the loss reduction application is considered the utility has to pay for the DC link, and the payback is 19.20 years. Therefore, the installation of this device only for this purpose is not interesting at all. However, when the maximization of DG penetration is analyzed, the payback of adding VSCs is increased just 0.17 years with respect to that of the base case. Hence, the installation of DC links is quite interesting for the DG promoter in the analyzed scenarios.

Conclusion

This work has analyzed the use of asynchronous DC links based on self-commutated power electronic devices to create controllable "loops" between a number of feeders. It has been demonstrated that these devices constitute a promising solution to solve current congestion problems arising in radial distribution networks, particularly in the presence of DG. The main objective of the chapter is to quantify those benefits, as a qualitative justification is surely not enough to persuade the involved actors (utilities, DG owners) to use this new technology. For this purpose, a general optimization framework has been described to help planning engineers numerically evaluate the benefits that can be achieved with this new network asset. The proposed generic framework allows several figures of merit to be systematically computed, through a single objective function, just by defining the appropriate set of independent variables. In this way, the planner may determine the maximum penetration of DG, the maximum load growth, or the minimum power losses. Comparing the resulting values with those obtained in absence of DC links will directly show the improvements associated with such devices.

Test results for the distribution network proposed by the CIGRE Task Force C6.04.02 are shown illustrating the application of the proposed optimization framework. The installation of a three-terminal VSC-based DC link is assessed, showing that this technology is an interesting choice for the different agents involved in the distribution business. On the one hand, the DG owners can make profit from the use of DC links because of the significant increase in penetration levels. On the other hand, distribution utilities can take advantage of the higher levels of system load growth, therefore delaying investments in network reinforcement.

In summary, DC links may be helpful to conciliate the somewhat contradictory interests of utilities and network users, such as DG. Like any other new technology, it will take a while before DC links are fully introduced in the electrical distribution business, but this work provides simple planning guidelines to partly overcome some of the existing barriers.

Acknowledgment

This work was supported by ENDESA, the Spanish Ministry of Economy and Competitiveness and Junta de Andalucía under research projects SMARTIE, ENE2011-24137 and P09-TEP-5170.

References

[1] Jiyuan Fan and Stuart Borlase, *The Evolution of Distribution To Meet New Challenges, Smart Grids Need Advanced Distribution Management Systems*, IEEE Power and Energy Magazine, Vol. 7 No. 2, March/April 2009.

[2] D. Aggeler,*Bidirectional Galvanically Isolated 25 kW 50 khz 5 kV/700V Si-SiC SuperCascode/Si-IGBT DC-DC Converter*, Ph. Dissertation, ETH ZURICH, 2010.

[3] H. Akagi, *The next-generation medium-voltage power conversion systems*, Journal of the Chinese Institute of Enginerrs, Vol. 30, No. 7, pp. 1117–1135, 2007.

[4] S. Inoue and H. Akagi, *A Bi-Directional Isolated DC/DC Converter as a Core Circuit of the Next-Generation Medium-Voltage Power Conversion System*, IEEE Conference on Power Electronics Specialists Conference 2006, 18–22 June 2006.

[5] N. Okada. *Verification of Control Method for a Loop Distribution System using Loop Power Flow Controller*, IEEE Power Systems Conference and Exposition, pp. 2116–2123, October 2006.

[6] M. A. Sayed, T. Takeshita, *Load voltage regulation and line loss minimization of loop distribution systems using UPFC*, 13th Power Electronics and Motion Control Conference, pp. 542–549, 1–3 September 2008.

[7] R. Simanjorang, Y. Miura and T. Ise, *Controlling Voltage Profile in Loop Distribution System with Distributed Generation Using Series Type BTB Converter*, 7th International Conference on Power Electronics, pp. 1167–1172, Daegu (Korea), 22–26 October 2007.

[8] R. Simanjorang, Y. Miura, T. Ise, S. Sugimoto and H. Fujita,*Application of Series Type BTB Converter for Minimizing Circulating Current and Balancing Power Transformers in Loop Distribution Lines*, Power Conversion Conference, pp. 997–1004, Nagoya (Japan) April 2007.

[9] A. Orths, Z.A. Styczynski and O. Ruhle, *Dimensioning of distribution networks with dispersed energy resources*, IEEE Power Systems Conference and Exposition, pp. 1354–1358, 10–13 October 2004.

[10] R.A.A. de Graaff, J.M.A. Myrzik, W.L. Kling and J.H.R. Enslin, *Series Controllers in Distribution Systems- Facilitaing Increased Loading and Hiher DG penetration*, IEEE Power Systems Conference and Exposition, pp. 1926–1930, USA, 2006.

[11] R.A.A. de Graaff, J.M.A. Myrzik, W.L. Kling and J.H.R. Enslin, *Intelligent Nodes in Distribution Systems- Optimizing Staedy State Settings*, IEEE Power Tech, pp. 391–395, Lausanne (Switzerland), 1–5 July 2007.

[12] Okada N, Takasaki M, Sakai H, Katoh S. Development of a 6.6 kV - 1 MVA Transformerless Loop Balance Controller. In: Proc. IEEE Power Electronics Specialists Conference, 2007.

[13] J.M. Mauricio, J.M. Maza-Ortega, and A. Gomez-Exposito, *Considering Power Losses of Switching Devices in Transient Simulations through a Simplified Circuit Model*, International Conference on Power System Transients, Paper No. 281, Session 6B - Power Electronics, Kyoto (Japan), 3–6 June 2009.

[14] E. Romero-Ramos, A. Gomez-Exposito, A. Marano-Marcolini, J.M. Maza-Ortega and J.L. Martínez-Ramos, *Assessing the loadability of active distribution networks in the presence of DC controllable links*, IET Generation Transmission and Distribution, Vol. 5, Iss. 11, pp. 1106–113, November 2011.

[15] K. Rudion, A. Orths, Z. Styczynski, and K. Strunz, *Design of benchmark of medium voltage distribution network for investigation of dg integration*, in IEEE Power Engineering Society General Meeting, 2006.

CHAPTER 8

Voltage-Based Control of DG Units and Active Loads in Smart Microgrids

Tine L. Vandoorn, Lieven Vandevelde
Electrical Energy Laboratory, Department of Electrical Energy, Systems and Automation, Ghent University

Introduction

Some of the reasons for the smart grid development are to cope with the increasing share of distributed generation (DG) units and the aging electric power system infrastructure. It is not expected that the emergence of the smart grid would mean a revolution in the electric power systems, but a more evolutionary change of the system is expected [1]. The smart grid is expected to emerge from smart microgrids (Fig. 8.1). Microgrids cluster (controllable) loads, storage, and DG units, often consisting renewable energy sources, as depicted in Fig. 8.2. They offer a coordinated approach to integrate the DG units with power-electronic interfaces in the electrical power system. Microgrids can operate both in grid-connected and islanded mode by using a point of common coupling (PCC) switch. In the islanded operating condition, the microgrid has to maintain the power balance independently of the main grid. One of the technical challenges is to

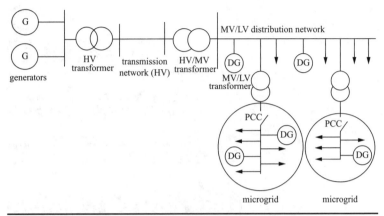

FIGURE 8.1 Microgrids connected to the distribution network. Smart microgrids lead the way to a full-scale smart grid.

develop advanced control strategies to live up to the expectations of microgrids, e.g., microgrids

- Can present themselves to the utility network as controllable entities, offering scale effects from the microgrid point of view and a reduced complexity for the utility's point of view
- Facilitate the integration of large amounts of DG units, especially with renewable energy sources
- Can increase the reliability of the system
- Support the integration of the smart grid concept in a coordinated manner

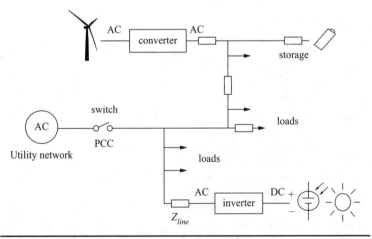

FIGURE 8.2 Microgrids as a cluster of (controllable) loads, energy storage devices, and DG units.

This chapter focuses on the islanded mode of the microgrid, with control strategies for both the DG units and the active loads. The DG units use voltage-based droop controllers for the power balancing and sharing between multiple units. These controllers provide the primary control in the islanded microgrid without communication, which increases the reliability of the system. The droop controllers are based on the mainly resistive line parameters of the low-voltage microgrid and the lack of rotating inertia. The power of the dispatchable DG units is changed more frequently than that of the renewable energy sources, which have power changes that are delayed to situations where the dispatchable DG units alone cannot guarantee a stable microgrid operation. This enables an optimized integration of the renewable sources without the need for communication. However, because of the small size of the islanded microgrid and the high share of renewables with an intermittent character, new means of flexibility in power balancing are required to ensure a stable operation. An active load control strategy integrated in the microgrid can offer the needed flexibility and change the current load-following control to a more generation-following strategy, which burdens the DG units less and avoids frequently changing the power of the renewable energy sources from their optimal operating point.

Active load control can be divided into primary and secondary active load control. Analogously to the current grid frequency control, the primary active load control focuses on increasing the reliability of the system. Therefore, it should be automatic and on a short time-scale, without need for communication. In this chapter, the active loads use a voltage-based control strategy analogous to that of the DG units, which is triggered by the microgrid voltage level as well. The combination of the voltage-based droop control strategy of both the generators and the active loads allows a reliable power supply without inter unit communication for the primary control. It leads to a more efficient usage of the renewable energy and can even lead to an increased share of renewables in the islanded microgrid. For active load control involving secondary objectives, such as further economic optimization and emergency load response, communication-based secondary active load control can be included next to the communication-less primary active load control for reliability issues. To include the secondary active load control strategy, incentives or time-differentiated pricing schemes can be provided to the participating customers. These advanced opportunities to encourage load participation are enabled by the smart grid concept and can be introduced in a smart microgrid.

Control Strategies for the DG Units

Microgrids can manage their energy production, storage, and consumption such that in islanded mode they become self-sufficient, which can increase the reliability of the system. In grid-connected mode, they enable an effective management of the renewable generation, become controllable units of the distribution system, and offer redundancy in the system.

To guarantee a stable operation of microgrids in islanded mode, several control strategies for the DG units have been developed. As most DG units are power-electronically interfaced to the electric network, the control of the DG units concerns the control of these interfaces, namely, the converter control. The control strategies for islanded microgrids can be classified in controllers with and without communication.

Communication-Based DG Unit Control

The communication-based control schemes achieve good voltage regulation and power sharing. However, these control strategies require expensive and vulnerable communication lines, which are especially restrictive in case of long distances. Although, in geographically small microgrids, this is less problematic, these communication links could reduce the system reliability and expandability, thus, limiting the flexibility of the system. Some of the best-known, effective, and intuitive ways for active power sharing and voltage control in islanded microgrids based on communication are central control, distributed control, and master/slave control.

In the central control method, a central controller coordinates the power-electronic interfaces in the microgrid to maintain the balance in active power P and reactive power Q in steady-state conditions [2, 3]. A central controller defines the set-value of the current for each module and is responsible for the power distribution. The central control unit measures the total load current i_L. The total load current i_L is divided between N modules in order to define the reference current for each sub-unit: i_L/N. The local controllers of each module control their output current to the reference current received from the central controller.

The distributed control method is a variant of central control. In the distributed control method, a central controller provides fundamental frequency power sharing between the different converters by distributing a low bandwidth signal to all converters [4, 5]. Power quality aspects are dealt within the local controllers by means of higher-frequency signals. The main advantage is that the signals can be transmitted via a limited bandwidth communication link.

In the master/slave control strategy, the master provides voltage regulation and specifies the reference current for the slaves [6, 7] as depicted in Fig. 8.3. The slave units operate as current controllers tracking the current command i_{ref} provided by the master. Master/slave control can also be combined with a central control unit that distributes the load current to all slave units.

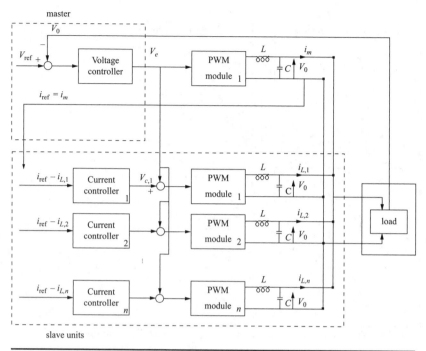

FIGURE 8.3   Schematic overview of master/slave control.

DG Unit Controllers without Communication

To avoid inter-unit communication links that can decrease the reliability of the system and form single points of failure, the droop-based control method is widely known. In case of inductive lines, often P/f droops are used, analogous to the conventional grid control [8–11]. However, in resistive microgrids, a linkage between active power and grid voltage instead of phase angle (which is dynamically determined by the frequency) exists. Therefore, P/V droop control, also called the reversed droop control, has been presented in [12, 13].

Some variants to improve the droop control strategy are the virtual output impedance method of [14, 15] and the usage of derivative instead of purely proportional controllers in [16–18]. In order to obtain an optimized integration of renewable energy sources in the microgrids, the voltage-based droop control for active power balancing and sharing between multiple DG units has been developed in [19]. As shown in Fig. 8.4, the reference voltage amplitude is determined by the active power controller, while the phase angle is determined by a Q/f droop controller. The voltage-based droop controller consists of a cascade of two controllers for the active power sharing as depicted in Fig. 8.4. First, the V_g/V_{dc} droop controller takes care of the power balancing in the network. Analogous to grid

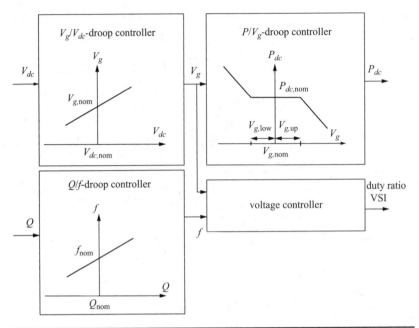

FIGURE 8.4 Voltage-based droop control consisting of V_g/V_{dc} and P_{dc}/V_g droop control.

frequency in conventional networks with rotating inertia, changes of the DC-link voltage of the inverters represent changes of the balance in production and consumption of the network. This control strategy changes the grid voltage V_g proportional to the DC-link voltage V_{dc} of the DG unit. Even a slight change of V_g leads to a change of the power delivered to the electrical network by the inverter. This effect is realized by a natural balancing due to the resistive loads and microgrid lines, and by intelligent loads that use voltage as a trigger for the active load control as presented in [20] and discussed further in this chapter. In order to avoid voltage limit violation, the P_{dc}/V_g droop controller in turn changes the DC-power of the DG unit dependent on V_g. Again a proportional controller with a negative slope is used. The controllers are based on the lack of inertia and the resistive lines of the microgrid.

In conventional networks, the penetration of renewable energy sources (RES) is limited in a conservative manner, e.g., to avoid over-voltages. With the voltage-based droop control, a higher penetration of RES can be achieved by including power curtailment (or even increase the injected power temporarily by including combined storage-generation solutions) in a distributed manner. The power changes by the P_{dc}/V_g droop controller of the RES are delayed compared to those of the dispatchable DG units. This is achieved by including a constant-power band with a width dependent on the nature of the energy source as depicted in Fig. 8.5. For RES, for example,

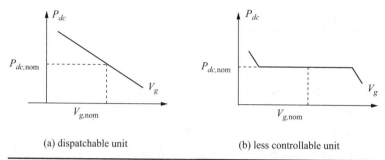

(a) dispatchable unit   (b) less controllable unit

Figure 8.5 Dispatchable versus less controllable units (a less controllable unit often uses a renewable source, such as photovoltaic panel, wind turbine, or combined heat and power with heat as primary driver).

a wide constant-power band ($2b = V_{g,low} + V_{g,up}$ in Fig. 8.4) is used. In case of near-nominal voltages, the output power of these units is not determined by the state of the network, but by the energy source, e.g., with maximum power point (MPP) tracking in case of photovoltaic panels. In case the voltage exceeds this constant-power band, these units also control their output power (storage or non-MPP condition) to ensure a stable operation of the microgrid. The dispatchable DG units on the other hand have small constant-power bands and react on small deviations of voltage compared to the nominal value. Units with a mediocre $2b$ can be combined heat and power (CHP) units, where in nominal conditions heat is the primary driver and electric power is the by-product. But in case the constant-power band is exceeded, the electric power can become the driver, thus, their generation becomes dependent on the state of the electrical network.

The difference in control strategy is dependent on the terminal voltage and is not triggered by communicated signals. Therefore, an optimized integration and possibly even a higher penetration of RES in the microgrid can be obtained.

Control Strategies for the Active Loads

The grid control is currently based on a "load following" strategy, where the generators follow the load changes without influencing the loads. In the load following strategy, the loads are quasi blind to the state of the network. The advent of large amounts of DG units leads to a relative decrease of centralized power plants compared to units with an intermittent variable character, which are often DG units (e.g., wind, solar). Therefore, the control flexibility of the generators to face the variability of the loads decreases in case of high amounts of DG in the network. Both the production and demand become variable and to adjust them such that production exactly matches demand becomes very complex. As the available storage capacity is mostly limited and/or expensive, a solution often presented is the usage of active load control to force the loads to react on the state of the electric power system. Some advantages of active load control are

- It can reduce the need for future utility investments and generation assets
- It can reduce/avoid congestion problems
- It reduces peak loads
- It can reduce the stress on the network
- It may increase the penetration limit for DG units while avoiding large system upgrades

If included in a high degree, the load following control strategy could reverse to a more "generation following" strategy in which the loads accomplish for the extra rigidity introduced at the supply side. Typical loads that might be controlled, e.g., in households are loads with a large electrical or thermal capacity and/or inertia like electrical-vehicle batteries, refrigerators, electrical boilers and freezers, or loads with less stringent timing requirements like dishwashers, washing machines, and dryers.

In general, the introduction of the smart (micro)grids offers high potential for load response, which is enabled by new smart grid features such as advanced metering and an embedded communication infrastructure that makes real-time pricing possible. Smart microgrids are especially suitable for load response. In order to force the loads to change their consumption based on external parameters, the drivers for the load participation should be considered. These drivers can be

1. Obligatory: General responsibility to maintain the stability of the system. This driver should be prevented if possible, but can achieve customer acceptance if it is absolutely necessary for the stability of the system, e.g., in small-scale systems. If this kind of driver is necessary to ensure the stability of the system, fast customer response is required (certainly in small-scale systems). Therefore, in this case, an automatic load response should be obtained.

2. Cost advantages: Customers acquire financial benefits when participating in the active load control.

Another classification can be based on the communication requirements of the active load control strategy.

Communication-Based Active Load Control

The first category of active load control uses communication to send direct (obligatory) control signals to the loads or to give the loads information about the financial benefits they can achieve when responding to the load control. The drivers for the loads to respond to this information generally come from the pricing strategy. Therefore, the pricing aspects are briefly dealt with in this section.

In general, there are two ways of electricity pricing to promote active load control: incentive-based and time-based pricing. In the incentive-based strategy, dedicated control systems are able to shed loads (obligatory driver) in response to a request from the utility. Incentive-based pricing is generally present to deal with emergency/event-based active load control, avoiding outages.

A second category of load response pricing uses time-based pricing. The transmission network and generators are sized to correspond to the peak demand, which involves a significant capital cost. Especially time-based active load control is often included to decrease this peak demand, i.e., peak shaving, by delivering cost benefits to customers that contribute to this peak reduction. Nowadays, for small consumers (that can be connected to the low-voltage microgrids), a basic time-based active load control program exists in the form of different meters.

The single-tariff meter, which is often used, involves a fixed price per kWh, independent of the production cost at the time of consumption. Therefore, consumers do not face real costs and have no incentives to reduce their consumption during peak times that face high costs of production. The loads are not sensitive to costs of production in the short term. For differences in day versus night consumption, a solution is brought by the two-tariff meter to lower the costs during nights and weekends compared to the day times. This is implemented in Belgium, for example. The switch between both tariffs is based on remote control. Another meter implemented in Belgium is a distinct meter for night tariff. This meter deals with a separate electrical circuit with heating appliances solely (heat accumulation and warm water boilers) to charge the heating appliances during night. The three-tariff meter, abolished in Belgium, charges higher tariffs during some hours of the day, e.g., to turn on the heating. Although these meters deliver differentiation in the prices over a day, still, real-time pricing is not up and running yet.

A newer concept is the smart meter by means of advanced metering infrastructures (AMI). The main characteristic of this meter is its ability for remote communication. It makes pricing based on real-time prices possible. This may lead to a more market-based pricing strategy.

Several real-time tariff schemes already exist, such as

- Time-of-use (TOU) pricing with different prices defined over a day, with a fixed number of timeslots. TOU reflects the average cost of generating and delivering power during those time periods. It encourages consumers to shift their consumption away from periods with high total load to periods with a lower demand with different prices in these periods as depicted in Fig 8.6. The prices are typically fixed on a monthly/seasonal basis. This is analogous to the two-tariff meter, but more differentiation is possible, e.g., to cope with seasonal effects.

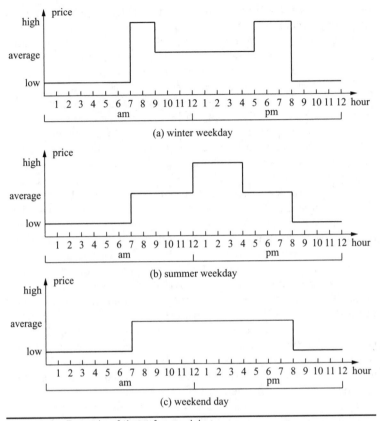

Figure 8.6 Example of time-of-use pricing.

- Real-time pricing (RTP) with typically hourly varying prices, related to the wholesale market price. Prices are typically fixed on a day-ahead or hour-ahead basis. RTP allows more gradation in the price compared to TOU pricing. As opposed to TOU pricing, next to seasonal effects, also intraday weather effects, for example, can play a major role.
- Critical peak pricing (CPP) uses trigger conditions that can change the price. It is often combined with TOU pricing.

In the smart grid, the number of time slots with different prices and the different price levels can be increased and also, the time the price is communicated in advance can be lowered compared to presently running systems. However, there will always be a trade-off between communication burden and number of time slots/different prices/difference between the information exchange and the real time-of-usage. Therefore, it is difficult to implement these kinds of communication-based active load control strategies if they are crucial for the stability of the system, e.g., in small-scale systems. Consequently, they cannot be used as primary load control algorithms but merely contribute to the secondary load control. For the primary load control, the real-time feature and a huge number of possible time slots is important, thus, communication should be avoided. Primary active load control should, thus, involve an automatic load response.

Active Load Control without Communication

In islanded microgrids, the current two-tariff meter for electricity pricing could have adverse effects on the stability of the network. For example, in microgrids with a high penetration of photovoltaic panels, relatively much power is injected during the daytimes. In such case, the consumption should be shifted to these times instead of to the nights. In conclusion, the load should be shifted in a dynamic manner. For this, communication can be used (such as the variable prices provided by a central control center). However, communication always involves some time delays (e.g., computational and aggregation delays in central control units) that can be dramatic for the microgrid robustness if this network is dependent on the load response for its stability. In microgrids, next to the distributed generation units and energy storage elements providing the control flexibility to deal with the variable nature of the loads, the microgrid could also depend on fast-acting controllable loads for its stability. The flexibility of the loads can be an important factor to ensure a stable operation because of the possible high share of renewable energy sources, the small scale of microgrids, and in case of emergency microgrids, the rare occurrence of islanded mode. Therefore, in [20], we have developed an active load program especially for islanded microgrids to allow customer response that does not depend on communication.

A schematic overview of this voltage-based active load strategy is shown in Fig. 8.7. This strategy complies with the voltage-based

FIGURE 8.7
Schematic overview of this voltage-based active load strategy (RE is renewable energy).

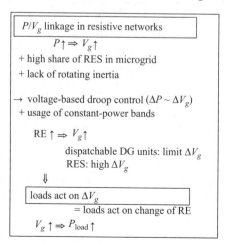

P/V_g linkage in resistive networks

$$P\uparrow \Rightarrow V_g\uparrow$$

+ high share of RES in microgrid
+ lack of rotating inertia

$\rightarrow$ voltage-based droop control ($\Delta P \sim \Delta V_g$)
+ usage of constant-power bands

$$RE\uparrow \Rightarrow V_g\uparrow$$

dispatchable DG units: limit ΔV_g
RES: high ΔV_g

$\Downarrow$

loads act on ΔV_g
= loads act on change of RE

$$V_g\uparrow \Rightarrow P_{load}\uparrow$$

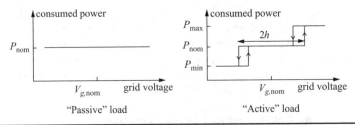

Figure 8.8   Active load control strategy based on the grid voltage.

droop control of the generators and does not require inter-unit communication. It is installed to deal with reliability issues, hence it is used as a primary active load control strategy, which will be clarified in the next section.

The parameter on which this active load control strategy can be based is the grid voltage. As shown in Fig. 8.8, the load shifts its consumption from low-voltage to high-voltage times. For this, a relay function can be used. The reason to use voltage as a control parameter is first, because of the resistive network lines and second, the usage of voltage-based droop control of the generators. First, in low-voltage electrical networks, the lines are mainly resistive, thus, there is mainly a linkage between the injected active power P and the grid voltage V_g. This follows from the power flow equation derived from Fig. 8.9:

$$P = \frac{V_g}{R_{line}}(V_g - V \cos \delta). \tag{8.1}$$

Hence, high voltages are present in case of high power injection or low load burden and vice versa. Second, in islanded microgrids exploited by the voltage-based droop control and with a high share of renewables, the renewable injection is mainly visible in the voltage. The usage of constant-power bands with a width $2b$ dependent on the nature of the energy source is the main condition for this. The dispatchable DG units have a low constant-power band width $2b$ such that they change their power to force the grid voltage near the nominal voltage. Renewable DG units on the other hand have a high constant-power band width. In nominal conditions, they operate as nondispatchable DG units. The power they deliver to the electrical network is not influenced by the state of the network, but by the state of their energy source, for example, maximum power point tracking in wind turbines and solar panels. Following from Eq. 8.1, this has a

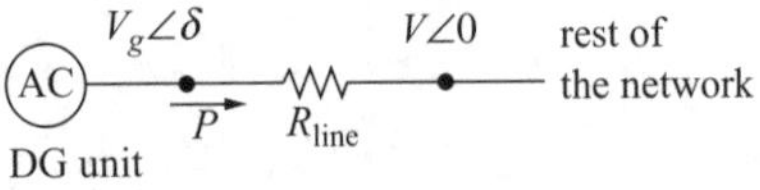

Figure 8.9
Power from DG unit (V_g, δ) to the rest of the microgrid $(V, 0)$ through resistive line R_{line}.

significant influence on the grid voltage. If the voltage exceeds the constant power band, these DG units change the power they deliver to the electrical network, for example, by using storage or by leaving the maximum power point. Hence, because of the usage of different $2b$ dependent on the energy source, extreme voltages are generally obtained in case of a high power input from renewable generation compared to the generation of the dispatchable DG units combined with a low load burden. In this way, the voltage is an important parameter for active load control. For example, if the loads shift their consumption to the high voltage times, they shift their consumption to the periods of high renewable generation as well. If they shift away from the low-voltage periods, they shift away from periods where the dispatchable, often not-renewable, units are heavily loaded.

The voltage-based droop control enables automatic power curtailment of the renewables in case of high voltages. However, loss of renewable energy should be delayed compared to load shifting. Therefore, the active load control is included with a constant-power band $2h$ as well, and generally the constant-power band width $2b$ of the renewable energy sources should be larger than that of the active loads. Again, the width of the constant power band is dependent on the nature of the load. Also, loads that allow a lower h should have more financial benefits than those with a higher h. The pricing for this load response should thus depend on

- The width h of the constant power band
- The number and duration of the load shifts

The smart grid features can deal with this pricing. The load response should be automatic and very fast, but the pricing involved may have some delay and can use bidirectional communication from the controllable load to an aggregation center.

In conclusion, by using the voltage as a parameter of load response, a voltage-based active load control strategy is obtained, which is analogous to the voltage-based control strategy of the generators. In this way, the consumption is automatically shifted from high load/low renewable injection times (low voltage) to low load/high renewable injection times (high voltage).

Primary versus Secondary Active Load Control

The primary and secondary active load control is based on the primary and secondary control of the generators in the transmission network ([UCTE Operation Handbook Appendix 1, 2004]), Fig. 8.10. In the transmission system, the primary control system is automated based on the rotating inertia of the system, which ensures a constant ratio between frequency changes and production changes and acts very fast. The primary control of the centralized generators is performed by an active power/frequency droop control strategy that

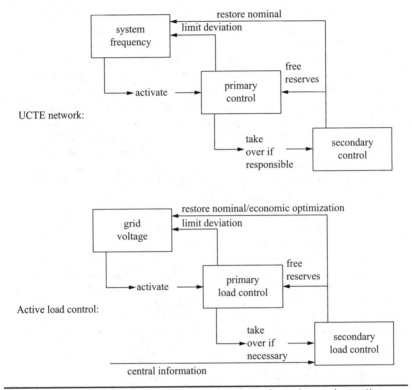

FIGURE 8.10   Primary/secondary grid control versus primary/secondary active load control.

measures the frequency and changes the input power accordingly. The action of primary control is to restore the power balance and to maintain the system frequency within specified limits. However, this generator adjustment results in deviations from agreed-upon power exchanges and a steady-state system frequency different from the nominal value. Therefore, secondary control is used to eliminate this frequency deviation and to correct errors in the power interchanges programmed between areas.

The primary and secondary active load control in smart microgrids can be considered as analogous to the primary and secondary control of the generators. The primary active load control is included for reliability issues, preferably without communication. Therefore, primary active load control should operate at a short time frame to ensure the stability of the microgrid, as is the primary generator control. The voltage-based active load control deals with these conditions. Together with the voltage-based droop control of the DG units, it can be considered as a new means of control flexibility of the renewable energy sources. The primary active load control is effective in case of

- Islanded microgrids with voltage-based droop control such that high voltages are mainly due to generation from non-dispatchable DG units (renewable energy sources) and low load times.

- Grid-connected microgrids: Nondispatchable DG units (renewable energy sources) can cause overvoltage during periods of low load and high injection. Generally, this is mitigated by conservatively limiting the number of nondispatchable DG units. The primary active load control can increase this maximum allowable number.

The secondary active load control can be included for optimization issues, and can operate on a larger time frame. The smart grid plays a major role in this secondary active load control as it offers bidirectional communication and smart sensors/devices. Some objectives of secondary active load control can be

- Communication about the availability of the load: bidirectional information between the smart appliances and the control centers

- Emergency actions (direct load control)

- Coordination of primary active load control actions

- Restoration of pre-agreed consumption patterns after a change by the primary active load control

Smart Microgrids

An approach to deal with the large increase of decentralized unpredictable power sources and the increasing (peak) consumption is to add intelligence to the power system. New communication and remote management abilities help to couple the grid elements via a fully interactive intelligent electricity network, the so-called smart grid, Fig. 8.11. The electricity generation, distribution, and consumption are overlayed with an additional metering and information system to save energy, reduce costs, increase reliability and transparency, and enable more customer choices for their energy management.

Essentially, the goals of microgrids and smart grids are the same: to minimize costs, meet the growing demand and integrate more sustainable generation resources, increase efficiency, and improve the reliability of the system in the face of increasing intermittency. The microgrid facilitates the penetration of DG into the utility grid as it delivers integration in a coordinated way by dealing with the unconventional behavior of DG. Microgrids offer significant benefits

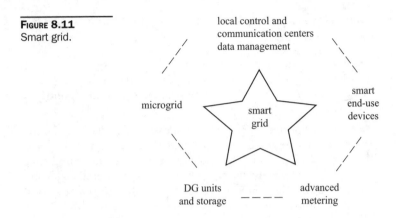

Figure 8.11
Smart grid.

in both the grid perspective and the consumer perspective. A key advantage from the grid point of view is that the microgrid appears to the power network as a single controllable unit, enabling it to deliver the cost benefits of large units as shown in Fig. 8.12. From the customer's point of view, the impact of the microgrid on the reliability of the distribution network is relevant, certainly in the future, with more unpredictable generation and higher consumption (peaks).

A smart microgrid is a microgrid with intelligent software to monitor, manage, and optimize energy supply, storage, and demand, which can

- Contain energy management systems
- Monitor the system and actively intervene in the consumption/generation
- Identify and maximize energy efficiency opportunities

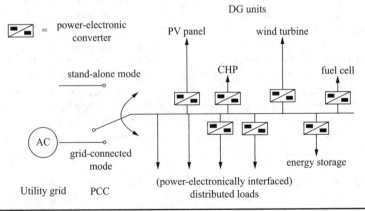

Figure 8.12 Microgrid with (power-electronically interfaced) loads, storage, and DG units in stand-alone or grid-connected mode.

- Use an extra communication and sensor layer to maximize cost savings and reduce carbon emissions
- Generate a more active participation of the consumers

It is not expected that the smart grid will emerge as a revolution; it will rather experience a gradual evolution with the smart grid emerging as a system of integrated smart microgrids [1]. Because of the flexibility and scalability (modular approach), microgrids have often been pointed out as a main actor in the development of the smart grid. The smart microgrid makes intelligent decisions for clean energy, smart appliances, etc. It can be seen as a microgrid, which can ensure a stable operation in the short term and is fast-acting, with an overlaying intelligence scheme, that operates slower to restrict the data burden and exploits the options of using communication, e.g., for economic optimization. Smart microgrids are especially interesting as they can be considered as small pilot versions of the future power system.

A balance between cost to incorporate intelligence and its benefits needs to be made. The infrastructure and control centers for smart grids can be very hulking to implement on a large scale and this may take many years. Rather than investing in incorporating intelligence on a large scale, investing in small smart microgrids can be done at a lower cost and more quickly. This makes smart microgrids very attractive for the implementation of smart grid features. Smart microgrids are, thus, a simpler alternative and could play a leading role in the deployment of smart networks. Also, a high level of intelligence built into the system is not necessary everywhere: Different areas require different levels of smartness. Areas with merely small residential units require/permit less interference of intelligent systems, while larger units benefit more from becoming controllable, and areas with high penetration of DG units allow higher levels of intelligence to increase the maximum DG penetration limit. These areas that allow more and benefit most from high levels of smartness of the system fit directly into the microgrid concept. In these microgrids, smart features can be installed faster than in the rest of the network with higher levels of intelligence than average. Therefore, the smart microgrid requires locally more intelligence compared to the rest of the system. As the level of intelligence is coupled with the cost, the smart microgrids can enable a new energy strategy while restraining the cost of making the whole system smart.

Conclusion

Smart microgrids enable a coordinated control of DG units, storage, and active loads, optimized by incorporating more intelligence in the network. The voltage-based droop control of the DG units operates without communication to ensure a proper voltage control, power sharing and balancing, and a stable and reliable operation of the microgrid. Therefore, it is a primary control strategy of the DG units in the islanded microgrid. This droop control is based on the specific characteristics of the low-voltage islanded microgrids, namely the resistive line impedances, high share of renewables, and lack of rotating inertia. Constant-power bands are used to delay the power changes of the renewable energy sources compared to those of the dispatchable DG units without inter unit communication. This enables an optimized integration of renewables in the microgrid. It also enables the renewable injection to become visible in the voltage of the network. Therefore, primary active load control, analogous to the voltage-based droop control of the DG units, is included in the network as well. The time scale of this active load control is slower than the power changes of the dispatchable units, but faster than those of the renewables, with faster and slower dependent on the terminal voltage. As this load control is voltage-based, it does not require communication and is very fast. Therefore, it deals with increasing the reliability of the microgrid and is used as a primary control strategy.

Smart grid features, such as more computational ability and an increased usage of communication and sensors, can help to optimize the microgrid. In this sense, secondary active load control can be used, e.g., for economic optimization, in case of emergency situations, or to support the primary control. The drivers for the secondary load control can be incentive-based or time-based. Secondary DG control can be included in an analogous way, e.g., as proposed in [21].

Acknowledgment

The work of T. Vandoorn is financially supported by the FWO-Vlaanderen (Research Foundation - Flanders, Belgium).

References

[1] H. Farhangi, "The path of the smart grid," in *IEEE Power & Energy Magazine*, vol. 8, no. 1, pp. 18–28, Jan./Feb. 2010.

[2] K. Siri, C. Q. Lee, and T. F. Wu, "Current distribution control for parallel connected converters part ii," *IEEE Trans. Aerosp. Electron. Syst.*, vol. 28, no. 3, pp. 841–851, July 1992.

[3] J. Banda and K. Siri, "Improved central-limit control for parallel-operation of dc-dc power converters," in *IEEE Power Electronics Specialists Conference (PESC 95)*, Atlanta, USA, Jun. 18–22, 1995, pp. 1104–1110.

[4] M. Prodanović, "Power quality and control aspects of parallel connected inverters in distributed generation," Ph.D. dissertation, University of London, Imperial College, 2004.

[5] M. Prodanovic and T. C. Green, "High-quality power generation through distributed control of a power park microgrid," *IEEE Trans. Ind. Electron.*, vol. 53, no. 5, pp. 1471–1482, Oct. 2006.

[6] K. Siri, C. Q. Lee, and T. F. Wu, "Current distribution control for parallel connected converters part i," *IEEE Trans. Aerosp. Electron. Syst.*, vol. 28, no. 3, pp. 829–840, July 1992.

[7] J.-F. Chen and C.-L. Chu, "Combination voltage-controlled and current-controlled PWM inverters for UPS parallel operation," *IEEE Trans. Ind. Electron.*, vol. 10, no. 5, pp. 547–558, Sept. 1995.

[8] M. C. Chandorkar, D. M. Divan, and R. Adapa, "Control of parallel connected inverters in standalone ac supply systems," *IEEE Trans. Ind. Appl.*, vol. 29, no. 1, pp. 136–143, Jan./Feb. 1993.

[9] J. M. Guerrero, J. Matas, L. García de Vicuña, M. Castilla, and J. Miret, "Wireless-control strategy for parallel operation of distributed-generation inverters," *IEEE Trans. Ind. Electron.*, vol. 53, no. 5, pp. 1461–1470, Oct. 2006.

[10] R. H. Lasseter and P. Paigi, "Microgrid: A conceptual solution," in Proc. *IEEE Power Electron. Spec. Conf. (PESC 2004)*, Aachen, Germany, 2004.

[11] J. A. Peças Lopes, C. L. Moreira, and A. G. Madureira, "Defining control stategies for microgrids in islanded operation," *IEEE Trans. Power Syst.*, vol. 21, no. 2, pp. 916–924, 2006.

[12] H. Laaksonen, P. Saari, and R. Komulainen, "Voltage and frequency control of inverter based weak LV network microgrid," in *2005 International Conference on Future Power Systems*, Amsterdam, Nov. 18, 2005.

[13] A. Engler, O. Osika, M. Barnes, and N. Hatziargyriou, *DB2 Evaluation of the local controller strategies.* www.microgrids.eu/micro2000, Jan. 2005.

[14] J. M. Guerrero, L. García de Vicuña, J. Matas, M. Castilla, and J. Miret, "Output impedance design of parallel-connected ups inverters with wireless load-sharing control," *IEEE Trans. Ind. Electron.*, vol. 52, no. 4, pp. 1126–1135, Aug. 2005.

[15] W. Yao, M. Chen, J. M. Guerrero, and Z.-M. Qian, "Design and analysis of the droop control method for parallel inverters considering the impact of the complex impedance on the power sharing," *IEEE Trans. Ind. Electron.*, vol. 58, no. 2, pp. 576–588, Feb. 2011.

[16] J. M. Guerrero, J. Matas, L. García de Vicuña, M. Castilla, and J. Miret, "Decentralized control for parallel operation of distributed generation inverters using resistive output impedance," *IEEE Trans. Ind. Electron.*, vol. 54, no. 2, pp. 994–1004, Apr. 2007.

[17] J. M. Guerrero, J. C. Vásquez, J. Matas, M. Castilla, and L. García de Vicuña, "Control strategy for flexible microgrid based on parallel line-interactive UPS systems," *IEEE Trans. Ind. Electron.*, vol. 56, no. 3, pp. 726–736, Mar. 2009.

[18] Y. Mohamed and E. F. El-Saadany, "Adaptive decentralized droop controller to preserve power sharing stability for paralleled inverters in distributed generation microgrids," *IEEE Trans. Power Electron.*, vol. 23, no. 6, pp. 2806–2816, Nov. 2008.

[19] T. L. Vandoorn, B. Meersman, L. Degroote, B. Renders, and L. Vandevelde, "A control strategy for islanded microgrids with dc-link voltage control," *IEEE Trans. Power Del.*, vol. 26, no. 2, pp. 703–713, Apr. 2011.

[20] T. L. Vandoorn, B. Renders, L. Degroote, B. Meersman, and L. Vandevelde, "Active load control in islanded microgrids based on the grid voltage," *IEEE Trans. on Smart Grid*, vol. 2, no. 1, pp. 139–151, Mar. 2011.

[21] J. M. Guerrero, J. C. Vásquez, J. Matas, L. Garcia de Vicuña, and M. Castilla, "Hierarchical control of droop-controlled AC and DC microgrids - A general approach towards standardization," *IEEE Trans. Ind. Electron.*, vol. 58, no. 1, pp. 158–172, Jan. 2011.

CHAPTER 9

Electric Vehicles in a Smart Grid Environment

David Dallinger, Daniel Krampe, Benjamin Pfluger
Fraunhofer Institute for Systems and Innovation Research

Smart grids and an increased use of information and communication technologies in the electricity sector are topics that have been discussed for many years, but there has only been slow progress in their practical realization. Most smart grid applications are still too complex and too costly when compared with their potential revenue. In addition, the main stakeholders in the electricity sector are not really interested in promoting flexible demand, which reduces the margins of current generation units. In the following chapter, we describe two main drivers that could spark a change and facilitate the implementation of smart grid technology.

The first driver is the transition from a controllable power plant park to a generation mix dominated by fluctuating generation from wind power and photovoltaic. If the worldwide reduction targets for CO_2 emissions are strictly enforced and technology learning further reduces the generation costs of renewable power generation, the electricity sector will change completely. This development can already be investigated in Denmark, Spain, or Germany, where renewable energy generation is playing an increasing role due to the early state subsidies in these countries.

The second driver is the renaissance of the electric vehicle. Compared to other mass applications in smart homes, the expected annual electricity consumption is very high. Later on, charging power electronics will allow advanced grid services. As automotive companies face new regulations to reduce CO_2, fleet emissions and utilities look for ways to integrate more fluctuating wind and photovoltaic power plants into the electricity grid, electric vehicles have good prospects of playing a key role in smart grid development. They enable a decoupling of electricity use and demand, and can balance fluctuating generation. In the following chapter, we analyze electric vehicles as a smart grid application and their interaction with fluctuating generation as well as possible revenues from reserve and spot markets.

Introduction

Load Shifting Using Electric Vehicles

A smart grid is a vague concept with still no clear definitions, and there are many different visions of an advanced distribution grid using communication technology. Important components of such a system include power electronics, or flexible AC transmission systems (FACTS), advanced metering systems, and appliances, as well as industrial processes that allow load shifting or distributed generation. Grid-connected electric vehicles (EVs), which include pure battery electric vehicles (BEVs) and plug-in hybrid electric vehicles (PHEVs), are one possible appliance that theoretically allows a decoupling of demand and supply (load shifting) and for energy to be fed back into the grid (vehicle-to-grid). Unlike other home appliances being discussed as smart grid devices, EVs use electric battery storage. Electric storage enables vehicle-to-grid services (V2G) and long grid management times with low storage losses, but comes with significantly higher costs compared to the thermal storages used by smart grid devices such as freezers, air conditioning systems, or heat pumps.

The main purpose of EVs is to fulfill mobility needs at costs equivalent to those of conventional vehicles. Hence, mobility behavior and the available grid management time are key values that have to be taken into account. The average standing time of vehicles is very long (95% to 98%) and theoretically offers good grid management potential. Average values, however, do not usually consider the interaction between driving [state of charge (SOC) reduction], charging and parking periods, and the required SOC before the start of a trip. Figure 9.1 shows the probability of starting the first trip of the day and returning from the last trip as well as driving, grid management, and charging time on a typical Monday for full-time employees.

In Germany, the average grid management time[1] during the day is 5 hours and 15 minutes (between the start of the first and the return

---

[1]The grid management time is defined as the time period between the start time of a trip and the return time of the next, minus the driving time and the charging time.

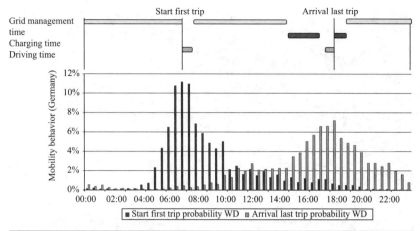

Figure 9.1 Principle of plug-in vehicle grid management time. WD: weekday.
Source: Own calculation; data basis [17].

from the last trip) and 12 hours and 17 minutes during the night. The possible EV shifting period is very long compared to other residential smart grid appliances. An average house in California with air conditioning, for instance, only allows load shifting for several minutes because of the generally poor level of insulation and low thermal building mass. The grid management potential of a heat pump with thermal storage is dependent on the ambient temperature and is mainly available in winter.

EVs users should maximize the distance driven electrically in order to recoup the higher initial investment.[2] Hence, load shifting here is restricted to shorter periods since regular charging is necessary. Frequent charging is particularly important for PHEVs with their smaller batteries (4–15 kWh). Analyses of different EV types indicate that the limited range of BEVs is a major sales restriction for those purchasing only one main car. In contrast, PHEVs, with smaller batteries, can ensure a high electric driving share without range concerns. In terms of overall costs, BEVs with larger batteries in combination with battery exchange stations or fast-charging infrastructure are less attractive than PHEVs. For these reasons, it is expected that most EVs in the future will be hybrids that can be characterized as a short-term grid management option with a storage capacity smaller than 20 kWh [1]. Load shifting over several days, which implies detailed trip planning by the customers and larger batteries, is therefore unlikely.

Another important aspect of EVs as smart grid devices is the parking location of the vehicles. Figure 9.2 illustrates four parking location categories of a German mobility survey [2].

---

[2]Costs per kilometer driving electric are smaller than costs per kilometer using gasoline or diesel.

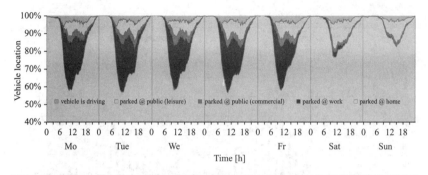

FIGURE 9.2 Frequency distribution of parking location.
Source: Own calculation; data basis [2].

Parking at home and at work occurs most often, and the parking location is relatively easy to predict. Public parking for shopping or leisure activities is characterized by a high degree of diversity. It is therefore more complex and cost-intensive to connect vehicles parking in these spots to the grid. In addition, the average public parking time is the shortest, which reduces the grid management time. Consequently, from a smart grid perspective, home and work charging seems to be of primary interest.

Control Equipment

We differentiate between direct and indirect control of smart grid devices. Direct control, or centralized optimal charging, implies that a service provider can shut down or reduce loads and directly control decentralized generation units. Examples include the direct load control of residential water heaters [3] and air conditioning loads in California, or of virtual power plants such as the real-time system of Iberdrola Renewables[3] to control wind turbine generation output. The advantages of direct control are prompt and predictable reactions in order to control signals. Drawbacks arise from reduced consumer acceptance, in the case of controlling loads in private homes or vehicles, and the communication and optimization efforts involved in controlling a large number of small storage or generation devices with varying consumer needs.

Indirect control uses price signals to manage loads or generation units. In this case, the service provider sends price signals and the consumer (or an automatically controlled device programmed by the consumer) decides to either reduce or shift the load when the price is high, or to pay the higher price. In this case, the decision to participate in a specific event remains on the consumer side. Disadvantages arise from the possibility of avalanche effects or simultaneous reactions to the signal [4] and inherent forecasting errors due to the necessity to predict the reaction of consumers to different price signals.

---

[3]Iberdrola Renewables is a utility based in Spain.

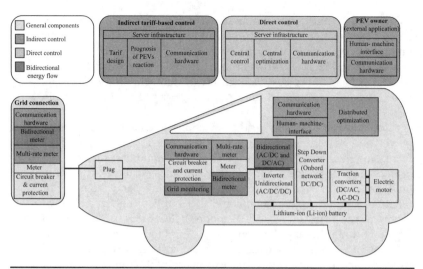

Figure 9.3 Components for automated control of electric vehicles.

A communication unit and a meter are the minimum requirements for a vehicle that is capable of automatically controlled smart charging (see Fig. 9.3). In principle, these components can be installed in the car or directly at the grid connection. Today, the billing point of electricity is a meter installed at the grid connection, with the exception of railway metering. Hence, each grid connection point is coupled to a contract with an energy supplier. As shown in Fig. 9.2, charging at home is the most likely option for EVs. Therefore, a smart meter at home offers a specific consumer the biggest benefits. To extend the smart grid connection time, other meters could be installed at work or at different commercial and leisure-related locations. In unbundled electricity markets it is possible that each of these grid connections has a different supplier contract. The moving vehicle needs to adopt the grid connection contract and the electricity rate or smart grid service agreement at each connection. Another contract is necessary to share the revenues of the smart grid services at public grid connections. Recent research projects have shown that billing customers at public charging stations with relatively low utilization costs more than the revenue from the electricity consumed. Further infrastructure at the grid connection and in the vehicle should provide the same functionality. For example, if the vehicle provides V2G but the grid connection only allows unidirectional metering, only a reduced service can be offered and the utilization of the bidirectional components would be lower.

Installing the meter and the communication unit in the EVs makes the vehicle more independent of the available infrastructure. But in this case the electricity is metered twice and it is necessary to deduct the

EV's electricity consumption from the regular electricity demand. From today's perspective, off-board metering without special EV billing is the easiest option. In the long term, however, onboard metering can guarantee a more efficient expansion of the infrastructure. Shorter vehicle life cycles also facilitate faster technology adoption. The components required for automated control are illustrated in Fig. 9.3.

Looking at the costs for grid connections and smart grid services, power and real-time capability play an especially important role. Increasing the power increases the costs[4] for circuit breakers, current protection, plugs, wiring, and power electronics. Real-time capability only comes with permanent communication links, and the need for reliable communication technology also implies higher costs. For example, the price for a 22 kW mode 3 IEC 61851 plug is about 300 euros today.[5] A regular household plug (about 3 kW), which is sufficient to charge a vehicle overnight, costs less than 5 euros.[6] Especially in the near term, we expect PHEVs to use low-power standard technology. In the medium term, it will be necessary to exploit synergies between components already available in vehicle and smart grid applications. Car-PCs or smart phone processors are able to schedule charging depending on the electricity rate (in Fig. 9.3, distributed optimization), or record meter data. Communication hardware will probably be available in 2020 in vehicles. Transaction and charging inverters could be integrated in one device using the same power electronics (e.g., silicon carbide switches). In this case, a bidirectional grid connection would come at a lower extra cost [5]. In contrast, installing additional charging infrastructure with its own communications, processor, meter, circuit breaker, and current protection does not seem to be the best approach in a world with rapidly evolving technologies and consumer needs.

For analyzing regulation reserve, cost data for the infrastructure (meter and communication system) are taken from [6].[7] For the bidirectional power electronics (power inverter, buck-boost converter, and grid monitoring), the prices of power inverters used in photovoltaic systems can be used as a guideline.[8] These do not account for the possibility of using the same power electronics for motor control

---

[4]Especially the step from 1-phase to 3-phase increases the costs.

[5]Economy of scale could reduce the price in the future. Costs are expected to be about 100 euros.

[6]A wall box (excluding smart charging) is offered by Daimler with plug and installation for 1550 euros.

[7]The assumed exchange rate is $1.40 = €1.

[8]According to a study of the costs of photovoltaic electricity generation, the prices for power inverters dropped by 70% down to 0.36 €/W between 1991 and 2007. It is assumed that by 2020 the price can be further reduced to 0.15 – 0.20 €/W due to economies of scale [7].

	Negative Regulation	Positive Regulation
Meter for invoicing	29 €[1]	29 €
Communication system	71 €	71 €
Bidirectional electronics	–	0.15 €/W

[1]This assumption seems optimistic. Buying an officially calibrated meter with SO-weighted impulse signal (DIN 43864) costs about 250 euros in Germany today for private consumers.
Source: Data basis [6] and [7].

TABLE 9.1 Necessary Investments in the Infrastructure[9]

because vehicles available today do not provide this function. The assumed prices are shown in Table 9.1.

To calculate the annuities, an interest rate (d) of 5% and a lifespan (n) of 12 years are assumed for the electronics and the battery. The costs of creating a pool or providing a control signal to the vehicles participating in the pool are still unclear and therefore not taken into account in this study.

Battery Degradation

A special feature of EVs as a smart grid device is the possibility of V2G. This enables not only load shifting but also feeding back power into the grid. Despite the fact that EVs can then be regarded as more flexible players in the smart grid, V2G comes with additional costs for battery degradation and power electronics. The uncertainty about the battery wearing out is a major concern with using the V2G opportunities. Figure 9.4 quantifies the battery degradation costs per energy unit.

This approach [8] uses cycle-life data of a Saft cell [9], U.S. Advanced Battery Consortium (USABC) goals, and A123 assuming an investment of 300 euros per kWh. With the depth of discharge (DoD)–based model [10], the cost function rises from low to full DoD rates. For USABC and Saft assumptions, the costs per kWh are between 2 ct. and 22 ct. per kWh. The model based on the energy proceeded [11] with the A123 battery performance results in DoD-independent costs of about 5 ct. per kWh. The cost for a full cycle with the Saft cell is about 22 ct. per kWh. This simplified approach with different estimations of degradation and cycle life shows that, even for optimistic estimations, V2G remains an expensive option. Only a technical breakthrough and extreme cost reductions, as well as rising base peak spreads in the electricity market, could bring V2G to a mass market.

[9]New charging circuits to the EV parking spot are not included. These costs are highly variable and expensive.

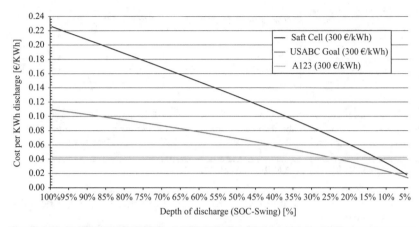

FIGURE 9.4   Battery degradation of lithium batteries.
Source: Own calculation [8], data basis [9], [10], and [11].

Regulation Reserve Market

One way to include electric vehicles in a smart grid environment is for them to provide regulation reserve to the utility. Especially at night, the vehicle is connected to the charging infrastructure for significantly longer than the time actually required for charging. This enables utilities or a central service provider to delay charging during hours of peak energy demand and to activate charging if the energy supply is higher than the actual demand.

This section starts with a brief description of the German market for ancillary services (for regulation reserve), introduces a dynamic simulation model of a fleet of electric vehicles that participate in such markets, and closes with a summary of the findings from our simulation and analysis (for details of the work see [12]).

German Markets for Ancillary Services

The European Network of Transmission System Operators for Electricity (ENTSO-E) is responsible for frequency control in Central Europe. Control is performed in a series of three independent control steps:

1. *Primary control* starts only seconds after a frequency deviation as a joint action of all the power system plants. This type of regulation capacity is mainly supplied by conventional power stations, which are operated slightly below their maximum capacity. Primary balancing power has to be deployed within 30 seconds and provided for up to 15 minutes.

2. *Secondary control* replaces primary control and restores the frequency to its nominal level. Adjustments of secondary control are realized in the time frame of seconds up to 15 minutes

Regulation Reserve			Capacity	Normalized Capacity Price	Dispatch	Energy Price	Dispatch Probability
			[MW]	[€/MW·h]	[MWh/ Month]	[€Cent/ kWh]	
Primary Control			667	20.51	–	–	n.s.
Secondary Control	HZ	Positive	3,081	22.05	120,163	11.16	14.9%
		Negative	2,451	4.04	106,521	0.1	16.6%
	NZ	Positive	3,050	7.41	116,290	6.91	8.1%
		Negative	2,413	8.23	270,227	0.01	23.8%
Tertiary Control	HZ	Positive	3,263	10.4	9,332	21.43	1.1%
		Negative	1,949	0.31	11,681	0.04	2.3%
	NZ	Positive	3,205	2.73	3,181	16.73	0.2%
		Negative	1,919	3.92	18,770	0.00	2.1%

TABLE 9.2 Average Market Prices in 2008 for Different Ancillary Services of the Four German TSO Areas

after an incident. The transmission system operator (TSO) in the control area is responsible for the activation of secondary control if there is an imbalance between generation and load. Secondary control is based on continuous automatic generation control.

3. If necessary, *tertiary control* is activated by the responsible TSO. *Tertiary control* reserves are activated manually in the framework of 15 minutes to one or two hours. These are primarily used to free up the secondary reserve in an imbalanced situation and as a supplement to the other reserves in case of large incidents (for detailed information see [13]).

Table 9.2 summarizes the market capacities, capacity, and energy prices as well as the monthly dispatch and the dispatch probability (dispatch to contract ratio) for the three German ancillary service markets and the four German TSOs[10] during 2008.

Prime time (Hauptzeit: HZ); secondary time (Nebenzeit: NZ); data basis: German Transmission System Operators 2009.[11]

In all three markets, an actor offers an exclusive bid for a specific capacity. Furthermore, for secondary and tertiary control, a distinction is made between positive and negative control as well as prime and

---

[10]50 Hertz Transmission GmbH (E.ON), Amprion GmbH (RWE), Transpower Stromüber-tragungs GmbH (Vattenfall) and EnBW Transportnetze AG.
[11]Dispatch probability for primary control is not specified (n.s.). A value of 10% is taken for the calculation.

secondary time. Prime time is defined as the time period between 8 A.M. and 8 P.M. on weekdays. Secondary time covers the remaining time on weekdays and the whole day at weekends. Positive regulation reserve means that additional power is fed into the grid in the case of an energy shortage, whereas negative regulation reserve means reducing the electricity generation or drawing electricity from the grid in the case of surplus energy (e.g., activating pump storage power stations). Beside the capacity price, a price for positive and negative energy is paid in the case of secondary and tertiary control. The dispatch probability describes how often capacity is retrieved and therefore the energy an actor has to provide or reduce in a certain time period. The operating availability is defined as the time a specific capacity has to be provided by a control unit (maximal energy an actor has to provide) to prequalify and is therefore essential for the bidding capacity of EVs. Since there are no defined requirements for battery storage, it is assumed that the operating availability in the secondary regulation and tertiary markets equals 4 hours. This corresponds to the rules for pump storage power stations. Since no published figures are available on the dispatched regulation for primary balancing power capacity, the dispatch probability is taken from [14].

Providing Positive Regulation Capacity with Electric Vehicles

As discussed in the previous section there are many objections to V2G, such as battery degradation fears and the uncertainty about the battery wearing out. Analyzing these factors and the potential benefits showed that providing positive regulation capacity currently does not seem to make economic sense ([12] and see Fig. 9.4). In the market for positive secondary regulation capacity, the high-dispatch probability results in very high variable costs. Approximately one-third of these costs comprise those for battery degradation and two-thirds those for energy costs.[12] In the market for positive spinning reserves, dispatches are so seldom that the income from providing regulation capacity and the fixed costs are decisive. The capacity price is too low, and the rare dispatch occurrences result in it not being economical to make the additional investment in the bidirectional power electronics.

The only profitable way to feed energy back into the grid is to participate in the primary control market. The profits are still relatively small at today's prices for regulation energy, but this option could become more relevant in the light of the strong price increases in the past (compare [15]) and the presumed upward trend in demand due to the expansion of renewable energies. At the moment participation seems to be ruled out by the regulatory requirements. The prequalification requirements are very high and since they do not allow resources to be pooled, each generation unit has to be able to provide at least 10 MW capacity.

---

[12]The vehicles need to charge the battery before they can feed back electric energy into the grid.

Positive and negative controls were analyzed separately to reduce the complexity and reveal the different secondary and tertiary markets. In general, either negative or positive regulation is needed within one regulating zone. Therefore, it is possible to bid for positive and negative control at the same time. Especially in the secondary market, it seems promising to realize further benefits by providing positive control after loading the battery with negative control services. Moreover, pooling vehicles provides new options for advanced bidding strategies. A vehicle pool can provide positive control simply due to the reduction of the load. Hence, the pool can participate on the positive control market without a bidirectional grid connection. Overall, this could result in a in an economic benefit since there are no costs for battery degradation or the bidirectional grid connection.

Providing Negative Regulation Capacity with Electric Vehicles

The study revealed that the biggest profits can be made in the market for negative secondary regulation capacity [12]. The relatively high-dispatch probability means that the energy costs of conventional charging can be avoided. In this way, drivers are able to draw some of their power practically free of charge. The technical effort and the investments in the infrastructure are relatively small. Battery degradation does not occur since the batteries are not additionally discharged. The tertiary control market is less attractive. The necessary investments are identical, but less income can be earned due to the lower dispatch probability.

As many factors (standing vehicles, prices, load curve) change dynamically over the course of the day, we examined this option in more detail in a dynamic simulation.

Dynamic Simulation Approach

We used a Monte Carlo simulation approach, simulating a pool of vehicles on a certain weekday and repeating this experiment 500 times in order to obtain insights into the variance of the results.

For the 1-day simulation, the approach is based on two steps:

1. First, the driving behavior of BEV and PHEV users is simulated. The vehicles enter the system after their last trip of the day and they leave it with the first trip on the next day. The battery of each vehicle and its state of charge are combined in a virtual pool battery. The simulation result is the energy that could be charged to the pool battery at each point in time on that specific day (negative regulation).

2. Second, the power that could be offered by the vehicle pool that day is computed. The bid is subject to the regulations for the providers of ancillary services.

The 1-day simulation is repeated 500 times.

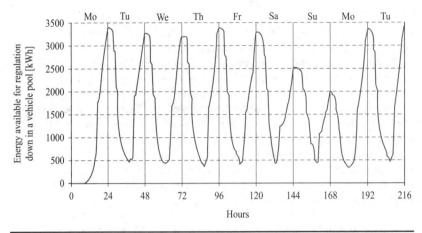

Figure 9.5 State of charge in a vehicle pool battery of 1000 vehicles. Vehicle pool consists of 10% city-BEVs (20 kWh) and 90% PHEVs (16 kWh) (assumptions about driving behavior based on [16] and [17]).

Step 1: Simulation output

Changing the simulation time from 1 day to 9 days gives an overview of the characteristics of each weekday. Figure 9.5 shows the result of the first step in a long-term simulation. The large variation in the pool battery across the 9 days indicates that considering the characteristics of different weekdays and the variation throughout the day yields significantly different results.

Step 2: Computation of the power for regulation

The power for regulation can be computed using the results from Step 1. The required dispatch time for supplying power t_{disp} is assumed to be 4 hours, which corresponds to the rules for pump storage power stations. For each point in time throughout the day it is assumed that the energy is constant and the possible power for ancillary services is computed. Weekdays are divided into a prime-time period ("Hauptzeit": HZ from 8 A.M. to 8 P.M.) and a secondary time period ("Nebenzeit": NZ from 8 P.M. to 8 A.M.). A bid is valid for one of the two time periods. The computation assumes that the pool only needs to provide power until the end of the time period, although t_{disp} may be larger. Therefore, the power increases at 8 A.M. and 8 P.M. in the example shown in Fig. 9.6.

Since the bid is valid for the whole time period, the minimum power available throughout the period determines the amount of regulation power that could be offered by the pool on that specific day. The example in Fig. 9.6 results in 90 kW in the prime time and 462.5 kW in the secondary time period. As most vehicles are used throughout the day and are not able to provide ancillary services, we focus on the secondary time period for providing regulation power.

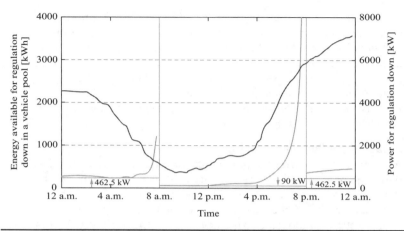

Figure 9.6 Available regulation power and energy for a vehicle pool with 1000 vehicles on a Monday (assumptions about behavior based [16] and [17]).

Results of the Dynamic Approach

In order to get an insight into the variance of the results, the 1-day simulation was repeated 500 times and the results evaluated statistically.

Impact of the Pool Size

Figure 9.7 shows the variation in regulation power (across the 500 iterations) that a pool of a certain size could provide per vehicle. It can be observed that the power converges toward a fixed value with increasing pool size. A large number of vehicles can level out the variation in the driving behavior of each individual and therefore provide more regulation power per vehicle.

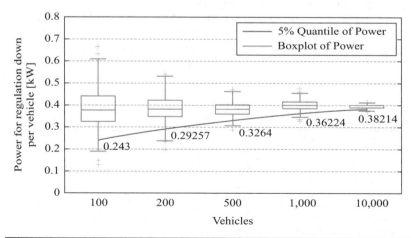

Figure 9.7 Regulation power for one vehicle on a Monday in the secondary operation time (assumptions about behavior based on [16] and [17]).

It is postulated that the pool needs to be able to provide the offered regulation power for 95% of all days (iterations). This provides additional security, since it is unlikely that ancillary power would be demanded at the weakest point in time on one of the 5% uncertain days. Therefore, the capacity that a pool of a certain size could offer is assumed to be the 5% quantile of the sample.

A pool with 10,000 vehicles can already determine the power per vehicle with a high degree of certainty.

Impact of the Duration of an Offer

According to the current requirements for the providers of ancillary power, an offer placed in the secondary control market is valid for the prime or secondary time period of 1 month. Since driving behavior depends mainly on the weekday concerned, this requirement is a strong restriction and leads to inefficient use of the pool's capabilities. Table 9.3 shows the high correlation between weekday and regulation power per vehicle. The offer for 1 month is limited by the relatively low power available at the weekend. For example, a pool of 100 cars could only offer 34 W per vehicle at the weekend, although it would be able to provide more than 10 times this amount from Tuesday to Friday. Changing the requirements would enable the pool operator to make more efficient use of the pool's capacity.

If the offers could be differentiated by weekday, the average power across the week could be increased by 360% to 900%. Smaller pools would profit more from a weekday-dependent offer than larger ones.

Impact of the Required Dispatch Time

Calculating the power for regulation in Step 2 is based on the currently required dispatch time of 4 hours. Reducing the dispatch time for a vehicle pool could increase the regulation power and facilitate participation in the regulation markets.

A decrease of the dispatch time t_{disp} by factor a yields higher power. The relation between the dispatch time and power is not reciprocally proportional as might be expected. Figure 9.8 illustrates

Pool Size	Mo	Tu – Th	Fr	Sa	Su	Minimum of All Weekdays
[Veh.]	[W/Veh.]	[W/Veh.]	[W/Veh.]	[W/Veh.]	[W/Veh.]	[W/Veh.]
100	243	508.8	501.7	34	64	34
1,000	362.2	629	612.1	77.9	112.1	77.9
10,000	382.1	663.2	644.7	100	135.9	100

Note: Secondary time period using a dispatch time of 4 hours.

TABLE 9.3   Regulation Power per Vehicle

Pool size	[Veh.]	100	1,000	10,000
Original regulation power	[W/Veh.]	34	77.9	100
Average regulation power after differentiation of weekday	[W/Veh.]	338.4	435.9	464.6
Increase of power		895%	460%	365%

Note: Secondary time period using a dispatch time of four hours.

TABLE 9.4   Increase of Power after Differentiation of the Offers Depending on the Weekday

the effect of decreasing the dispatch time from 4 hours to 1 hour. In the secondary time period, the minimum power is located in section (II) after the decrease in the dispatch time. In section (II), the difference in power for regulation is not reciprocally proportional. Therefore, the power increase is 4 times smaller than the previous power. In the prime-time period, the minimum before and after the decrease is located in section (I). In this section, the power difference is reciprocally proportional and is, therefore, 4 times higher than for a dispatch time of 4 hours.

Table 9.5 shows the power per vehicle for each weekday after reducing the dispatch time to 1 hour. The relative increase compared to Table 9.4, which shows the results based on 4 hours, is given in brackets.

On Saturdays and Sundays, the minimum capacity providing negative regulation reserves is located in section (I) (see Fig. 9.8) and the power could be increased by 300%. The weekend is the limiting period for the entire monthly offer. If the offers were distinguished by weekdays, the pool could provide four times the amount of power. If both changes were realized at the same time, i.e., differentiation by

Pool Size	Mo	Tu – Th	Fr	Sa	Su	Minimum of All Weekdays
[Veh.]	[W/Veh.]	[W/Veh.]	[W/Veh.]	[W/Veh.]	[W/Veh.]	[W/Veh.]
100	312 (28%)	696 (37%)	752 (50%)	136 (300%)	256 (300%)	136 (300%)
1,000	555.2 (53%)	965.6 (54%)	1017.2 (66%)	311.6 (300%)	448.4 (300%)	311.6 (300%)
10,000	626.2 (64%)	1110.6 (67%)	1169.6 (81%)	400 (300%)	543.7 (300%)	400 (300%)

Note: Secondary time period.

TABLE 9.5   Regulation Power per Vehicle Using a Dispatch Time of 1 Hour

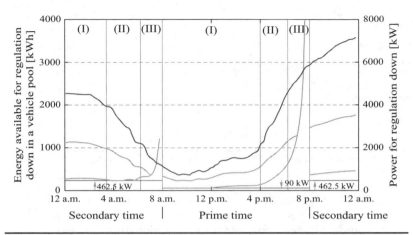

FIGURE 9.8   Available regulation power of 1000 vehicles by reducing the dispatch time on a Monday. The lower gray line represents the power for a dispatch time of 4 hours and the upper gray line the power for a dispatch time of 1 hour.

weekday and decrease in the dispatch time, the average capacity per offer would still increase, but by less than 4 times because the minimum power on weekdays is in section (II) (see Fig. 9.8).

Value of Vehicle-to-Grid Power for Regulation

The value of vehicle-to-grid power supplied by a vehicle pool strongly depends on the pool size and the requirements on the markets for ancillary power.

Table 9.6 shows the potential profit per vehicle and year excluding the administration costs of the pool operator under different conditions and pool sizes. It is assumed that a pool consists of 90% PHEV and 10% BEV.[13] The impact of the vehicle technology is relatively low because the maximum range is rarely exceeded. The result shows that it is not economical to provide ancillary power from electric vehicles under today's circumstances.[14]

If the user already had a contract for his vehicle with an energy supplier prepared to install a smart meter and provide the monthly accounting, the additional costs for providing ancillary services may be negligible. In this case, or if energy prices increase significantly (and "free charging" via negative regulation becomes very attractive), participating in the markets for regulation could already be economical, even if the suggested requirement changes were not fully implemented. The corresponding scenarios are marked gray in Table 9.6.

---

[13] The Fraunhofer ISI evaluated different scenarios for the diffusion of electric vehicles in Germany [18]. The "ISI Dominance Scenario" postulates that 98% of the electric vehicles in 2020 will be PHEV. This fraction will decrease to 86% in 2030. A different study of the first users of electric vehicles assumes that the fraction of PHEV will be between 64 and 86% in 2020 [16]. Since there is large uncertainty about which technology will dominate in the future, this study assumes a share of 90% PHEV and 10% BEV.

[14] No differentiation of weekdays and a required dispatch time of 4 hours.

		Differentiation of Offers Depending on the Weekday			
		No		Yes	
		100 Veh.	10,000 Veh.	100 Veh.	10,000 Veh.
Decrease of the required dispatch time from four hours to one hour	No	− 4.25 € (34 W)	9.40 € (100 W)	58.62 € (338.4 W)	84.88 € (464.6 W)
	Yes	16.85 € (136 W)	71.44 € (400 W)	93.36 € (506.3 W)	168.01 € (867.3 W)

Color codes indicate profitability: Dark gray: not profitable; gray: may be profitable in the future; white: profitable.

TABLE 9.6   Potential Power and Value of V2G per Vehicle and Year Under Different Conditions and Pool Sizes

Generally, it is favorable to integrate many vehicles in one pool in order to even out the stochastic behavior of individuals and thus allow better forecasts of the possible regulation power. If the suggested changes of decreasing the dispatch time and integrating a weekday-based differentiation of the offers were implemented, a large vehicle pool could already be economical even at today's energy prices. A differentiation between weekday and weekend is also a reasonable improvement for EVs.

Electricity Markets

Besides regulation markets, EVs can participate in electricity markets by using price differences between base and peak load generation. In a hypothetically perfect electricity market, prices are represented by marginal electricity generation costs. The marginal generation costs of power plants are mainly determined by the efficiency and fuel costs. In Europe, the opportunity costs for CO_2 emission certificates also increase the marginal generation costs. In reality, an additional margin or a markup might be added in certain hours to cover fixed costs, such as capital costs or operational expenditures, and to generate profits. Power plants with a higher specific investment and higher efficiency or lower fuel costs are used as baseload power plants.[15] Gas turbines and, in some cases, oil power plants with lower investment and efficiency are used to provide peak power. With general load fluctuation and a power plant park with mostly dispatchable generation, peaks traditionally appear around noon and during the early evening. During nighttime hours, when demand is the lowest, primarily baseload power plants operate.[16] The spread of

---

[15]For example, coal power plants and combined-cycle gas power plants.
[16]From an environmental perspective, load shifting can result in higher utilization of coal power plants that are often used to provide baseload power.

marginal generation costs between a typical baseload power plant[17] and a peak power plant[18] is approximately 45 euros/MWh. In this simplified case, the maximal possible revenue is 90 euros[19] for an EV owner with a demand of 2000 kWh per year. Including investment costs for automated load shifting and operation shows that the realizable profit is very low (see Fig. 9.10). Other smart grid devices in the residential sector are characterized by a shorter load management time and a lower demand, which further reduces the revenue potential. Managing all smart grid devices in a household could bring higher revenues, but profits for a single household remain low. An additional issue in this context is that consumers have become used to a high level of personal flexibility in their consumption and could perceive demand-side management as a restriction of this freedom. If consumers are viewed as "homi economici," even a low potential profit should result in a behavioral change, but in reality higher financial incentives will probably be necessary to change ingrained consumer habits. On a utility level, consumer profits due to greater demand flexibility mean substantially reduced revenues, since profits are mostly generated during peak hours. All these issues can be considered as major barriers to smart grids.

Effects of Fluctuating Electricity Generation

The recent high growth of renewable energy sources (RES) in countries like Denmark, Spain, and Germany has mostly been triggered by political incentives. But the generation costs of RES will also decrease due to high learning rates, whereas fuel prices and the investment risks for conventional power plants are rising.[20] Including external costs shows that RES are the most efficient solution for a sustainable energy supply [19].

A higher share of RES in the electricity system has a direct effect on the energy produced by conventional power plants, as in most countries, the dispatch of electricity from RES is prioritized. The generation full load hours of base power plants and therefore the basis of utility profits will drop if high generation from RES forces the plant operator to throttle or shut down the plant. On the other hand, peak capacity has to increase in this scenario in order to keep system security on the same level. In the future, it will no longer be possible to make a clear distinction between baseload during nighttime hours and peak load periods during the day, and very flexible power plant parks will be needed (see Fig. 9.9).

---

[17]For example, a coal power plant with 45% efficiency and a fuel price of 13 euros/MWh.
[18]For example, a gas turbine with 35% efficiency and a fuel price of 26 euros/MWh.
[19]Spread: 4.5 ct/kWh * Demand: 2000 kWh = Revenue: 90 euros.
[20]Natural gas prices currently decrease in the United States because of new exploration technologies and transport restrictions. But also in the United States price increases are likely in the long term.

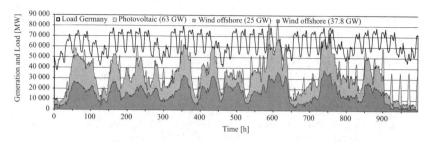

FIGURE 9.9   Fluctuation of renewable generation. Source: Own calculation, data basis [20], 2030 scenario, installed capacity: wind onshore 37.8 GW, wind offshore 25 GW, photovoltaic 63 GW.

Marginal cost-based electricity prices follow the residual load,[21] which leads to a higher price volatility if a high share of wind and photovoltaic capacity is installed. Theoretically, the marginal generation costs of wind and photovoltaic are zero, which affects the functionality of electricity markets. The merit-order-effect reduces the price on the electricity exchange markets if a high amount of fluctuating generation is available by replacing plants with high marginal costs, thus reducing the price bid for all generators [21]. Average gross profits per MWh will then drop, and utilities will tend to include a higher markup during peak times to cover fixed costs (Fig. 9.10). In hours with low residual load, baseload power plants will bid prices below marginal generation costs to avoid cycling, which results in lower or even negative prices [22]. As a result, the electricity price spread between base and peak residual load will

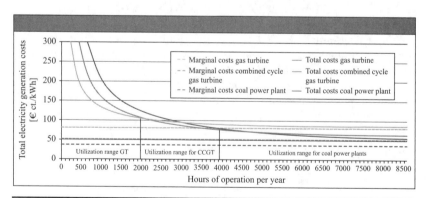

FIGURE 9.10   Marginal versus total costs of different power plant options. Assumptions: Gas turbine (GT)—specific investment, 350 euros/kW; efficiency, 37.5%. Combined cycle gas turbine (CCGT)—specific investment, 750 euros/kW; efficiency, 59%; gas price, 27.1 euros/MWh$_{therm}$. Coal power plant—specific investment, 20 euros/kW; efficiency, 49%; coal price, 12 euros/MWh$_{therm}$; CO_2 price, 15 euros/t; interest rate, 10%.

---

[21]Residual load: Total system load minus fluctuating generation from renewable energy sources.

increase. This growing price spread leads to market opportunities for flexible demand and storage technologies.

Earning Potential of Smart Grid Devices

The earning potential of smart grid devices on the electricity markets is determined by the spread between low and high prices over the load management time.[22] When analyzing the price volatility caused by fluctuating renewable energies, the selected generation time series have a very strong influence. These RES time series depend on different weather conditions and the RES technology mix, which makes it very difficult to derive general statements. For example, a 3-day period with very high wind generation can result in low electricity prices for the entire period. Hence, the usable spread over the grid management time is low. On the other hand, a 2-hour solar generation peak or a shorter period of high wind generation can result in a high price spread over the grid management time period of EVs.

Figure 9.10 illustrates possible spreads in electricity market prices and compares the marginal and total electricity generation costs depending on the operating hours. The price spread between base (coal power plant) and peak (gas turbine) for the example shown would be in the range of 40 to 50 euros/MWh. For electricity systems with a high share of fluctuating RES generation, the total generation costs of a power plant become more relevant for the pricing. The difference between the total costs of a coal power plant with 6000 hours of operation and a gas turbine with 500 hours of operation is about 130 euros/MWh. The total cost curves rise sharply with low operating hours. This shows the tendency for price peaks in electricity systems with a high capacity used to provide back-up power for fluctuating RES generation.[23]

On a residential level, electricity prices also include additional components such as grid fees and service costs. Making these fixed-price components more flexible could give rise to greater price spreads. Today, flat retail rates are most common, which do not provide load-shifting incentives. Time-of-use (TOU) rates are also available and offered in combination with heat pumps, night storage heating, and in some cases for EVs.[24] However, often very high prices are charged when a higher demand of nonshiftable appliances (cooking, lighting, etc.) is probable. Hence, the low rate in a specific time period results in low

---

[22]EVs are able to shift loads over 5 to 24 hours (see the section "Load Shifting Using Electric Vehicles").

[23]Another aspect affecting the spread is the fuel price development. If the prices for gas rise faster than those for coal, higher spreads are more likely, whereas a rising CO_2 price has the opposite effect.

[24]For example: San Diego Gas and Electric offer super off-peak: Midnight to 5 A.M. 14.4 ct./kWh; off-peak: 5 A.M. to 12 P.M., and 6 P.M. to midnight 16.7 ct./kWh; peak 12 P.M. to 6 P.M. 25.7 ct./kWh [23]

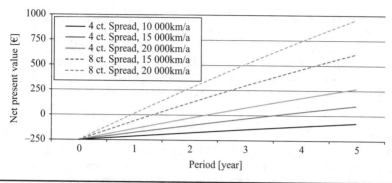

FIGURE 9.11   Earning potential for EV users depending on price spread and mileage. Assumptions: Interest rate 5%; energy usage 0.2 kWh/km, driving 250 days per year, investment 250 euros, operating expenses 40 euros per year.

costs for demand-side management appliances but higher costs for regular electricity consumption.

Figure 9.11 shows the resulting net present value for an assumed investment in automated load shifting of 250 euros and yearly operation costs of 40 euros with a price spread of 4 and 8 euro cents.

As discussed earlier, today's available price spreads of 4 ct. or lower result in a very low or negative net present value after five years of operation. Only higher price spreads in combination with a high shiftable demand result in reasonable incentives for consumers.

A current exception that enables high price spreads is the German feed-in tariff for power generated from photovoltaic. In Germany, the subsidies for consuming self-produced electricity from photovoltaic allow a spread of 12 to 16 ct. per kWh. Companies like Conergy or SMA are already selling systems to consume self produced PV generation. In this case high revenues are possible if the EVs can be used to consume electricity generated by photovoltaic that is fed into the grid and cannot be used locally. Also older PV modules that are still operating after the guaranteed subsidy period of 20 years can create a market for own PV consumption. Looking at driving patterns indicates that most highly utilized cars do not park at home during the early afternoon and therefore cannot be used to consume local PV generation from their own PV system. However, PV systems located at workplaces could be a possible option. At first glance, load shifting into peak hours does not seem to be a sustainable approach and it is probably only a matter of time until these subsidies are stopped. Nevertheless, PV is expected to be the first RES technology causing grid integration problems because of the short peaks in power output (two to three hours during the early afternoon). In some distributed grids with high PV generation in Germany the load flow is already being affected today (PV generation is greater than local demand).

For consumers without feed-in tariffs, own generation could still be lucrative because the fees for grid infrastructure and taxes included in retail electricity prices are not levied if electricity is self-generated. Grid parity, or the point at which own generation costs the same as electricity obtained from the grid, will be reached within the next five years or is even being achieved today.[25] This development could lead to a fundamental change of the electricity system and act as an enabler for load shifting and smart grid devices.

Conclusion

The analysis of grid-connected EVs in a smart grid environment reveals that income can be generated in regulation reserve markets and regular electricity markets. Compared to other home appliances, EVs are characterized by a high shiftable demand and a larger grid management period. In contrast to the U.S. studies [6], the delivery of regulation reserve to the grid is not economic in the German case under today's conditions. This is mainly due to the longer dispatch time (operating availability) necessary to prequalify as a regulation service supplier and the reduced power a vehicle can then provide for regulation. When real-life driving patterns are taken into account for a certain time period, the potential income from participating on the regulation markets is significantly reduced in comparison to approaches based on average values. The conclusions in detail are

Regulation reserve markets:

- *Participating in regulation reserve markets comes with higher control complexity.* Regulation reserve is used to balance the mismatch between planned demand and generation. Therefore, a fast reaction to control signals is required. An EV control mechanism with a response period of minutes, or even in real time, requires a complex grid connection. It is also necessary to have a separate bill for the regulation reserve service, metering to prove the service as well as pooling of several EVs. All this comes at a significantly higher cost than a timer that can be used to react to time-of-use tariffs. A grid code requiring a frequency or voltage measurement with an interruption of the charging process under certain conditions seems more likely than vehicles participating in regulation reserve markets. The control mechanism for EVs must be as simple as possible, and the technology should be available in the vehicles as much as possible.

- *Because of battery degradation, V2G is not economic.* The uncertainty about battery degradation and the possible need to change the

---

[25]In terms of module prices, grid parity is already reached in areas with high radiation and high retail electricity prices. However, because of individual and locally different costs for construction, installation, and permit, it is difficult to make an exact prediction of grid parity.

battery within the vehicle's lifetime are the main barriers to EVs. Analyzing available battery aging data and cost expectations reveals that providing V2G usually costs more than the expected revenues. The switch from gasoline to electricity as a fuel saves about 4 ct./km; load shifting with a spread of 4 ct./kWh accounts for less than an additional cent[26] if no additional technology costs are included. Arbitrage electricity trading with a 4 ct. price spread results in negative revenues even under optimistic assumptions about battery degradation.

- *A dynamic approach is required, since driving behavior has a strong impact on participation in the regulation markets.* For acceptance reasons, the vehicle owner's mobility should not be constrained when offering V2G services. This is an essential difference to the current technologies for ancillary power. Pump storage systems and gas turbines are stationary systems, whose major purpose is to generate electrical power. Electric vehicles primarily provide mobility and V2G services only as a by-product. Considering the dynamic driving behavior when estimating the V2G value leads to significantly different results compared to a static approach that focuses on average values [12].

- *The market for negative secondary control in the secondary time period offers the best potential for electric vehicles.* The market for negative secondary control represents the best offer for electric vehicles [12]. The simulation conducted provides evidence that a pool can offer more ancillary power in the secondary than in the prime-time period because most cars are connected to the grid at night. Furthermore, the demand for negative regulation reserves is larger during this secondary period, which offers the highest potential for "free charging." A combined offer in both prime and secondary periods would not necessarily improve the results since energy that was charged during the day cannot be charged again during the secondary time period and the possible regulation reserve power offered would decrease. Participation should, therefore, focus on the secondary time period.

Regular electricity markets:

- *Base and peak load spread: Fluctuating renewable energies are the activator of the smart grid.* The idea of smart grids and demand response is not new. But the progress made over the last years has been relatively small, even though there have been significant improvements in communication technologies. One explanation for the slow progress is that the (monetary)

---

[26]Spread 4 ct./kWh * consumption 0.2 kWh/km = 0.8 ct./km.

incentives for individual homes to participate are too low. EVs can shift the load and thus make use of the spread between base and peak-load power generation prices. Rising energy prices and a higher penetration of fluctuating renewable energies could widen the gap between base and peak load prices and hence increase consumer incentives.

- *Private photovoltaic generation: Own PV consumption is the first niche market for smart grid devices in Germany.* The difference between retail and wholesale prices, as well as the additional incentive for self-consumption of the generated power, could make it lucrative to consume self-generated electricity[27] even if it is not beneficial in terms of total system operation. As a result, consumers with PV home systems will start to shift their electricity demand. Products are already available that allow the synchronization of PV generation and the demand of washing machines, dishwashers, and even battery storage systems. This could also be profitable for those commercial companies without access to wholesale prices that have the possibility to manage the electricity demand of employees or company vehicles.

This study has shown that electric vehicles could play an increasingly important role in a smart grid. Their earning potentials depend on the price spread between regulation reserve and residential electricity prices and on the hourly price variations of wholesale electricity prices. Modern information and communication technologies, which are being increasingly integrated into the grid, enable the coordination of distributed energy producers and consumers. These new technologies form the basis for integrating electric vehicles. The vehicle owners have the opportunity to reduce their energy costs without limiting their mobility or degrading the battery. Thus, over time smart grid services can facilitate the diffusion of electric vehicles and improve their economic efficiency compared to conventional vehicles.

Acknowledgment

This work has been cofinanced under a grant from the German Federal Ministry of Economics and Technology (BMWI) as part of the project "MeRegioMobil" and under a grant from the German Federal Ministry of Education and Research (BMBF) within the project "Fraunhofer Systemforschung Elektromobilität." We thank Martin Wietschel, Danilo J. Santini, Jakob Zwick, Tomas Gómez, and Gillian Bowman-Köhler for discussions and critical reading of the manuscript.

---

[27]In the next 5 years, the costs of generating electricity from photovoltaic are expected to drop below retail prices in many sunny parts of the world.

References

[1] Mock, P., Schmid, S. (2009), Market prospects of electric passenger vehicles and their effect on CO_2 emissions up to the year 2030—A model based approach, German Aerospace Center (DLR), Institute of Vehicle Concepts, Plug-In Hybrid and Electric Vehicles (PHEV'09), Montreal Canada.

[2] German Mobility Panel (2002–2008). Project handling: Institute for Transport Studies of the University of Karlsruhe, Retrieved: 11 July 2011, URL: http://www.dlr.de/cs/en/desktopdefault.aspx/tabid-704/1238_read-2294

[3] Ericson, T. (2009), Direct load control of residential water heaters, Energy Policy, vol. 37.

[4] Dallinger, D. and Wietschel, M. (2012), Grid integration of intermittent renewable energy sources using price-responsive plug-in electric vehicles. *Renewable and Sustainable Energy Review* 16 (5): 3370–3382.

[5] AC Propulsion (2003), Development and Evaluation of a Plug-in HEV with Vehicle-to-Grid Power Flow, Retrieved: 11. November 2011, URL: http://www.acpropulsion.com/icat01-2_v2gplugin.pdf

[6] Tomić, J., Kempton, W. (2007). Using fleets of electric-drive vehicles for grid support. Journal of Power Sources 168, 459–468.

[7] Meinhardt, M., Burger, B., Engler, A. (2007). PV-Systemtechnik – Motor der Kostenreduktion für die photovoltaische Stromerzeugung. FVS BSW-Solar, Retrieved: 11. November 2011, URL: http://www.fvee.de/fileadmin/publikationen/tmp_vortraege_jt2007/th2007_15_meinhardt.pdf

[8] Link, J., Büttner, M., Dallinger, D. and Richter, J. (2010). Optimisation Algorithms for the Charge Dispatch of Plug-in Vehicles based on Variable Tariffs," Working Paper Sustainability and Innovation, 2010. Retrieved: 11. November 2011, URL: http://ideas.repec.org/p/zbw/fisisi/s32010.html

[9] Kalhammer, F. R., Kopf, B. M., Swan, D. H., Roan, V. P., and Walsh, M. P. (2007). Status and Prospects for Zero Emission Vehicle Technology. Retrieved: 11. November 2011, URL: http://www.arb.ca.gov/msprog/zevprog/zevreview/zev_panel_report.pdf

[10] Rosenkranz, K., (2003) Deep-Cycle Batteries for Plug-in Hybrid Application, EVS20 Plug-In Hybrid Vehicle Workshop, Long Beach, CA.

[11] Peterson, S., Apt, J. and Whitacre, J. (2009). Lithium-ion battery cell degradation resulting from realistic vehicle and vehicle-to-grid utilization, Journal of Power Sources. Retrieved: 11 July 2011, URL: http://dx.doi.org/10.1016/j.jpowsour.2009.10.010

[12] Dallinger, D., Krampe, D. and Wietschel, M. (2011). Vehicle-to-Grid Regulation Reserves Based on a Dynamic Simulation of Mobility Behavior, IEEE Transactions on Smart Grid, vol. 2, no. 2, pp. 302–313.

[13] European network of transmission system operators for electricity, (Entsoe). (2011) Web page, URL: https://www.entsoe.eu/home/

[14] Tomić, J. and Kempton, W. (2007). Using fleets of electric-drive vehicles for grid support. *Journal of Power Sources* 168, 459–468.

[15] Holttinen, H., Meibom, P., Orths, A., Lange, B., O'Malley, M., Tande, J.O., Estanqueiro, A., Gomez, E., Söder, L., Strbac, G., Smith, J C., Van Hulle, F. (2008). Impacts of large amounts of wind power on design and operation of power systems, results of IEA collaboration. Results of IEA collaboration: Brian K Parsons; National Renewable Energy Laboratory (U.S.)

[16] Biere, D., Dallinger, D. and Wietschel, M. (2009). Ökonomische Analyse der Erstnutzer von Elektrofahrzeugen. Zeitschrift für Energiewirtschaft, Bd. 33, S. 173–181.

[17] MiD-2002. Mobilität in Deutschland. www.clearingstelle-verkehr.de: DIW Berlin & DLR-Institut für Verkehrsforschung.

[18] Wietschel, M., Dallinger, D., Peyrat, B., Noack, J., Tübke, J., Schnettler, A., et al. (2008). *Marktwirtschaftliche Analysen für Plug-In-Hybrid Fahrzeugkonzepte*. Studie im Auftrag der RWE Energy AG.

[19] Awerbuch, S. (2008). Energy Economics, Finance & Technology, Retrieved: 11 November 2011, URL: http://www.awerbuch.com/

[20] Nitsch, J., Pregger, T., Scholz, Y., Naegler, T., Sterner, M., Gerhardt, N., Von Oehsen, A., Pape, C., Saint-Drenan, Y. and Wenzel, B. (2010). Langfristszenarien und Strategien für den Ausbau der erneuerbaren Energien in Deutschland bei Berücksichtigung der Entwicklung in Europa und global, Deutsches Zentrum für Luft- und Raumfahrt, Fraunhofer Institut für Windenergie und Energiesystemtechnik, Ingenieurbüro für neue Energien, vol. BMU - FKZ0, Retrieved: 11 November 2011, URL: http://www.erneuerbare-energien.de/inhalt/47034/40870/

[21] Sensfuß, F. (2007). Assessment of the impact of renewable electricity generation on the German electricity sector - An agent-based simulation approach, University of Karlsruhe (TH), 2007.

[22] Genoese, F., Genoese, M. and Wietschel, M., (2010). In Occurrence of negative prices on the German spot market for electricity and their influence on balancing power markets. EEM2010 - 7TH INTERNATIONAL CONFERENCE ON THE EUROPEAN ENERGY MARKET

[23] San Diego Gas and Electric (2011). Time of use electricity rate for electric vehicles, Retrieved: 11 November 2011, URL: http://regarchive.sdge.com/environment/cleantransportation/evRates.shtml

CHAPTER **10**

Low-Voltage, DC Grid—Powered LED Lighting System with Smart Ambient Sensor Control for Energy Conservation in Green Building

Yen Kheng Tan
Energy Research Institute @ Nanyang Technological University, Singapore

King Jet Tseng
Nanyang Technological University, Singapore

Introduction

Many opportunities to save waste energy exist in building environments. Besides the general offer of high efficacy that significantly reduced power consumption, extended operating life, and reduced maintenance are two other important factors that favor the eventual conversion of most lighting uses to solid-state sources, most particularly light-emitting diodes (LEDs). In addition, the key challenge in our drive to energy sustainability lies in cutting energy consumption, without negatively impacting our lifestyle, is the approximately one-third of our electricity consumption today for lighting applications [1]. To underscore this contention, after three years of relatively slow growth, the market use of LEDs has virtually exploded into double-digit growth in the past 2 years, particularly in fixed-position lighting [2] found in application areas like buildings, i.e., common areas, office space, etc. According to statistics presented in a U.S. Department of Energy (DOE) report [2], there is a common consensus on energy-efficient LED lighting systems for strong penetration into various lighting application areas, like green building.

However, one of the subtle obstacles facing LEDs, despite their otherwise compelling benefits, is that driver circuits for these devices must include power conversion capability to transform alternating current (AC) branch distribution voltage (typically 240 Vac) to low-voltage direct current (DC) power. While this process is fairly simple, it adds cost and can reduce the otherwise extraordinary power conversion efficiency of the LEDs themselves. The AC-to-DC power conversion process is not just lossy, but also greatly unwanted and unnecessary. Given today's typical building AC power distribution infrastructure, however, there is not much choice.

To overcome the drawbacks of needing to use the conventional AC-DC power conversion approach, a low-voltage (LV) DC grid is proposed. In typical commercial and residential buildings, a 110/230 Vac grid is incorporated to provide electrical power to all kinds of electrical loads, regardless of AC or DC loads, in the buildings. However, a large portion of these electrical loads are DC in nature and they are generally low-voltage, low-power electronic devices, like LED lighting systems. Rather than using the AC power from the existing AC grid converted into DC power to support these DC devices, it makes more sense to have a separate LV DC grid be integrated in the buildings instead. The LV DC grid opens up many opportunities for other advanced technologies, like DC renewable energy sources, i.e., solar and thermal, energy storage, electric vehicles, i.e., vehicle to grid (V2G) or grid to vehicle (G2V), etc., to be introduced into the green smart building context.

When compared with the conventional AC power grid, the LV DC grid is a more efficient way to provide DC power for the electrical LED lighting system. Of course, LEDs are not the first devices to benefit from the LV DC grid; expandable for all types of electronic devices used in the buildings have the same—they are DC devices, trying to exist in an AC environment, like computers, printers, cell phone chargers, and assorted other personal-use devices. These low-power electronic loads, running on standard voltage of 3 V, 5 V, 9 V, or 12 V, are greatly benefited by the low-voltage feature of the 24V DC grid. On top of that, the LV DC grid allows the electronic loads to do away with their bulky and lossy power adaptors and replaces them with a direct plug-and-play DC power main. Instead of having many low-power-rating AC-DC and DC-DC power converters, a single power conversion unit connecting the AC power source and the LV DC grid together is sufficient. As such, with this LV DC grid embedded into the building, the electronic devices can now easily tap their electrical power from the DC main to operate.

This chapter will also discuss achieving energy conservation from adequate control of the LED lighting system. The brightness of the room, hence, the electrical power consumed by the LED lightings, is controllable according to the needs of the occupants as well as their presences. This is really the smartness brought about by the miniaturized ambient wireless sensors deployed around a building. Unlike the conventional sensors deployed at fixed and unrealistic positions for sensing, the effectiveness of the proposed ambient wireless sensors is high and reliable. On top of that, this wireless feature removes the additional cost incurred in the conventional approach where the sensed signals are channeled back to the control station via long and lossy cables. In this chapter, an energy-efficient LED lighting system powered directly from a low-voltage DC grid will be discussed. More details of the design and development of this integration will be provided. Furthermore, the lighting system is controlled by a set of ambient wireless sensors to conserve any unwanted waste of energy incurred from the negligence of the building's occupants. The ambient wireless sensor mainly comprises an energy source, i.e., a battery, a power management circuit, a microcontroller, a radio frequency (RF) transmitter, and some sensory devices, i.e., light sensor. The light irradiance level is first sensed to determine the ambient brightness and then compared with the desired lighting condition. The digital controller will send the actuating signal, in a wireless manner, to control the LED lighting system.

Low-Voltage DC Microgrid

Looking back a century at the struggle for dominance in the business of electricity, a great battle was fought over which paradigm would hold sway, AC or DC [3]. A great deal of business history was written about these attacks and counterattacks, as well as a few torrid battlefield accounts, which all boil down to, for our purposes, four important points: (1) wholesale power production in large plants was cheaper than many distributed small ones; (2) AC could travel long distances with low losses, unlike DC; (3) incandescent lamps were the majority of the load and they operated on AC or DC; and (4) semiconductors had not yet been invented. Is the AC or DC power system the future for our next generation? Pros and cons? Let's explore this in the chapter with respect to building context.

Motivation: AC Grid versus DC Grid Technology

In this post-modern era, our conventional electric power system is still designed and structured to move centralized station AC power, via high-voltage transmission lines and lower voltage distribution lines, to households and businesses that use the power in incandescent lights, AC motors, and other AC equipment. Ironically to the source side, today's consumer equipment and tomorrow's distributed renewable generation, which is the demand side, requires us to rethink this conventional AC model. Electronic devices (such as computers, fluorescent lights, variable-speed drives, and many other household and business appliances and equipment) need DC input. Yet, all of these DC devices require conversion of the building's AC power into DC for use, as illustrated in Fig. 10.1, and that conversion typically uses inefficient rectifiers.

Referring to Fig. 10.1, it can be seen that the number of power conversion stages of a DC grid as compared to an AC grid has been reduced from four steps to two steps where the additional DC/AC and AC/DC converters have been removed. Moreover, distributed renewable generation (such as rooftop solar) produces DC power but must be converted to AC to tie into the building's electric system, only later to be reconverted to DC for many end uses. These AC-DC conversions (or DC-AC-DC in the case of rooftop solar) result in substantial energy losses. There are more than these reasons where it makes sense to explore an alternative to the present AC system, especially for the building context of this chapter.

One possible solution is a DC microgrid, which is a DC grid within a building (or serving several buildings) that minimizes or eliminates entirely these conversion losses. In the DC microgrid system, AC power converts to DC when entering the DC grid, using a high-efficiency rectifier, which then distributes the power directly to DC equipment served by the DC grid. On average, this system reduces AC-to-DC conversion losses from an average of about 32% down to 10% [3]. In addition, as seen in Fig. 10.2, rooftop photovoltaic (PV) and other distributed DC generation can be fed directly to DC equipment, via the DC microgrid, without the double conversion loss (DC to AC to DC), which would be required if the DC generation output was fed into an AC system. The DC grid is easier and more efficient for renewables to be integrated with it.

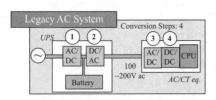 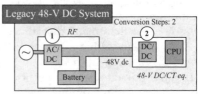

FIGURE 10.1   Architecture of AC and DC systems [4].

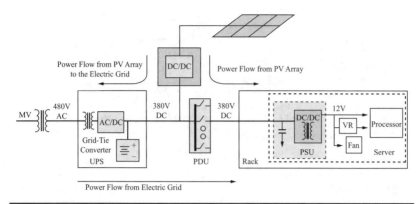

Figure 10.2 DC grid–tied photovoltaic (PV) system hybrid with AC grid system [5].

Emerging Standards for LVDC

EMerge Alliance, a nonprofit organization, was formed to address the convergence of the DC microgrids and the smart grid work of more than 50 international companies brought together to promote low-voltage DC power standards for device manufacturers and systems integrators, as illustrated in Fig. 10.3. This group expects the momentum of LEDs as a light source for common lighting applications to continue and eventually dominate the lighting market. LEDs typically plug into a 110-volt or 230-volt AC power supply that converts that power to 24-volt DC, which is what the light source consumes to make visible

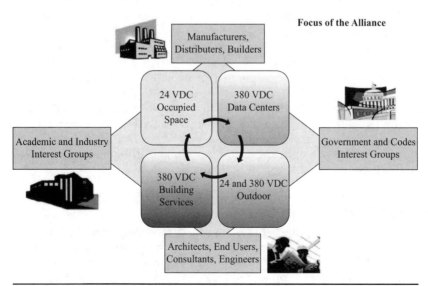

Figure 10.3 Focus of EMerge Alliance; DC microgrids throughout buildings, [5].

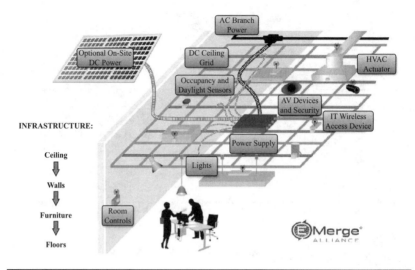

FIGURE 10.4 EMerge Alliance first standard ceiling view [7].

light. Not coincidently, 24-volt DC is the first DC power standard promulgated by the EMerge Alliance [6] (see Fig. 10.4).

Several companies that participate in the EMerge Alliance have already developed products compatible with this DC power standard that enable a new kind of suspended ceiling that distributes low-voltage 24-volt DC power through the metal grid support structure in which ceiling tiles sit. This innovation in DC power distribution through the ceiling provides a new, highly efficient channel for DC power generators to serve DC loads, like electronically ballasted or LED lighting. According to [3], if all of the new ceilings that are installed in the United States every year were specified to distribute DC in this new way, with solar inputs, these systems would accommodate over a gigawatt of solar PV in the first 2 years. Similarly, rooftop solar could be incorporated in its native DC form at 99% efficiency to a portion of offset air-handling loads, potentially providing over 50 TWh of annual avoided peak load in the United States per annum [3]. This is energy that, as in the lighting example, brings both the user and the grid consumer base benefits that the AC paradigm does not, avoiding transmission and distribution losses as well as conversion losses at the building site.

LED Solid-State Lighting System

Nowadays, the adoption of LED lamps in daily lighting instead of the common lamps like incandescent, fluorescent, and halogen lamps, as shown in Fig. 10.5, has given rise to the opportunity of saving even more energy in many application areas, in particularly in buildings. Recently, a study has been carried out on an LED light source as a

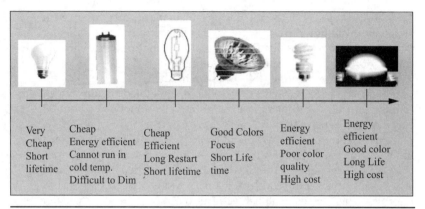

| Very Cheap Short lifetime | Cheap Energy efficient Cannot run in cold temp. Difficult to Dim | Cheap Efficient Long Restart Short lifetime | Good Colors Focus Short Life time | Energy efficient Poor color quality High cost | Energy efficient Good color Long Life High cost |

FIGURE 10.5 Different light sources in commercial use and their characteristics.

study lamp with the visual perception of the user, comparing its performance with compact fluorescent lamp (CFL) [8]. There was no substantial difference found in the visual performance using the two types of lamps. Another comparison [9] has been made between LED lamps and incandescent lamps, and again no significant difference was found. The current trends in the lighting industry show that LED lamps are going to be used vastly in domestic lighting [10–11] in the near future. This drastic change is coming to the domestic lighting market as a result of concern for energy savings, extra-long operating life, and a product that is environmentally safe.

For the general lighting in a building, particularly where the direct replacement of lamps such as MR-16, E-27, and general luminaires is concerned, important issues to be addressed include

- Directed versus broad dispersion lighting. LEDs are generally better in directed light applications.
- LED color rendition and lighting flux in required areas.
- Retrofit versus green-field applications. Retrofit of luminaire versus lamp provides different opportunity for packaging.
- Ability to disperse heat from fixtures.
- Life of the LED lamp in typical operating conditions.
- Electrical input characteristics and grid interface of the lamp.

Typical applications require the replacement of 40, 60, and 100 watt incandescent lamps, either as stand-alone lamps or inside a fixture. Furthermore, replacement of MR-16 halogen lamps with 50 watt consumption, with equivalent LEDs that consume 6 watts or less is now becoming feasible. Unlike CFLs, the ease with which the intensity can be varied provides further opportunity for control, enabling comfort and further energy savings. Finally, LED lamps are now available in a range of styles and fixture designs that allow integration with architectural lighting. In such cases, the ability to mix colors using red, green, and blue (RGB) LEDs to realize a continuously variable color rendition provides unique capability.

In solid-state lighting (SSL) research, there is a well-known statement similar to Moore's Law for silicon-integrated circuits, which claims that LEDs will increase in brightness by a factor of about 30 every decade while their costs will decrease by a factor of about 10. This statement proposed by Roland Haitz [12–13] of Hewlett Packard has held true since the 1960s, with the exception that the light output from the latest generation of LEDs has already exceeded the achievable long-term lumen per watt prediction. Moreover, further major improvements are achievable. Electrical-to-optical energy conversion efficiencies over 50% have been achieved in infrared light-emitting devices [14]. If similar efficiencies were achieved in visible light-emitting devices, the result would well exceed 150 lm/W for a white-light source. This would be nearly two times more efficient than fluorescent lamps, and ten times more efficient than incandescent lamps. So energy savings possible when using SSL devices for general lighting applications are enormous [15–16].

Intelligent Wireless Sensors Network System for Smart-Induced Green Building

Smart buildings are a field closely linked to smart grids. Smart buildings rely on a set of technologies that enhance energy efficiency, energy saving, and user comfort, as well as the monitoring and safety of the buildings. Technologies include new, efficient building materials as well as information and communication technologies (ICTs). An example of newly integrated materials is a second façade for glass skyscrapers. The headquarters of the New York Times Company has advanced ICT applications as well as a ceramic sunscreen consisting of ceramic tubes that reflect daylight and thus prevent the skyscraper from collecting heat [17].

ICTs can be used in several areas of a green building, and they include: (1) building management systems that monitor heating, lighting, and ventilation; (2) software packages that automatically switch off devices such as computers and monitors when offices are empty; and (3) security and access systems. These ICT systems can be found at both the household and office level. According to Siemens, sensors and sensor networks in smart building systems significantly contribute to energy reduction. They estimate the energy savings due to more precise climate, air quality, and occupancy sensors at 30% compared to buildings with traditional automation technology. Among these ICT green building technologies, lighting control systems that deliver the correct amount of light, where you want it, and when you want it is one of the more prominent and promising technologies. Lights can automatically turn on, off, or dim at set times or under set conditions, plus users can have control over their own lighting levels to provide optimal working conditions [6]. Lighting control helps to reduce costs and conserve energy by turning off or dimming lights when they are not required. As opposed to the conventional fixed-position sensory devices, which are neither robust, flexible, nor intelligent enough to monitor the environment conditions (ambient intelligence) for adequate control of the lights, the wireless sensor network and its sensors are introduced into the lighting control process to achieve a truly smart lighting system.

Wireless Sensor Network and Its Sensors: A Technology Overview

The term "smart lighting system" refers a system where multiple lighting fixtures and sensors collecting environmental information, hence ambient intelligence, are connected and they cooperate, forming a network [5]. With good knowledge about the ambient condition of lighting areas in the building, it is so much easier and efficient for the control system to manage the energy consumption of each LED light and interact with its low-voltage DC grid. Other benefits of having the sensory devices connected in network form in the smart lighting system include wide coverage of fault monitoring, coherent protection and remedial abilities, a reduction in wiring (hence its related power losses and monetary costs), and easy implementation and maintenance (self-sustainable with energy-harvesting technology [18]).

The main objectives of such a smart and coherent lighting system are to provide energy saving on the one hand and user satisfaction on the other hand through the cooperation of the individual nodes communicating in a wireless manner. Each of the sensor nodes, coupled with their respective LED lights, has the means to speak with one another to communicate and relay information about the ambient conditions back to the base station for post-processing and control. Several researchers are working toward developing a smart lighting system [19–20]. Thus far, research in this field has been focused mostly on utilizing traditional light sources such as incandescent and fluorescent. Energy savings of up 40% have been achieved by adopting lighting preferences such as daylight harvesting, occupancy sensing, scheduling, and load shedding [21].

To achieve the said smart lighting system, a wireless sensor network (WSN) with many sensor nodes meshed together in a wireless network fashion is introduced to sense and potentially also control its deployed environment. WSN communicates the information through wireless links "enabling interaction between people or computers and the surrounding environment" [22]. The data gathered by the different nodes is sent to a sink, which either uses the data locally, through, e.g., actuators, or which "is connected to other networks (e.g., the Internet) through a gateway [22]. Figure 10.6 illustrates a typical WSN. Within a

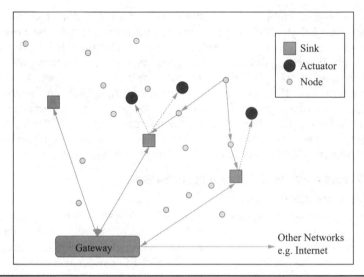

FIGURE 10.6   A typical wireless sensor network.

WSN, sensor nodes are the simplest devices in the network. As their number is usually larger than the number of actuators or sinks, they have to consume low power, be small in size, be light in weight, and be relatively cheap. The other communication devices in the network are more complex because of the energy-hungry functionalities they have to provide [22].

A sensor node, as shown in Fig. 10.7, typically consists of five main parts: One or more sensors gather data from the environment. The central unit in the form of a microprocessor manages the tasks. A transceiver (included in the communication module in Fig. 10.7)

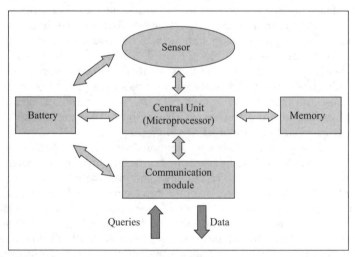

FIGURE 10.7   Architecture of a sensor node.

communicates with the environment, and memory is used to store temporary data or data generated during processing. The battery supplies all parts with energy (see Fig. 10.7). To assure a sufficiently long network lifetime, as energy efficiency in all parts of the network is crucial. Due to this need, data-processing tasks are often spread over the network, i.e., nodes cooperate in transmitting data to the sinks [22]. Although most sensors have a traditional battery there is some early-stage research on the production of sensors without batteries, using similar technologies to passive radio frequency identification (RFID) chips without batteries.

Sensors measure multiple physical properties, and include electronic sensors, biosensors, and chemical sensors. This chapter deals mainly with sensor devices, which convert a signal detected by these devices into an electrical signal, although other kinds of sensors exist. These sensors can thus be regarded as "the interface between the physical world and the world of electrical devices, such as computers" [17]. The counterpart is represented by actuators that function the other way round, i.e., whose tasks consist in converting the electrical signal into a physical phenomenon (e.g., displays for quantities measures by sensors, e.g., speedometers, temperature reading for thermostats). Table 10.1 provides examples of the main sensor types and their outputs. Further sensors include chemical sensors and biosensors, but these are not dealt with in this chapter. Outputs are mainly voltages, resistance changes, or currents. Table 10.1

Physical Property	Sensor	Output
Temperature	Thermocouple	Voltage
	Silicon	Voltage/current
	Resistance temperature detector (RTD)	Resistance
	Thermistor	Resistance
Force/Pressure	Strain gauge	Resistance
	Piezoelectric	Voltage
Acceleration	Accelerometer	Capacitance
Flow	Transducer	Voltage
	Transmitter	Voltage/current
Position	Linear variable differential transformer (LVDT)	AC voltage
Light Intensity	Photodiode	Current

Souce: OECD based on Wilson, 2008.

TABLE 10.1 Examples of Sensor Types and Their Outputs [22]

shows that sensors that measure different properties can have the same form of electrical output.

The simplest lighting control system turns off or dims lights at a specified time when the building is assumed to be empty, and turns lights back on again before people arrive for work the next day. This is a start, but with today's offices where people are increasingly working longer, more flexible hours, additional controls are needed. Occupancy sensors are useful not only to address flexible working hours, but also to control lights in areas with irregular usage patterns. For example, lights could be dimmed by default in a large room like a laboratory or warehouse. When the sensors detect that someone has entered, the lights corresponding to the location in which the person is detected can be brightened to provide sufficient illumination. Occupancy sensors can also be used to create "corridors of light" to follow people like security guards and cleaners as they move through a building [7].

With daylighting photosensors, when sunlight comes streaming in through windows, the electric lighting can be dimmed or even turned off accordingly. And as the natural light fades, the lights can automatically come back on again. This helps not only to conserve lighting energy, but also to reduce the amount of heat being emitted by the electric lights, which in turn, can help save money on air conditioning costs. In addition to the automated control provided by the timers and sensors, lighting control systems can also place control in the hands of individuals. People often require different levels of lighting depending on factors such as their age and the type of work they are doing. Lighting systems can provide the ability for office workers to adjust personal lighting levels directly from the PCs on their desks. All these smart lighting features are going to be an important and indispensable part of the smart-induced green buildings. In the next section, a case study of the low-voltage, DC grid–powered LED lighting system with smart ambient sensor control for energy conversation in a green building will be provided.

Case Study: Low-Voltage, DC Grid–Powered LED Lighting System with Smart Ambient Sensor Control for Energy Conservation in a Green Building

In this section, a case study of an existing test bedding project is discussed. The objective of this test bedding project is to employ solid-state devices like LED light engines and lighting fixtures, coupled to a low-voltage DC (LVDC) grid and controlled by the ambient wireless sensors. Taking the form of a hybrid distribution layer of low-voltage (typically 24 Vdc) power, it does not replace AC in a building, but complements it. After all, there are still some high-current loads that favor the use of AC power. The goal of this test bedding project is to efficiently aggregate or eliminate the multiple AC-to-DC conversions, thereby making devices simpler, safer, and more flexible in use. The project intends to achieve the following features, namely:

1. *Efficiency* Both alternative power generation and device consumption become more efficient with the consolidation or elimination of poor and highly fractionated power conversion.

2. *Cost effective* More and more devices, like LEDs, are native users of DC power, and therefore can be easier to build and are smaller in size when directly connected to DC power.

3. *Safe to use* Low-voltage DC power allows the use of greatly simplified and less expensive low-voltage wiring and device protection, greatly reduces spark and fire hazards, and eliminates shock/startle hazards.

4. *Flexible to deploy* Low-voltage DC power allows for "hot-swap" plug-and-play connectivity that essentially can be embedded into an existing building structure and elements, i.e., suspended ceiling grid, modular furniture, etc.

5. *Smart control* Adding wireless ambient sensors as closed-loop feedback for smart control. This provides energy saving when only the required amount of lumen from the LED lighting system is applied. When daylight is available, the power consumption of LEDs will reduce automatically. Intelligent human detection wireless sensors will be employed to switch off lightings when there is no occupant.

The test bedding project is designed to cover an office floor space of around 200 m² and is implemented and retrofitted into an existing building. The project deliverables are as follows:

1. Simulate the light intensity of the floor area (~300–500 lux) and design the LED lighting system.

2. Direct-powered 24 Vdc driven 2' × 2' LED luminaries to illuminate the office space.

3. Low-voltage DC grid (24 Vdc power supply) with complete overload and overcurrent safety protection.

4. DC power, current, and voltage measurement sensors at source with data acquisition via PC-based system for continuous power trending. This will allow continuous monitoring and adjustment of sensor setpoint, sensors, and other external conditions to achieve energy efficiency.

5. Design and deployment of ad hoc and distributed ambient wireless sensors in a noninvasive fashion within the office environment.

6. Design and implement smart control protocols to optimize the energy consumption of the LED lighting system.

LED Lighting Simulation

For the target office space of around 200 m², the LED lighting system is designed with the help of the lighting vendor, Philips, and simulated the light intensity of the floor area (~ 300–500 lux) with lighting software, DIALux. With kind cooperation from Philips, two of their LED products are modified to meet our defined 24 V LVDC grid requirement. The technical details of the LED lights are given in Fig. 10.8 and they are used in the LED lighting simulation. Figures 10.9 and 10.10 show the lighting setup designed for two of

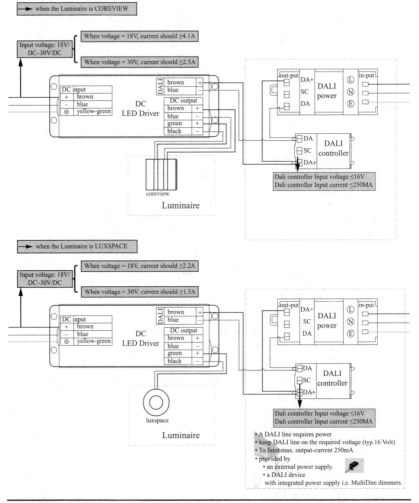

FIGURE 10.8   Technical details of the LED lights.

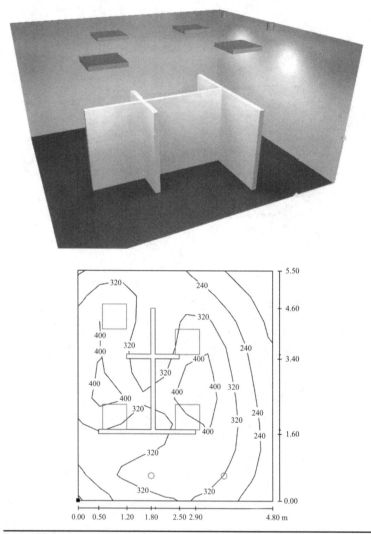

Figure 10.9 Lighting simulation of a smaller workplace.

the workspaces in the office. The partitions of the work desks are also included in the simulation to make sure that their shadowing and light blockage effects are taken in consideration. For the smaller workspace seen in Fig. 10.9, four 2' × 2' COREVIEW luminaries and two downlight LUXSPACE luminaries are used to brighten up the room. Referring to Fig. 10.9, it can be seen from the image contrast that most parts of the room are relatively bright (light gray color) and a small portion of the room is darker in dull gray color. Similarly,

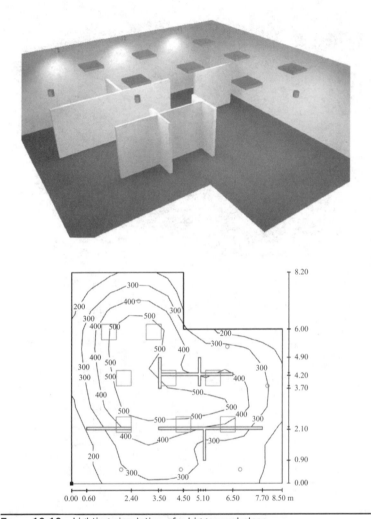

Figure 10.10 Lighting simulation of a bigger workplace.

for the bigger workspace seen in Fig. 10.10, the same observation can be seen. Generally, the ambient brightness at work desks, where the partitions are located, is good and consistent. It is an important objective of this project to have a good and undisturbed lighting condition for the office users.

In order to better understand the adequateness of the lighting condition, it is necessary to quantify the lux values at each point of the rooms illustrated in Figs. 10.9 and 10.10. The simulation results show that the lux level of both the small and big rooms ranges from 300 lux to 500 lux, which is well within the standard lighting requirement for an office workspace. This positive outcome verifies that the positioning

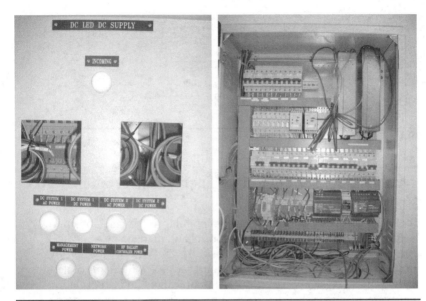

Figure 10.11   Dashboard of the LVDC power supply for the LED lightings.

of the LED lights is proper and acceptable such that the lighting condition in each of the rooms is sufficient for the users.

Implementation of a Low-Voltage, DC Grid–Powered LED Lighting System

Based on the lighting design discussed earlier, the test bedding project has been brought one step further into its development stage. Figures 10.11, 10.12, and 10.13 show some of the photographs of the low-voltage DC grid LED lighting system used in an office area, and the smart control sensor nodes, respectively.

Figure 10.12   LED lighting system in an office.

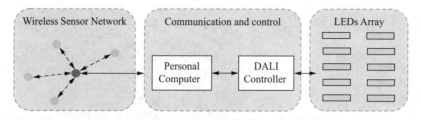

FIGURE **10.13**   Smart wireless sensor node.

Smart Ambient Sensor Control for Energy Conservation in a Green Building

The WSN is becoming more and more popular in the fields of smart buildings, home security, industrial control and maintenance, medical assistance, and traffic monitoring. Basically, the WSN is formed by multiadvanced technologies such as smart sensors, MEMS (mircoelectromechanical systems), wireless communication and information management, and embedded computing. In this chapter, the proposed architecture of the smart wireless sensor network for conserving the energy wasted in the LED lightings is depicted in Fig. 10.14.

Referring to Fig. 10.14, the ambient sensors of the smart lighting system are wirelessly networked together in a star topology. The existing WSN platform based on Texas Instruments' SimpliciTI wireless communication protocol was used to set up a star communication network; the access point (AP) sensor node functions as a data collection hub of the end devices (ED), and each ED is programmed to function as a sensor node in the WSN. According to the greenhouse criteria, SimpliciTI technology is an efficient solution for greenhouse applications because of its low cost, low power

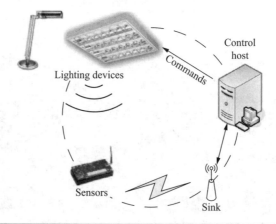

FIGURE **10.14**   Closed-loop smart lighting device control procedure.

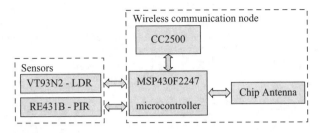

Figure 10.15   Block diagram of a wireless sensor node.

consumption, and high network capacity. The platform of sensor nodes used in this test bedding project is the eZ430-RF2500 wireless development tool manufactured by Texas Instruments, which is composed of the MSP430F2274 microcontroller and CC2500-2.4GHz wireless transceiver. The former is mainly digital processing, while the latter is used to perform the sleep-mode task to lessen the operating current and save the consuming energy.

The hardware structure of a sensor node is shown in Fig. 10.15. A sensor node device is constituted by integrating multisensors together such as pyroelectric infrared (PIR) sensors and light sensors (light-dependent resistor, or LDR). In the first module, the LDR sensors have been used to control the LEDs system to meet the reference value. The LEDs system will be turned on/off or dimmed automatically during office hours. After office hours, as usual, the lighting system is automatically turned off by an LDR sensor-based control system; however, the lighting system at the locations where the office staffs are still working needs the on state to be maintained based on PIR sensors. These PIR sensors detect human movement and give back the sensing signal to let the control system know that the staffs are still at work and the lighting needs to be on even after office hours.

In this application, the address information of each LED light and the sensor nodes are encoded to ensure that each sensor node will control the particular LEDs accordingly. Besides that, the reference value for each sensor, varying between 300 lux and 500 lux, can be changed flexibly versus the specific location of that sensor node. This lighting adjustment feature not only helps to alter the brightness of the lights to suit the individual user but also helps to conserve the energy consumed by the lights themselves. This is how energy conservation is achieved, through smart lighting systems, in a smart-induced green building. The LED lighting system is controlled by the proposed smart wireless sensor network as summarized in the following steps:

- Lux sensing and transmitting by ED back to AP and then calculate the actual room lighting lux value by the computer connected via RS485 to the AP.

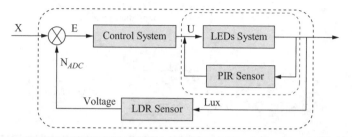

Figure 10.16 Proposed closed-loop smart lighting control scheme with the consideration of human activity.

- Determine the time to change between various sensing modes—LDR sensor or PIR sensor—so as to gain better ambient intelligence and at the same time, conserve energy consumption of the EDs for prolonged operational lifetime.

- Receive and classify sensed signals from wireless sensor nodes, compare feedback signal with reference control signal, and then convey the desired actuating signal to the digital addressable lighting interface (DALI) controller interfaced with the electronic ballast of the LED lights.

The principle of operation of the smart lighting system is illustrated as follows: Each sensor node of the smart WSN is set to receive the light lux value from the LDR sensor periodically every two seconds. The sensed lux value is then converted into a digital number (N_{ADC}) performed by its onboard 10-bit analog-to-digital converter (ADC). This digital output is calculated by the closed-loop control approach, with the reference value (X) set by users as illustrated in Fig. 10.16, where the control signal is obtained by:

$$E = X - N_{ADC}$$
$$U = k_p * E + k_i * \text{Sum } E$$

(10.1)

As soon as the control signal, U, is calculated, the control signal and other relevant information are tied together into a string of messages sent to the AP, which is connected with the PC via serial communication. This message contains information like the address of the sensor node, energy level, and the calculated U value from the sensor node (ED).

After the sensor node sends the string of messages, the access point determines the heading number and area number of the particular LED light for control. Note that the address of each LED is encoded so that the sensor node can only control its assigned LEDs. In this application, the group of LEDs can be controlled by one sensor node. The access point is capable of determining the time to control the sensing frequency of the sensor node as well as to change the mode between PIR sensors and LDR sensors when possible. For example, LDR sensors control the LDEs system during office hours from 8:00 A.M. to 6:30 P.M., and the control mode is selected by PIR sensors.

The next step is to control the LED system. A personal computer has been used to send an 8-byte message to the DALI controller via serial communication to turn on/off or dim up/down the LEDs. The message comprises the Optcode together with the U value from the sensor node.

Conclusion

The chapter presents a whole new approach to how electrical power is distributed in a decentralized situation in the context of a building: the concept of the DC microgrid, which is nothing new, but simply the LVDC grid. The LVDC grid is a more direct and efficient approach, without all the extra power conversion stages, to powering up many of the electrical loads (DC in nature) in a building. LED lighting is one of the most immediate beneficiaries of this LVDC grid introduction. In this chapter, a case study of the low-voltage, DC grid—powered LED lighting system with smart ambient sensor control for energy conversation in a green building is provided and illustrated with design details. The test bedding project has successfully been implemented, and future research outcomes will be reported subsequently.

References

[1] R. Mehta, D. Deshpande, K. Kulkarni, S. Sharma and D. Divan, "LEDs - A Competitive Solution for General Lighting Applications," *IEEE Energy2030*, 2008.

[2] Building Technologies Program, Office of Energy Efficiency and Renewable Energy, "Energy Savings Estimates of Light Emitting Diodes in Niche Lighting Applications," *U.S. Department of Energy*, October 2008.

[3] P. Savage, R.R. Nordhaus, and S.P. Jamieson, "DC Microgrids: Benefits and Barriers," *Yale School of Forestry & Environmental Studies*, April 2010.

[4] Kaoru Asakura, "Development of Higher-Voltage Direct Current Power Feeding System in Datacenters," *NTT Energy and Environment Systems Laboratories*, DC Building Power Asia, December 2010.

[5] Jeff Shepard, "Improving the Future of Power," *Darnell Group*, DC Building Power Asia, December 2010.

[6] Executive Summary of 1.0 version the standard, http://www.emergealliance. org accessed on March 16, 2011.

[7] Karen Lee, "A Hybrid AC/DC Power Platform for Commercial Buildings" *EMerge Alliance-Smart Power Standards*, July 2010.

[8] S. Varadharajan, K. Srinivasan, S. Srivatsav, A. Cherian, S. Police, R.K. Kumar, "Effect of LED-based study-lamp on visual functions," in proceedings of Experiencing Light, Eindhoven, The Netherlands, 2009.

[9] S. Newel and B. Albert, "Factors in the perception of brightness for LED and incandescent lamps," *Society of Automotive Engineers Transactions*, vol. 114, pp. 908–920, 2005.

[10] M. Richards and D. Carter, "Good lighting with less energy: Where next?," *Lighting Research and Technology*, vol. 41, no. 3, pp. 285–286, 2009.

[11] T. M. Goodman, "Measurement and specification of lighting: A look at the future," *Lighting Research and Technology*, vol. 41, no. 3, pp. 229–243, 2009.

[12] R Haitz, F. Kish, J.Y. Tsao, and J. Nelson, "The Case for a National Research Program on Semiconductor Lighting," *Optoelectronics Industry Development Association (OIDA) forum*, Washington D.C., 1999.

[13] A. Bergh, G. Craford, A. Duggal, and R. Haitz, "The Promise and Challenge of Solid-State Lighting," *Physics Today*, vol. 54, no. 12, pp. 42–47, December 2001.

[14] M. Wendt and J.W. Andriesse, "LEDs in Real Lighting Applications: from Niche Markets to General Lighting," 41st IAS Annual Meeting on IEEE Industry Applications Conference, pp. 2601–2603, 2006.

[15] M. Kendall and M. Scholand, "Energy Savings Potential of Solid State Lighting in General Lighting Applications," *U.S. Department of Energy*, Washington, DC, April 2001.

[16] D.A. Steigerwald, J.C. Bhat, D. Collins, R.M. Fletcher, M. Ochiai Holcomb, M.J. Ludowise, P.S. Martin, and S.L. Rudaz, "Illumination With Solid State Lighting Technology," *IEEE Journal On Selected Topics In Quantum Electronics*, vol. 8, no. 2, pp. 310–320, March/April 2002.

[17] Working Party on the Information Economy (WPIE), "Smart Sensor Networks: Technologies and Applications for Green Growth," *Organisation for Economic Co-Operation and Development (OECD)*, December 2009.

[18] Y.K. Tan and S.K. Panda, "Review of Energy Harvesting Technologies for Sustainable WSN," *Sustainable Wireless Sensor Networks*, Tan Yen Kheng (Editor), ISBN 978-953-307-297-5, INTECH, 2010.

[19] W. Yao-Jung and AM. Agogino, "Wireless networked lighting systems for optimizing energy savings and user satisfaction," *IEEE Wireless Hive Networks Conference*, pp. 1–7, Texas, August 2008.

[20] M.S. Pan, L.W. Yeh, Y.A. Chen, Y.H. Lin, and Y.C. Tseng, "A WSN-based intelligent light control system considering user activities and profiles," *IEEE Sensors Journal*, vol. 8, issue 10, pp. 1710–1721, October 2008.

[21] E. Mills, "Global lighting energy savings potential," *Light and Engineering*, vol. 10, pp. 5–10, 2009.

[22] J. Wilson, *Sensor Technology Handbook*, Newnes/Elsevier, Oxford, 2008.

[23] D.M. Han, J.H. Lim, "Smart home energy management system using IEEE 802.15.4 and zigbee," *IEEE Transactions on Consumer Electronics*, vol. 56, no. 3, pp. 1403–1410, August 2010.

CHAPTER 11

Multiple Distributed Smart Microgrids with a Self-Autonomous, Energy Harvesting Wireless Sensor Network

Josep M. Guerrero
Aalborg University, Denmark

Yen Kheng Tan
Energy Research Institute, Nanyang Technological University, Singapore

Introduction

Worldwide, electrical grids are expected to become smarter in the near future. In this sense, the increasing interest in intelligent microgrids able to operate in island mode or connected to the grid, which will be a key point to cope with new functionalities, as well as integration of renewable energy resources. A microgrid can be defined as a part of the grid with the following elements: prime energy movers, power electronics converters, distributed energy storage systems, and local loads, which can operate autonomously but also interact with the main grid. Some of the functionalities expected for these small grids include black-start operation, frequency and voltage stability, active and reactive power flow control, active power filter capabilities, and storage energy management. With these functionalities incorporated in the microgrid, electrical energy can be generated and stored near the consumption points, thus increasing the reliability and reducing the losses produced by the large power lines. These microgrid-associated advantages are brought about by the presence of many wireless sensor devices deployed in an ad hoc fashion in the microgrid. These wireless sensor devices, meshed together in network form, provide several key functionalities for achieving the proposed multiple distributed microgrids, namely wireless communication means, sensing of electrical parameters, and adequate intelligent power control. On top of the mentioned functionalities, the autonomous sensing capability of the wireless sensor network allows for real-time condition-based monitoring of the microgrids, wireless power metering, security, etc. Typically, these wireless sensor devices are battery-operated and they go into idle state after some time, say a few months or, at most, a year. To sustain the operations of these devices, the energy harvesting technology has been introduced. By harvesting the ambient energy available at the point of deployment of these wireless sensor devices, not only will the operational lifetime of the devices be lengthened, but also their capabilities can be more powerful to perform more energy-hungry tasks.

The electrical grid is tending to be more distributed, intelligent, and flexible. New power electronic equipment will dominate the electrical grid in the next decades. The trend of this new grid is to become more and more distributed, and hence the energy generation and consumption areas cannot be conceived separately. Nowadays, electrical and energy engineering have to face a new scenario in which small distributed power generators and dispersed energy storage devices have to be integrated together into the grid. The new electrical grid, also called a smart grid (SG), will deliver electricity from suppliers to consumers using digital technology to control appliances at consumers' homes to save energy, reducing cost and increasing reliability and transparency. In this sense, the expected whole energy system will be more interactive, intelligent, and distributed. The use of distributed generation (DG) of energy systems makes no sense without using distributed storage systems to cope with the energy balances.

A microgrid is a local electrical grid—for instance, inside a house—with small generators, energy storage systems, and load consumption. This house microgrid can work in isolation or connected to neighbors' microgrids, forming microgrid clusters inside a neighborhood. Thus, the microgrids can exchange energy inside the cluster instead of interchanging energy with the main grid, and thus utilize the power close to the source where generated.

In order to control the elements inside microgrids, power electronics systems are needed to connect generators, storage systems, and consumption and to enable grids to work in a cooperative way. Cooperative control systems are available for robotics, in which several robots have to cooperate for a common objective. Controllers for power electronics systems connected to the grid are available in the state of the art, but energy management and cooperative control for microgrid clusters needs a frontier approach.

Microgrids (MGs), also called minigrids, are becoming an important concept to integrate DG and energy storage systems. The concept has been developed to cope with the penetration of renewable energy systems, which can be realistic if the final user is able to generate, store, control, and manage part of the energy that will consume. This change of paradigm allows the final user to be not only a consumer but also a part of the grid.

DC- and AC-MGs have been proposed for different applications, and hybrid solutions have been developed [1–12]. Islanded MGs have been used in applications like avionic, automotive, marine, or rural areas. The interfaces between the prime movers and the MGs are often based on power electronics converters acting as voltage sources (voltage source inverters, VSI, in the case of AC-MGs). These power electronics converters are parallel-connected through the MG. In order to avoid circulating currents among the converters without using any critical communication between them, the droop control method is often applied.

In the case of paralleling inverters, the droop method consists of subtracting proportional parts of the output average active and reactive powers to the frequency and amplitude of each module to emulate virtual inertias. These control loops, also called P–ω and Q–E droops, have been applied to connect inverters in parallel in uninterruptible power systems (UPS) to avoid mutual control wires while obtaining good power sharing. However, although this technique achieves high reliability and flexibility, it has several drawbacks that limit its application.

For instance, the conventional droop method is not suitable when the paralleled system must share nonlinear loads, because the control units should take into account harmonic currents, and, at the same time, balance active and reactive power. Thus, harmonic current-sharing techniques have been proposed to avoid the circulating distortion power when sharing nonlinear loads. All of them entail distorting the voltage to enhance the harmonic current-sharing accuracy, resulting in a trade-off. Recently, novel control loops that adjust the output impedance of the units by adding output virtual reactors [5–6] or resistors [7] have been included into the droop method, with the purpose of sharing the harmonic current content properly.

Further, by using the droop method, the power sharing is affected by the output impedance of the units and the line impedances. Hence, those virtual output impedance loops can solve this problem. In this sense, the output impedance can be seen as another control variable.

Another important disadvantage of the droop method is its load-dependent frequency deviation, which implies a phase deviation between the output voltage frequency of the UPS system and the input voltage provided by the utility mains. This fact can lead to a loss of synchronization, since the bypass switch must connect the utility line directly to the MG bus. Consequently, this method, in its original version, can be only applied to islanded MGs [9]. Hence, this technique is not directly applicable to line-interactive MGs, since the transition between islanded and grid-connected modes will be difficult, due to the loss of synchronization. In addition, the inherent trade-off of this method between frequency and amplitude regulation in front of active and reactive power-sharing accuracy cannot be avoided in islanded mode [14–19]. Significant efforts have been made to improve the droop control method to avoid the aforementioned frequency deviation. In [12], a local controller shifts up the droop function to restore the initial frequency of the inverters by using an integrator to avoid frequency deviation. However, in practical situations when the inverters are not connected to the AC bus at the same time, the integrator initial conditions are different, and as a consequence the power sharing is degraded.

Another possibility is to wait for accidental synchronization, taking advantage of the fact that the mains and the MG frequencies are not exactly equal; hence, the phase of the MG drifts gradually toward the mains phase. Thus, the bypass switch can be closed when the two phases match. However, it is hazardous and can take a lot of time. Further, the transient at the time of reclosing can be uneven since the frequencies may not have matched. To overcome this problem, an inverter can be located physically near the bypass switch. This inverter then can be synchronized with the mains grid for seamless reconnection. However, during the synchronization process this inverter will be overloaded if there is a significant number of inverters supplying the MG. Thus, the reliability of those methods is considerably low and they are definitely not practical, so the use of communications seems unavoidable, it being desirable to implement together with the DG a low-bandwidth, noncritical communication system.

On the other hand, in the case of paralleling DC power converters, the droop method consists of subtracting a proportional part of the output current to the output voltage reference of each module. Thus, a virtual output resistance can be implemented through this control loop. This loop, also called adaptive voltage position (AVP), has been applied to improve the transient response of voltage regulation modules (VRMs) in low-voltage, high-current applications. However, the droop method also has an inherent trade-off between the voltage regulation and the current sharing between the converters [8–13].

To cope with this problem, an external control loop, called the secondary control, has been proposed to restore the nominal values of voltage inside the MG. Further, additional tertiary control can be used to bidirectionally control the power flowing when the MG is connected to a stiff power source or the mains (in case of AC MGs) [14]. Hierarchical control applied to power dispatching in AC power systems is well-known, and it has been used extensively for decades. Nowadays, these concepts are starting to be applied to wind power parks and were proposed for isolated photovoltaic systems. However, with the raise of power electronic-based MGs, which are able to operate both in grid connected fashion and in islanded mode, the hierarchical control and energy management systems are necessary. Some authors proposed secondary [33–36] and tertiary controllers [37–38]. The main problem to be solved in such works was the frequency control of the system. However, voltage stability and synchronization issues are also important to achieve enough flexibility to operate in both modes. Only few works conceived the MG as a whole problem, taking into account the different control levels.

MGs can be conceived to use DC or AC voltage in the local grid. Also, there are AC sources or MGs interconnected by means of power electronic interfaces to a DC MG. Thus, hybrid DC-AC MGs are often implemented, being necessary to control the power flow between DC and AC parts. In this sense, it seems reasonable that the DC-MG area can be connected to storage energy systems like batteries, supercapacitors, or hydrogen-based fuel cells. Although DC transmission and distribution systems for high-voltage applications are well-established and there is a notable increase of DC-MG projects, we cannot find so many studies about the overall control of these systems.

The chapter is organized as follows. In the next section, a general approach of the hierarchical control stem from the ISA-95 is adapted to MGs. In the third section, the hierarchical control is applied to AC-MGs, solving the trade-off of the droop method by implementing a secondary control loop and operating in grid-connected and islanded modes. In the fourth section the approach is applied over a DC-MG consisting of droop-controlled converters, which is able to share the load together with a stiff DC source, which can be a DC generator, a DC distribution grid, or rectified AC grid.

Then the chapter covers the smart wireless sensors for microgrids, as well as the energy harvesting technology used to sustain the operations of these sensors. Last, a case study on the multiple distributed smart microgrids with a self-autonomous, energy harvesting wireless sensor network is presented.

A General Approach of the Hierarchical Control of MGs

The need for standards in MG control is related to the new grid codes that are expected to appear in the near future. In this sense, ANSI/ISA-95, or ISA-95 as it is more commonly referred to, is an international standard for developing an automated interface between enterprise and control systems. This standard has been developed for global manufacturers to be applied in all industries and in all sorts of processes, like batch, continuous, and repetitive processes. The objectives of ISA-95 are to provide consistent terminology that is a foundation for supplier and manufacturer communications to provide consistent information models and to provide consistent operations models, which is a foundation for clarifying application functionality and how information is to be used. In this standard a multilevel hierarchical control is proposed, with the following levels [25–26]:

Level 5: Enterprise. The enterprise level comprises the superior management policies of a commercial entity. This level has operational and development responsibility for the entire enterprise, including all of its plants and their respective production lines.

Level 4: Campus/Plant. The campus or plant level comprises superior management policies of a branch or operational division of an enterprise, usually including the elements of the enterprise financials that are directly associated with that business entity.

Level 3: Building/Production. The building or production level comprises the management and control policies required to administer the states and behaviors of a building and its environmental and production systems

Level 2: Area/Line. The area or production line level comprises the management and control policies required to administer states and behaviors of a specific area or production line.

Level 1: Unit/Cell. The unit or cell level comprises the management and control policies required to govern the states and behaviors of a unit of automation or a manufacturing cell.

Level 0: Device. The device level comprises the set of field devices that sense and provide actuation of physical processes within the environmental and production systems.

Each level has the duty of the command level and provides supervisory control over lower-level systems. In this sense, it is necessary to ensure that the command and reference signals from one level to the lower levels will have low impact in the stability and robustness performance. Thus, the bandwidth must be decreased when increasing the control level.

In order to adapt ISA-95 to the control of an MG, zero to three levels can be adopted as follows [7–14]:

- *Level 3 (tertiary control)*   This energy production level controls the power flow between the MG and the grid.

- *Level 2 (secondary control)* Ensures that the electrical levels into the MG are inside the required values. In addition, it can include a synchronization control loop to seamlessly connect or disconnect the MG to the distribution system.

- *Level 1 (primary control)* The droop control method is often used in this level to emulate physical behaviors that make the system stable and more damped. It can include a virtual impedance control loop to emulate physical output impedance.

- *Level 0 (inner control loops)* Regulation issues of each module are integrated in this level. Current and voltage, feedback and feedforward, and linear and nonlinear control loops can be performed to regulate the output voltage and to control the current while keeping the system stable.

On the other hand, AC-MGs should be able to operate both in grid-connected and islanded modes [27]. The bypass switch is responsible for connecting the MG to the grid. This bypass switch is designed to meet grid interconnection standards, e.g., IEEE 1547 and UL 1541 in North America. Now the IEEE P1547.4 *Draft Guide for Design, Operation, and Integration of Distributed Resource Island Systems with Electric Power Systems* is in draft form [28]. It will cover MGs and intentional islands that contain distributed energy resources (DER) with utility electric power systems. The draft provides alternative approaches for the design, integration, and operation of MGs, and includes the ability to connect and disconnect to/from the grid.

In grid-connected mode, the MG operates according to IEEE 1547-2003. The transition to islanded mode is done by intentional or unintentional events, e.g., grid failures. Thus, proper islanding detection algorithms must be implemented. In islanded mode, the MG must supply the required active and reactive powers, as well as provide frequency stability and operate within the specified voltage ranges. The reconnection of the MG to the grid will be done when the grid voltage is within acceptable limits and phasing is correct. Active synchronization is required to match the voltage, frequency, and phase angle of the MG.

Hierarchical Control of AC-MGs

AC-MGs are now on the cutting edge of the state of the art [20]–[25]. However, the control and management of such a system needs still further investigation. MGs for stand-alone and grid-connected applications have been considered in the past as separate approaches. Nevertheless, nowadays it is necessary to conceive flexible MGs able to operate in both grid-connected and islanded modes. Thus, the study of topologies, architectures, planning, and configurations of MGs is necessary. This is a great challenge due to the need of integrating

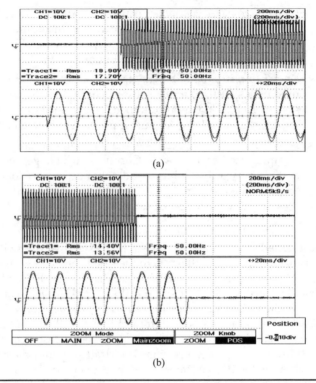

FIGURE 11.1 Current waveforms of two-inverter MGs for load step changes (a) from no-load to 7.5 kW and (b) back to no load.

different technologies of power electronics, telecommunications, generation, and storage energy systems, among others. In addition, islanding detection algorithms for MGs is necessary to ensure a smooth transition between grid-connected and islanded modes. Furthermore, security issues such as fault monitoring, predictive maintenance, or protection are very important regarding MGs' feasibility.

This section deals with the hierarchical control of AC-MGs, consisting of the same three control levels as presented in the last section. The UCTE (Union for the Coordination of Transmission of Electricity, Continental Europe) has defined a hierarchical control for large power systems, as shown in Fig. 11.1. In such a system, it is supposed to operate over large synchronous machines with high inertias and inductive networks. However, in power electronic-based MGs there are no inertias and the nature of the networks is mainly resistive. Consequently, there are important differences between both systems that we have to take into account when designing their control schemes. This three-level hierarchical control is organized as follows. The primary control deals with the inner control of the DG units, adding virtual inertias and controlling their output impedances. The secondary control is designed to restore the frequency and amplitude deviations produced inside the MG by the virtual inertias and output virtual impedances. The tertiary control regulates the power flows between the grid and the MG at the point of common coupling (PCC).

Inner Control Loops

The use of intelligent power interfaces between the electrical generation sources and the MG is mandatory. These interfaces have a final stage consisting of DC/AC inverters, which can be classified into current-source inverters (CSIs), which consist of an inner current loop and a phase-locked loop (PLL) to continuously stay synchronized with the grid, and voltage-source inverters (VSIs), consisting of an inner current loop and an external voltage loop. In order to inject current into the grid, CSIs are commonly used, while in island or autonomous operation, VSIs are needed to keep the voltage stable.

VSIs are very interesting for MG applications since they do not need any external reference to stay synchronized. Furthermore, VSIs are convenient since they can provide to distributed power generation systems features like ride-through capability and power quality enhancement. When these inverters are required to operate in grid-connected mode; they often change its behavior from voltage to current sources. Nevertheless, to achieve flexible MG, i.e., able to operate in both grid-connected and islanded modes, VSIs are required to control the exported or imported power to the mains grid and to stabilize the MG [14].

VSIs and CSIs can cooperate together in an MG. The VSIs are often connected to energy storage devices, fixing the frequency and voltage inside the MG. The CSIs are often connected to photovoltaic (PV) or small wind turbines (WT) that require maximum power point tracking (MPPT) algorithms, although those DG inverters could also work as VSIs if necessary. Thus, we can have a number of VSIs and CSIs, or only VSIs, connected in parallel forming an MG.

Primary Control

When connecting two or more VSIs in parallel, circulating active and reactive power can appear. This control level adjusts the frequency and amplitude of the voltage reference provided to the inner current and voltage control loops. The main idea of this control level is to mimic the behavior of a synchronous generator, which reduces the frequency when the active power increases [30]. This principle can be integrated in VSIs by using the well-known P/Q droop method [31]:

$$\omega = \omega^* - G_p(s) \cdot (P - P^*) \qquad (11.1)$$

$$E = E^* - G_Q(s) \cdot (Q - Q^*) \qquad (11.2)$$

with ω and E being the frequency and amplitude of the output voltage reference, ω^* and E^* their references, P and Q the active and reactive power, P^* and Q^* their references, and $G_p(s)$ and $G_Q(s)$ their corresponding transfer functions (typically proportional droop terms, i.e., $G_p(s) = m$ and $G_Q(s) = n$). Note that the use of pure integrators is not allowed when the MG is in island mode, since the total load will not coincide with the total injected power, but they can be useful in grid-connected mode to have a good accuracy of the injected active and reactive power [14]. This control objective will be achieved by the tertiary control level.

The design of $G_p(s)$ and $G_Q(s)$ compensators can be achieved by using different control synthesis techniques. However, the DC gain of such compensators (named m and n) provide for the static $\Delta P/\Delta \omega$ and $\Delta Q/\Delta V$ deviations, which are necessary to keep the system synchronized and inside the voltage stability limits. Those parameters can be designed as follows:

$$m = \Delta \omega / P_{max} \qquad (11.3)$$

$$n = \Delta V / 2Q_{max} \qquad (11.4)$$

with $\Delta \omega$ and ΔV being the maximum frequency and voltage allowed, and P_{max} and Q_{max} the maximum active and reactive power delivered by the inverter. If the inverter can absorb active power since it is able to charge batteries like a line-interactive UPS, then $m = \Delta \omega / 2P_{max}$.

Further, the primary control can be used to balance the energy between DG units with storage energy elements, e.g., batteries. In this situation, depending on the state of charge (SoC) of the batteries, the contribution of active power can be adjusted according to the availability of energy of each DG unit. Thus, the frequency droop function can be expressed as:

$$\omega = \omega^* - \frac{m}{\alpha} \cdot (P - P^*) \qquad (11.5)$$

with m being the frequency-droop coefficient, and α the p.u. level of the charge of the batteries ($\alpha = 1$ is fully charged and $\alpha = 0.01$ is empty). The coefficient α is saturated to prevent $G_p(s)$ from rising to an infinite value. This way, the DG units will provide the energy proportionally to the batteries' SoC.

In the conventional droop method used by large power systems, it is supposed that the output impedance of synchronous generators, as well as the line impedance, is mainly inductive. However, when using power electronics the output impedance will depend on the control strategy used by the inner control loops (level 0). Further, the line impedance in low-voltage applications is near to pure resistive. Thus, the control droops (Eqs. 11.1 and 11.2) can be modified according to the park transformation determined by the impedance angle θ:

$$\omega = \omega^* - G_p(s)\big[(P - P^*)\sin\theta - (Q - Q^*)\cos\theta\big] \qquad (11.6)$$

$$E = E^* - G_Q(s)\big[(P - P^*)\cos\theta + (Q - Q^*)\sin\theta\big] \qquad (11.7)$$

The primary control level can also include the virtual output impedance loop, in which the output voltage can be expressed as [17]:

$$v_o^* = v_{ref} - Z_D(s) \cdot i_o \qquad (11.8)$$

where v_{ref} is the voltage reference generated by Eqs. 11.6 and 11.7, with $v_{ref} = E \sin(\omega t)$, and $Z_D(s)$ is the virtual output impedance transfer function, which normally ensures inductive behavior at the line-frequency. Figure 11.2 depicts the virtual impedance loop in relation with the other control loops: inner current and voltage loops, and the droop control. The Thévenin equivalent circuit of an inverter with the virtual impedance [47], which consists of a controlled voltage source, $G(s)v_{ref}$, with $G(s)$ being the closed-loop voltage gain transfer function, connected to the MG through the closed-loop output impedance Z_o and the virtual impedance Z_D. Usually Z_D is designed to be bigger than Z_o; this way the total equivalent output impedance is mainly dominated by Z_D [17]. The virtual output impedance Z_D is equivalent to the series impedance of a synchronous generator. However, although the series impedance of a synchronous generator is mainly inductive, the virtual impedance can be chosen arbitrarily. In contrast with a physical impedance, this virtual output impedance has no power losses; thus, it is possible to implement resistance without efficiency losses.

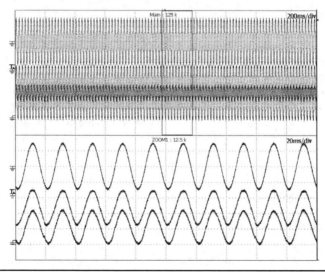

Figure 11.2 Waveforms of the parallel system sharing a linear load. Output voltage (top) and load currents (middle and bottom).

Notice that by using the virtual impedance control loop, the inverter output impedance becomes a new control variable. Thus, we can adjust the phase angle of Eqs. 11.6 and 11.7 according to the expected X/R ratio of the line impedance, $\theta = \tan^{-1} X/R$, and the angle of the output impedance at the line frequency. Furthermore, the virtual output impedance can provide additional features to the inverter, such as hot-swap operation and harmonic current sharing.

When connecting a DG unit to the MG, there are unavoidable small differences in phase and/or amplitude that result in current spikes, which can damage or overload the unit. For instance, in large wind turbines, the output impedance of the generator is increased by using external resistors and thyristors, and gradually reducing the output impedance [45]. In our case, we can implement a similar soft-starter by properly changing the value of the virtual impedance; hot-swap operation thus can be obtained, expressed by:

$$Z_D(t) = Z_f - \left(Z_f - Z_i\right)e^{-t/T} \tag{11.9}$$

with Z_i and Z_f being the initial and the final virtual impedance values, and T the time constant of the startup process.

On the other hand, by using a bank of band pass filters, we can independently adjust the output impedance "seen" by the fundamental and the current harmonics. This way, we can cope with the trade-off between the current harmonic sharing and the voltage total harmonic distortion (THD). Thus, the virtual impedance can be expressed as follows:

$$Z_D(s) = L_D \frac{2k_1 s^2}{s^2 + 2\xi\omega_1 s + w_1^2} + \sum_{\substack{i=1 \\ odd}}^{n} R_i \frac{2k_i s}{s^2 + 2\xi\omega_o s + \omega_o^2} \tag{11.10}$$

where L_D and R_i are the inductive and resistive impedance values, k_i is the coefficient of the filter for every harmonic-i term. The output impedance can be also designed taking into account, not only the power ratting of the inverter, but also the voltage dropped by this loop and the amplitude droop loop (Eq. 11.7).

These control loops allow the parallel operation of the inverters. However, there is an inherent trade-off between P/Q sharing and frequency/amplitude regulation [48–51]. Figures 11.3 and 11.4 show the experimental results of the current and voltage waveforms of two inverters sharing a common 7.5 kW resistive load. Notice the good current sharing obtained due to the use of the primary control.

Secondary Control

In order to compensate for the frequency and amplitude deviations, a secondary control is proposed. The secondary control ensures that the frequency and voltage deviations are regulated toward zero after every change of load or generation inside the MG [32–35]. The

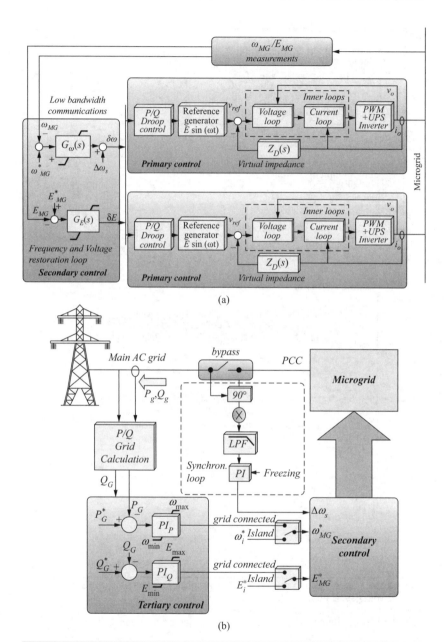

FIGURE 11.3 Block diagrams of the hierarchical control of an AC-MG. (a) Primary and secondary controls of an AC-MG. (b) Tertiary control and synchronization loop of an AC-MG.

frequency and amplitude levels in the MG ω_{MG} and E_{MG} are sensed and compared with the references ω^*_{MG} and E^*_{MG}; the errors processed through compensators ($\delta\omega$ and δE) are sent to all the units to restore the output voltage frequency and amplitude.

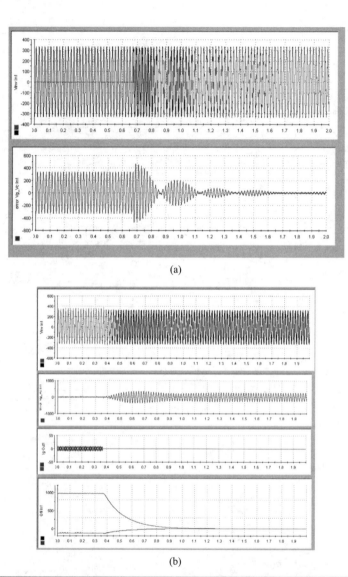

Figure 11.4 MG transition modes: (a) Synchronization process. Top: Grid and MG voltages. Bottom: Voltage difference. (b) Transfer from grid-connected to island mode. Top: Grid and MG voltages. Middle: Voltage difference and current grid. Bottom: Active and reactive power injected to the PCC.

Taking into account the grid exigencies [7], the secondary control should correct the frequency deviation within allowable limits, e.g., ±0.1 Hz in Nordel (North of Europe) or ±0.2 Hz in UCTE. It is defined as:

$$\delta P = -\beta \cdot G - \frac{1}{T_r}\int G dt \qquad (11.11)$$

where δP is the output set point of the secondary controller, β is the gain of the proportional controller, T_r is the time constant of the secondary controller, and G is the area control error (ACE), which is normally calculated in about 5- to 10-second intervals by computers in the dispatch center as:

$$G = P_{meas} - P_{sched} + K_{ri}\left(f_{meas} - f_0\right) \tag{11.12}$$

with P_{meas} being the sum of the instantaneous measured active power transferred at the PCC, P_{prog} being the resulting exchange program, K_{ri} being the proportional factor of the control area set on the secondary controller, and $f_{meas} - f_0$ being the difference between the instantaneous measured system frequency and the set-point frequency. From Eq. 11.11, note that the control action δP increases by integral formula, if the deviation of ACE remains constant (proportional-integral [PI]-type controller). This controller is also called load-frequency control (LFC) or automatic gain controller (AGC) in the United States.

In case of an AC-MG, the frequency and amplitude restoration controllers, G_ω and G_E shown in Fig. 11.5a, can be obtained similarly as follows [14]:

$$\delta\omega = k_{p\omega}\left(\omega_{MG}^* - \omega_{MG}\right) + k_{i\omega}\int\left(\omega_{MG}^* - \omega_{MG}\right)dt + \Delta\omega_s \tag{11.13}$$

$$\delta E = k_{pE}\left(E_{MG}^* - E_{MG}\right) + k_{iE}\int\left(E_{MG}^* - E_{MG}\right)dt \tag{11.14}$$

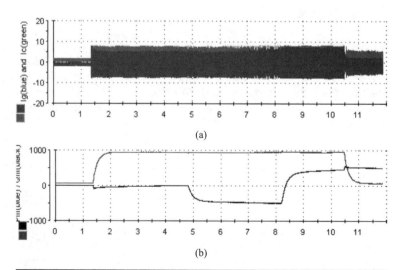

(a)

(b)

Figure 11.5 Experimental results (a) Currents at the PCC and at the MG load (b) Active and reactive power injected into the grid.

with $k_{p\omega}$, $k_{i\omega}$, k_{pE}, and k_{iE} being the control parameters of the secondary control compensator, and $\Delta\omega_s$ is a synchronization term that remains equal to zero when the grid is not present. In this case, $\delta\omega$ and δE must be limited in order to not exceed the maximum allowed frequency and amplitude deviations. Figure 11.5a shows the block diagram of the primary and the secondary control loops of an AC-MG. The primary control is only based on local measurements of the output voltage and current to calculate P and Q for the droop method and the virtual impedance control loop. The secondary control will be implemented by an external centralized controller that uses Eqs. 11.13 and 11.14 to restore the deviations produced by the primary control.

In order to connect the MG to the grid, we have to measure the frequency and voltage of the grid, and that will be the reference of the secondary control loop. The phase between the grid and the MG will be synchronized by means of the synchronization control loop shown in Fig. 11.5b, which can be seen as a conventional PLL. The output signal of the PLL, $\Delta\omega_s$, will be added to the secondary control (see Eq. 11.13) and sent to all the modules to synchronize the MG phase. After several line cycles, the synchronization process will have finished and the MG can be connected to the mains through the static bypass switch. At that moment, the MG does not have any exchange of power with the mains.

Figure 11.4a shows the synchronization process of the MG with the grid. It can be seen that when voltage error is little enough, we can swap to the grid-connected operation mode. Figure 11.4b depicts a nonplanned islanding scenario, leading the MG to operate out of synchronization.

Tertiary Control

When the MG is operating in grid-connected mode, the power flow can be controlled by adjusting the frequency (changing the phase-in steady state) and amplitude of the voltage inside the MG [36–38]. As can be seen in the tertiary control block diagram of Figure 11.5b, by measuring the P/Q through the static bypass switch, P_G and Q_G, they can be compared with the desired P_G^* and Q_G^*. The control laws, PI_P and PI_Q, shown in Fig. 11.3b, can be expressed as following [14]:

$$\omega_{MG}^* = k_{pP}\left(P_G^* - P_G\right) + k_{ip}\int\left(P_G^* - P_G\right)dt \qquad (11.15)$$

$$E_{MG}^* = k_{pQ}\left(Q_G^* - Q_G\right) + k_{iQ}\int\left(Q_G^* - Q_G\right)dt \qquad (11.16)$$

with k_{pP}, k_{iP}, k_{pP}, and k_{iQ} being the control parameters of the tertiary control compensator. Here, ω_{MG}^* and E_{MG}^* are also saturated in case they are outside the allowed limits. These variables are inner generated in island mode ($\omega_{MG}^* = \omega_i^*$ and $E_{MG}^* = E_{MG}^*$) by the secondary control. When the grid is present, the synchronization process can start, and ω^*_{MG} and E^*_{MG} can be equal to those measured in the grid. Thus, the frequency and amplitude references of the MG will be the frequency and amplitude of the mains' grid. After synchronization, these signals can be given by the tertiary control [15–16]. Notice that, depending on the sign of P^*_G and Q^*_G, the active and reactive power flows can be exported or imported independently.

Note that by making k_{iP} and k_{iQ} equal to zero, the tertiary control will act as a primary control of the MG, thus allowing the interconnection of multiple MGs, forming a cluster.

This control loop also can be used to improve the power quality at the PCC. In order to achieve voltage dips ride-through, the MG must inject reactive power to the grid, thus achieving inner voltage stability. Islanding detection is also necessary to disconnect the MG from the grid and disconnect both the tertiary control references as well as the integral terms of the reactive power PI controllers to avoid voltage instabilities.

Experimental results were done by using an MG lab with a DG generator working as a VSI. The MG was able to operate in island, as well as in grid-connected mode.

Figure 11.4a shows the synchronization process of the MG with the grid. After the MG is completely synchronized, it can be connected to the grid and command P and Q. Since the MG is based on VSIs, if there is some nonplanned grid disconnection, the MG can remain working in island mode. Figure 11.4b shows the transfer from grid-connected to island mode.

Figure 11.5 shows the experimental waveforms of the currents P and Q injected into the grid. At 1.2 s the MG was connected to the grid and the tertiary control started. At this point we changed the P^* and Q^* as follows. First we started with $P^* = 1$ kW and $Q^* = 0$ VAr, injecting real power to the grid and achieving unity power factor. At $t = 4.8$ s, we changed the Q^* from 0 to –500 VAr, and hence the MG was acting like a capacitor. Then, at 9.2 s, we suddenly changed Q^* from –500VAr to +500VAr, acting like an inductor. Finally, at $t = 10.5$ s, P^* was fixed to zero without changing the reactive power.

Results

Some simulation results from a three-inverter MG were performed. The inverters consisted of a full bridge with an LC filter, rated at 5 kVA. The local controller consisted of current and voltage loops, the P and Q calculations, and the droop control with a virtual output impedance of 50 μH. In this example, the AC MG hierarchical control of the bandwidth of the level 0 is 5 kHz for the voltage control loop and 20 kHz for the current control loop. The bandwidth for the levels 1 and 2 are 30 Hz and 3 Hz, respectively. The selected control parameters are listed in Table 11.1.

Figures 11.6a and 11.6b show, respectively, the active power and reactive sharing dynamics of the MG system. First, the three-inverter

Parameter	Symbol	Value	Units
Power Stage			
Grid voltage	V_g	311	V
Grid frequency	F	50	Hz
Grid inductance	L_g	1e-3	H
Grid resistance	R_g	1	Ω
Loss resistance inverter I	$R_{loss\,1}$	0.1	Ω
Loss resistance inverter II	$R_{loss\,2}$	0.11	Ω
Loss resistance inverter III	$R_{loss\,3}$	0.09	Ω
Inverter I inductance	L_1	50e-3	H
Inverter II inductance	L_2	55e-3	H
Inverter III inductance	L_3	45e-3	H
Inverter I resistance	r_1	0.1	Ω
Inverter II inductance	r_2	0.15	Ω
Inverter III inductance	r_3	0.05	Ω
Load	R_L	25	Ω
Primary Control			
Derivative frequency droop	m_d	0.0001	W/rd
Proportional frequency droop	m_p	0.0015	Ws/rd
Proportional amplitude droop	n_p	0.001	VAr/V
Secondary Control			
Proportional frequency droop	k_{pw}	1	Ws/rd
Integral frequency droop	k_{iw}	10	W/rd
Proportional amplitude droop	k_{pE}	1	VAr/V
Integral amplitude droop	k_{iE}	100	VAr·s/V
Tertiary Control			
Proportional phase term	k_{pP}	1e-5	Ws/rd
Integral phase term	k_{iP}	0.1	W/rd
Proportional amplitude term	k_{pQ}	1	W/rd·s
Integral amplitude term	k_{iQ}	100	VAr·s/V

TABLE 11.1 AC-MG Control System Parameters

system connected to the grid was started up. The grid reference was fixed from 0 to 1 kW at $t = 2.5$ s by the tertiary control, while the three inverters gave 650 W to each one. At $t = 2.5$ s, a preplanned islanding scenario occurs, and the grid is disconnected from the MG, operating in autonomous (islanded) operation. Afterward, at $t = 5$ s, one inverter

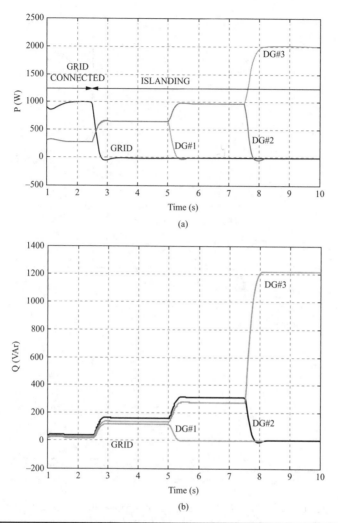

FIGURE 11.6 Transient response of a two-inverter AC-MG. (a) Active power and (b) reactive power.

(DG#2) is disconnected suddenly from the grid, and inverters DG#2 and DG#3 provide the power to the local loads. At $t = 7.5$ s the inverter DG#2 is disconnected; thus, the inverter DG#3 is supplying all the required power. Notice the flexible operation of the MG, being able to operate both in grid-connected and islanded modes.

During islanded mode, the primary control produces frequency and amplitude deviations, which can be compensated by the secondary control loops. Figures 11.7a and 11.7b depict, respectively, the frequency and amplitude restoration action done by the secondary control. Note that these control loops avoid the inherent steady-state error produced by the primary control (detailed waveforms are shown in Fig. 11.8).

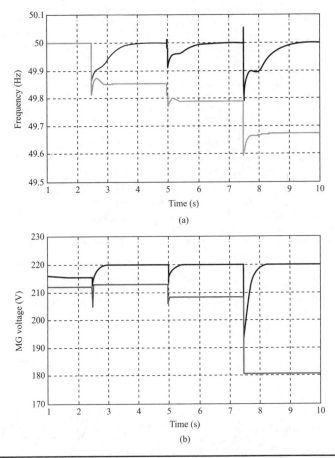

FIGURE 11.7   (a) Frequency and (b) voltage RMS transient response of the AC-MG without (grey line) and with (black line) the secondary control.

Figure 11.9 shows the transient response of the frequency and amplitude of the AC-MG for a nonplanned islanded scenario occurring at $t = 2.5$ s. After 1 second, the islanding operation is detected, the tertiary control loops are disabled, and the frequency and amplitude references for the secondary control are self-generated. Notice that the small transient frequency deviation can be used to detect that the AC-MG is operating in island mode. This transience barely affects the performance of the MG system.

Figure 11.10 shows the active and reactive power flow from the grid to the MG. In this case, the three inverters remained connected to the grid, and the tertiary control changed the reference of the active power while keeping reactive power equal to zero in steady state.

Finally, Fig. 11.11 depicts the P dynamics during different scenarios: at $t = 0$ s the MG is in island mode, after the synchronization process, at $t = 5$ s, the MG is connected to the grid; and at $t = 10$ s,

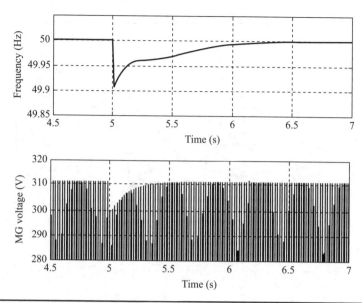

FIGURE 11.8 Detail of the frequency and amplitude restorations.

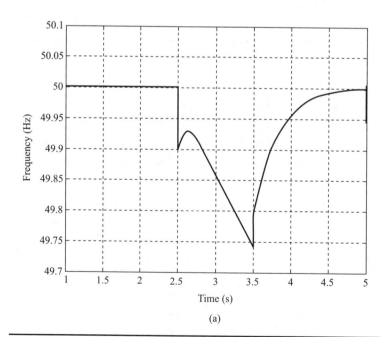

(a)

FIGURE 11.9 Islanding detection for a nonplanned islanded scenario. Transient response of the (a) frequency and (b) amplitude.

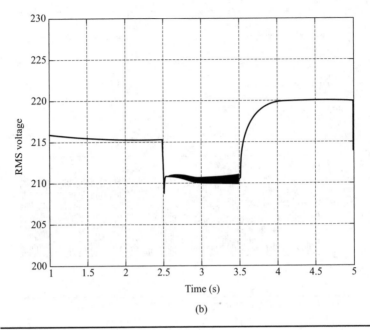

(b)

FIGURE 11.9   (Continued).

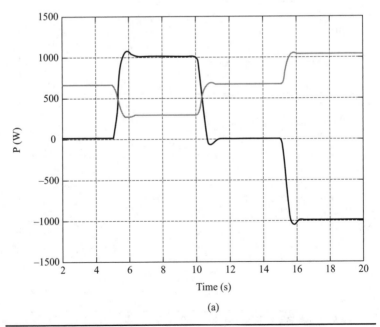

(a)

FIGURE 11.10   Tertiary control: (a) active power and (b) reactive power.

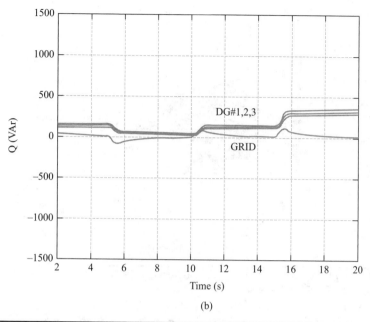

(b)

FIGURE **11.10** (Continued).

the reference of P is changed from 0 to 1 kW. Figure 11.12 shows detail of the voltage difference between the MG and the grid during the synchronization process, showing the action of the synchronization loop. Consequently, the operation during island and grid-connected modes, as well as the transitions between modes and the corresponding synchronization process, has been performed successfully.

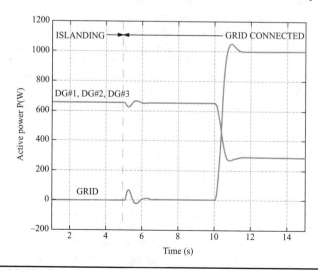

FIGURE **11.11** Transfer process from islanding to grid-connected mode.

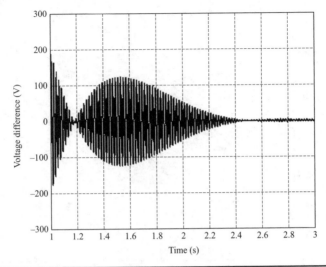

Figure 11.12 Detail of the voltage difference between the grid and the MG during the synchronization process.

Self-Powered Smart Wireless Sensors for Microgrids

Overview of Wireless Sensor Networks

Wireless sensor networks (WSNs) are networks of nodes that sense and potentially also control their environment. They communicate the information through wireless links "enabling interaction between people or computers and the surrounding environment" [54]. The data gathered by the different nodes is sent to a sink, which either uses the data locally, through, e.g., actuators, or which "is connected to other networks (e.g., the Internet) through a gateway. Sensor nodes are the simplest devices in the network. As their number is usually larger than the number of actuators or sinks, they have to be cheap. The other devices are more complex because of the functionalities they have to provide [55]. A sensor node typically consists of five main parts: One or more sensors gather data from the environment. The central unit in the form of a microprocessor manages the tasks. A transceiver included in the communication module communicates with the environment, and memory is used to store temporary data or data generated during processing. The battery supplies all parts with energy. To assure a sufficiently long network lifetime, energy efficiency in all parts of the network is crucial [56–58]. Due to this need, data processing tasks are often spread over the network, i.e., nodes cooperate in transmitting data to the sinks. Although most sensors have a traditional battery, there is some early-stage research on the production of sensors without batteries, using similar technologies to passive radio frequency identification (RFID) chips without batteries.

Problems in Powering the Sensor Nodes

As the network becomes dense with many wireless sensor nodes, the problem of powering the nodes becomes critical, even more so when one considers the prohibitive cost of providing power through wired cables to them or replacing batteries [56–58]. In order for the sensor nodes to be conveniently placed and used, these nodes must be extremely small, as tiny as several cubic centimeters. When the sensor nodes are small, severe limits would be imposed on the nodes' lifetime if powered by a battery that is meant to last the entire life of the device. State-of-the-art, nonrechargeable lithium batteries can provide energy up to 800 WH/L (watt hours per liter) or 2880 J/cm^3 [59]. If an electronic device with a 1 cm^3 coin-size battery is to consume 200 μW of power on average (this is a challenging average power consumption by the load), the device could last for 4000 hours or 167 days, which is equivalent to half a year. Clearly, a lifetime of half a year for the electronic device to operate is far from sufficient because the duration of the device's operation could last for several years. This implies that the battery supply of the electronic device has to be regularly maintained. The need to develop an alternative method for powering the wireless sensor and actuator nodes is acute. Hence, the research direction is targeted to resolving the energy supply problems faced by the energy-hungry wireless sensor nodes.

The wireless sensor node harvests its own power to sustain its operation instead of relying on finite energy sources such as alkaline/rechargeable batteries. This is an alternative energy system for the WSN. The idea is that a node would convert renewable energy abundantly available in the environment into electrical energy using various conversion schemes and materials for use by the sensor nodes. This method is also known as "energy harvesting" because the node is harvesting or scavenging unused, freely available ambient energy. Energy harvesting is a very attractive option for powering the sensor nodes because the lifetime of the nodes would only be limited by failure of their own components. However, it is potentially the most difficult method to exploit because the renewable energy sources are made up of different forms of ambient energy, and therefore there is no one solution that would fit all applications. This option would be able to extend the lifetime of the sensor node to a larger extent compared to the other two possibilities, i.e., improvements on the existing finite energy sources and reduce the power consumption of sensor nodes. Due to the availability of existing power lines of the microgrids, in this chapter, the magnetic energy harvesting technique is introduced to sustain the monitoring operation of the self-powered wireless sensors.

Wind Energy Harvesting for Smart Wireless Sensor Networks

In recent years, the research work on renewable energy harvesting from the ambient environment to extend the lifetime of miniature low-power wireless sensor nodes/networks (WSN) is becoming very popular [60–62]. Even though there are great advancements in the low-power electronic circuits design, high-energy density storage devices, and optimized power-aware network protocols, the amount of energy provided by the finite energy storage element still constraints the autonomy of distributed embedded systems. In practical applications, longer lifetime is an important goal of many WSN systems. To achieve this objective, there is a need to make a paradigm shift from the battery-operated conventional WSN, which solely relies on batteries, toward a truly self-autonomous and sustainable energy harvesting wireless sensor network (EH-WSN) [63, 64]. For EH-WSN, the sensor nodes are integrated with some form of energy harvesting mechanisms that harvest energy such as wind, solar, vibration, etc., directly from the surrounding environment of the remote deployment site for recharging the onboard batteries/supercapacitors of the sensor nodes. Hence, very little maintenance of the nodes is required for an extended period of operation.

Like any of the common renewable energy sources, wind energy harvesting (WEH) has been widely researched for high-power applications where large wind turbine generators are used for supplying power to remote loads and grid-connected applications [65–66]. However, to the author's best knowledge, very few research works can be found in the literature that discuss the issue of small-scale WEH using micro-wind turbine generators [67–69], which are miniaturized in size and highly portable, to power small autonomous sensors deployed in remote locations for sensing and/or even to endure long-term exposure to hostile environments such as an outbreak of fire, sandstorm, etc.

For a space-efficient micro-WEH system operating at low wind speeds, the AC voltage, V_p (peak), generated by the wind turbine generator is in the range of 1 to 3 V, which is relatively smaller than that of the large wind turbine of Mega watt (MW) power rating and output voltage of hundreds of volts [65]. It is thus challenging for the AC-DC rectifier to use conventional diodes, which have a high on-state voltage drop, V_{on}, of 0.7 to 1 V, to rectify and convert the low-amplitude AC voltage into a form usable by the electronic circuits. Another challenging issue is that the electrical power harvested by the WEH system for powering the wireless sensor nodes is often very low, of the order of mW range or less. The situation is even worse if the wind turbine generator is not operating at its maximum power point. The primary concern is to develop a high-efficiency power converter and its micro-power associated electronic circuits with the MPPT algorithm incorporated to track and maintain maximum output power from the wind turbine generator to sustain the wireless sensor node operation over a wide range of operating conditions. MPPT techniques have been commonly used in large-scale WEH systems [66–68] for harvesting much higher amounts of energy

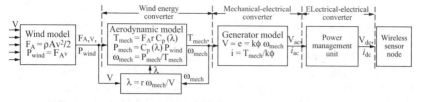

FIGURE 11.13 Functional block diagram and conversion stages of WEH wireless sensor node, where v is the wind speed, F_A, C_p, λ, and r are the aerodynamic force, power coefficient, tip-speed ratio, and blade radius of wind turbine generator; P_{wind} and P_{mech} are the wind and mechanical powers; e, k, Φ, V, and i are the back-EMF voltage, motor constant, flux, terminal voltage, and current of the electrical generator.

from the environment. However, these MPPT techniques require high computational power to fulfill their objective of precise and accurate MPP tracking. Implementation of such accurate MPPT techniques for small-scale WEH whereby the power consumed by the complex MPPT circuitry is much higher than the harvested power itself and therefore not desirable. A resistor emulation approach has been investigated for a micro-wind turbine generator. The rationale behind the resistor emulation approach is that the effective load resistance is controlled to emulate the internal source resistance of the wind turbine generator [69–71] so as to achieve good impedance matching between the source and load, and hence the harvested power is always at its maximum at any operating wind speed.

The system architecture of the self-autonomous WEH wireless sensor node for remote sensing of the target multiple distributed smart microgrid application is presented in Fig. 11.13. It consists of three main building blocks, i.e., wind turbine generator, power management unit, and wireless sensor node. To better understand how the principal components of this WEH system work and interact, the power conversion stages of the functional block diagram is depicted in Fig. 11.13: starting from the wind turbine generator source that generates an output single-phase AC power to the power management unit, a regulated DC power is generated to power the sensing and communicating operations of the wireless sensor node.

Wind Turbine Generator

Referring to the functional structure illustrated in Fig. 11.15, the governing relationship of the power conversion process from one stage to another, i.e., from wind to mechanical stage and mechanical to electrical stage, has been defined in each of the structural models to determine the amount of generated power and its conversion efficiency. Knowing the power generation of the stages, the overall output power harvested by the wind turbine generator can be designed to meet the power requirement of the wireless sensor node in remote sensing the wind-driven wildfire spread.

The wind model in Fig. 11.15 presents the relationship between the input wind speed variable, v, and the output power available in wind, P_{wind}, for a given windfront contact area, A. According to the wind model, the power in the wind is proportional three variables: (1) intercepting cross-sectional area, A, in m^2 of wind turbine being swept by the wind; (2) cube of the wind speed, v, in m/s; and (3) density of air, ρ, in kg/m^3, which is typically 1.225 kg/m^3 at sea level. In this chapter, a miniaturized wind turbine generator with a blade radius of 3 cm has been used, and its windfront contact area can be computed to be 28.3 cm^2. Based on these technical specifications, the amount of wind power, P_{wind}, available in the airflow can be determined. The higher the wind speed, v, the higher the wind power, P_{wind}, for harvesting as a function of the cube of its wind speed. Due to the aerodynamic effect of the wind turbine blades, some portion of the wind power is converted by the wind turbine into mechanical rotational power, P_{mech}. $C_p(\lambda)$. Also known as the power coefficient factor, this is the power conversion efficiency factor between P_{wind} and P_{mech} and it can be obtained either by the analytical functions [72] approach or by the direct data measurement and calculation approach. For simplicity, the direct data measurement and calculation approach is adopted in this chapter to determine the power coefficient, C_p, and the interference factor, a, of the wind turbine, which can be expressed as:

$$C_p(\lambda, \theta_{pitch}) = 4a(1-a)^2 \qquad (11.24)$$

$$a = \frac{u_0 - u_2}{2u_0} \qquad (11.25)$$

Some experimental measurements are conducted to collect the upstream, u_0, and downstream, u_2, wind speeds of the wind turbine to be substituted into Eqs. 11.24 and 11.25. At the average wind speed flow throughout the month in the sample remote area of 3.62 m/s [73], it can be read from [73] that 39% of the wind power, P_{wind}, of 82 mW is trapped in the wind turbine, and the rest is the mechanical power, P_{mech}, of 32 mW generated at the output shaft of the wind turbine, which is directly coupled to the rotor shaft of the electric generator of volumetric size of 1 cm^3. The mechanical power applied to the electric generator is then converted into electrical power for the output load, i.e., power management unit and wireless sensor node.

The electrical characteristic of the wind turbine generator is investigated for different incoming wind speeds applied to the input of the wind turbine generator, and different electrical loads are connected to its output. The investigation process examines the performance of the wind turbine generator under diverse operating conditions. By doing so, one can determine how much electrical power can be harvested by the wind turbine generator at each operating condition so as to meet the power requirement of the WEH wireless sensor node deployed for remote sensing of wildfire spread.

Referring to [73], the linear gradients of the $I-V$ curve represent the internal impedances of the wind turbine generator. Since these gradients are parallel to one another, there exists only optimal resistance value, R_{opt}, to match with the internal impedance of the wind turbine generator, which is given by:

$$R_{opt} = \frac{V_{mppt}}{I_{mppt}}$$

(11.26)

where V_{mppt} and I_{mppt} are the voltage and current at the MPP of the wind turbine generator, respectively. It can also be observed in [73] that the maximum electrical power, P_{mppt}, that can be harvested from the wind turbine generator is at the matching load resistance of 100 Ω, which is essentially the optimal resistance value, R_{opt}, defined in Eq. 11.26. At a wind speed of 3.62 m/s (the average wind speed of the target deployment area), the electrical output power generated by the wind turbine generator for different loadings is estimated as 3 mW @1.8 kΩ, 13 mW @100 Ω, and 6.5 mW @39 Ω. The results show that shifting away from the optimal output load of the wind turbine generator, either very light or heavy loads, results in a significant drop in the electrical output power being generated by the generator, i.e., 77% @1.8 kΩ and 46% @39 Ω drop in harvested power with respect to 13 mW @100 Ω. As such, it is required to incorporate an MPPT scheme in the power management unit of the WEH system to track its peak power points so that maximal power can be harvested to sustain the remote sensing operation of the wireless sensor node in a wildfire spread environment.

Optimal Power Management Unit for a Wind Energy Harvesting System

The power management unit provides a proper matching between the source (i.e., wind turbine generator) impedance and the load (energy storage, power management unit, and sensor node) impedance to achieve high-power conversion efficiency of the WEH system and more electrical energy is harvested. MPPT techniques are very commonly used in the world of large-scale energy sources for harvesting much higher amounts of energy from the environment. For smaller devices, the goal of MPPT is also to maximize the transfer efficiency as well as to minimize the MPPT overhead because in this case energy supply is scarce. These tiny pervasive wireless sensor nodes often need to be small in size; therefore, micro-wind turbines, which generate very limited energy, are used. The energy consumption and efficiency of the MPP tracker are, therefore, very important design criteria in wind energy harvesting systems for sensor nodes rather than the MPPT accuracy.

According to [74], the MPPT algorithms can be grouped into indirect and direct methods. The indirect methods are based on the use of a datasheet that includes parameters and data, such as curves typical of the solar panel for different irradiances and temperatures, or on the use of the mathematical functions obtained from empirical data to estimate the MPP. In contrast to the indirect method, where a priori knowledge of the solar panel characteristics is needed, the direct method measures the solar panel's voltage and current to compute and obtain the actual maximum power at a given operating point. For the WEH system described in this chapter, it can be deducted from [73] that the indirect method does not apply here because there is no single voltage or current point on the power curves that can be used to represent all the MPPT operating points. As for the direct method, it still applies for the WEH system; however, it gives rise to excessive energy loss in the iterative oscillating search, which is very undesirable in this case for small-scale wind energy harvesting. To overcome that, the MPPT technique based on the internal resistance of the energy harvester is explored instead of the generated voltage and current. The power curves plotted in [73] show that when the load resistance matches the source resistance, i.e., 100 Ω, the harvested power is always maximum for any incoming wind speed. By applying the essence of the direct method to iteratively search and compute for the evaluated source resistance, a quick and accurate way of achieving the MPPT points for various incoming wind speeds can be achieved. The proposed MPP tracking algorithm is based on the concept of emulating the load impedance to match the source impedance; this is known as resistor emulation or impedance matching.

In this chapter, a microcontroller-based resistance emulator with a closed-loop feedback resistance control scheme is designed as the MPP tracker of the WEH wireless sensor node for various dynamic conditions. The MPPT circuit depicted in Fig. 11.14 is essentially composed of three main building blocks, namely (1) a DC-DC boost converter seen in Fig. 11.14 (a) to manage the power transfer from the wind turbine to the load, i.e., supercapacitor, power management unit, and wireless sensor node; (2) a voltage and current sensing circuit [see Fig. 11.14(b)] to generate a feedback resistance signal for the MPP tracking algorithm; and (3) an MPP tracker and its control and pulse width modulation (PWM) generation circuits, as illustrated in Figs. 11.14 (c) and (d) respectively, to electronically adapt the wind turbine, which is coupled to the AC-DC rectifier (thus providing a voltage V_{rect}) with the supercapacitor (characterized by voltage $V_{supercap}$) to its maximum power points by adjusting the duty cycle of the PWM gate signal of the boost converter.

Several experimental testings have been conducted to evaluate the performance of the WEH system and its maximum power point tracking capability. Taking both the power losses in the boost converter and the control, sensing, and PWM generation circuits into consideration, the WEH performance analysis with and without MPPT are performed, and the experimental results are tabulated in

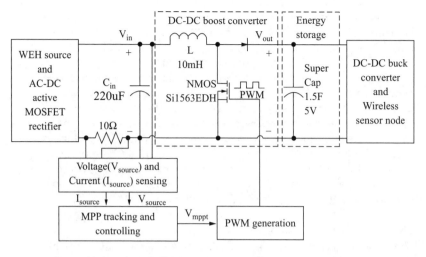

(a) Overview of wind energy harvesting wireless sensor node

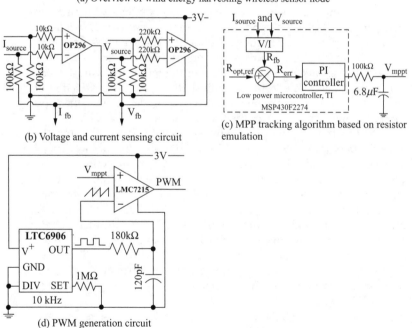

(b) Voltage and current sensing circuit

(c) MPP tracking algorithm based on resistor emulation

(d) PWM generation circuit

FIGURE 11.14 Wind energy harvesting wireless sensor node (a) Overview, (b) Voltage and current sensing, (c) MPP tracking and controlling, and (d) PWM generation.

the bar chart as shown in [73]. At low wind speed of around 3 m/s, the gross electrical power harvested by the wind turbine using the MPPT scheme is 7.7 mW, which is twice of that without the MPPT scheme. The difference in the harvested power is even wider when the wind speed goes as high as 8.5 m/s; up to four times more electrical power is harvested from the wind turbine generator. Considering the associated losses in the MPPT circuits, it is observed in [73] a small fraction of the total power harvested, i.e., around 10% to 20%, is consumed by the boost converter and the control, sensing, and PWM generation circuits. For all the wind speed measurement points shown in [73], it is observed that the performance of the WEH system with MPPT, including the converter's efficiency loss and the circuits' power loss, is more superior than the WEH system without MPPT, which is even more obvious for higher wind speeds. This exhibits the importance as well as the contribution of implementing MPPT in the WEH system to sustain the lifetime of the wireless wind speed sensor in remote monitoring of the wind-driven wildfire spread.

WEH Wireless Sensor Node

The designed WEH system is connected to a commercially available wireless sensor node supplied by Texas Instruments (TI). As the direction of the wind is unpredictable, it may happen that for a given operating condition the wind incident angle is not perpendicular to the wind-turbine plane resulting in nonoptimal wind energy extraction. However, for the wind turbine generator used in this work, this concern is well taken care of because of the tail of the wind turbine generator (WTG). The WTG tail adjusts the wind-front direction of the WTG to face the incoming wind at a right angle. As such, the wind incident angle is always perpendicular to the turbine plane. The eZ430-RF2500 is a complete wireless development tool for the MSP430 microcontroller and CC2500 RF transceiver that includes all the hardware and software required to develop an entire wireless project with the MSP430 in a convenient Universal Serial Bus (USB) stick. The eZ430-RF2500 uses the MSP430F2274 16-bit ultra-low-power microcontroller, which has 32 kB flash, 1K RAM, 10-bit ADC, and 2 op-amps and is paired with the CC2500 multichannel RF transceiver designed for low-power wireless applications. Since the microcontroller is part of the wireless sensor node, it is very convenient to make use of the onboard microcontroller to achieve a more accurate and faster MPPT scheme than a simple MPPT analog circuit [61–70].

The operations of the sensor node comprise (1) sensing some external analog signals and (2) communicating the sensed information to the gateway node in every one second time. The information collected at the base station is post-processed into important parameters like wind speed, temperature, power quality, etc., to facilitate the condition-based maintenance team in understanding the environmental as well as operational conditions of the distributed microgrids. The data transfer rate of 1 Hz should be fast enough for the monitoring team to perform real-time monitoring of the condition and also to react accordingly to various emergencies.

Magnetic Energy Harvesting for a Smart Wireless Sensor Network

Magnetic energy harvesting is another potential energy harvesting solution for powering the smart wireless sensor network used in this multiple distributed microgrid application. The research work on magnetic energy harvesting via inductive coupling utilizes induction as the energy harvesting technology. It is based on the combination of the famous Ampere's law and the Faraday's law of induction. Ampere's law describes the magnetic flux density of the stray magnetic energy source available for induction by the surge coil. Faraday's law of induction states the induced electromotive force, V_{emf}, in a surge coil is directly proportional to the time rate of change of magnetic flux, Φ, through the winding loop. The induced voltage, V_{emf}, generated at the output of the surge coil is processed by a power management unit and stored in an energy storage device, i.e., capacitor. This stored energy is then used to power up the operation of a wireless sensor node.

Magnetic Energy Source Generated from Power Lines

In the experimental tests, the characterization process of the magnetic energy harvester is divided into two parts, namely (1) the magnetic energy source, i.e., magnetic field containing the magnetic energy governed by Ampere's law and (2) the magnetic energy harvester, i.e., a toroid-based surge coil wound with many turns of wires, N, as per Faraday's law of induction. The magnetic energy source is first characterized. Since the magnetic flux density, B, along a current-carrying electrical power cable is a function of the current, I, flowing through the power cable and the radius distance, r_a, between the measurement point and the center of the conductor, it is possible to determine the magnetic field lines that best describe the magnetic energy generated by the current flowing in the power line.

The second part of the characterization process is to determine the voltage induced by the toroid-based magnetic energy harvester, constructed by physically winding N number of turns of copper wires on a circular ring-shaped ferrite magnetic core. When the current-carrying power cable is laid through the center of the ferrite core, magnetic field lines are generated. These magnetic field lines circulate around the ferrite core and its copper winding, and an AC voltage is induced. The induced voltage is proportional to the rate of change in the number of flux lines enclosed by the loop per unit time and the number of winding turns, N, in the loop. In other words, the induced voltage is related to the magnetic field, B; the loop area, A; the winding number of turns, N; and the frequency of the current, f.

The experimental setup measures the induced voltage, V_{emf} [$-\omega NBA \sin(\omega t)$ where the negative sign is due to Lenz's law], from the toroid-based magnetic energy harvester via the current-carrying

				B			
Measured and Calculated Induced EMF Voltage							
$V_{emf} = \omega NBA \sin(\omega t)$							
I, Current flowing in power line	μ_r	ω $2\pi f$ f at 50 Hz	N	$\frac{\mu_0 \mu_r I}{2\pi r_a}$ r_a at 1.5 cm	Area $(\pi r b^2)$ $r_b \sim 0.5$ cm	Calculated V_{emf} (V_{rms})	Measured V_{emf} (V_{rms})
4A	1500	100π	500	0.08T	$2.5 \times 10^{-5}\,\pi$	0.987	1.025
3A	1500	100π	500	0.06T	$2.5 \times 10^{-5}\,\pi$	0.740	0.748
2A	1500	100π	500	0.04T	$2.5 \times 10^{-5}\,\pi$	0.493	0.449
1A	1500	100π	500	0.02T	$2.5 \times 10^{-5}\,\pi$	0.247	0.194

TABLE 11.2   Measured and Calculated Induced EMF Voltage for Different Current Flowing in the Power Line

power cable. The current flowing in the circuit can vary from 1–4 A by adjusting the voltage knob of the AC power supply. Due to the high current along the circuitry, the high-wattage resistor load bank is utilized. A summary of the measured and calculated induced EMF voltage for difference current flowing in the primary-side power line is tabulated in Table 11.2.

It can be observed from Table 11.2 that as the current flowing in the mainstream power line increases from 1 to 4 A, the magnetic field, B, obtainable at 1.5 cm away from the center of the conductor also increases from 0.02 to 0.08 T. For that reason, the induced voltage generated at the output of the toroid magnetic energy harvester has been increased. During the characterization process, the r_a distance of 1.5 cm is set as the reference point based on the practical considerations of the physical diameter of the power cables and the space taken by 500 turns of copper windings.

Performance of Magnetic Energy Harvester

To study how the magnetic energy harvester performs under various operating conditions, experiments were carried out on the designed magnetic energy harvester. Referring back to Table 11.2, the open-circuit voltage of the harvester, which consists of one toroid coil, is quite low, ranging from 0.2 to 1 V. In order to achieve a higher output voltage, three sets of ferrite cores are connected in series. The improved version of the magnetic energy harvester is connected to different loading resistances, and the source current flowing in the power line varies between 1 and 4 A. This enables one to find out the performance of the harvester for various input and output operating conditions.

From the experimental IV curve, the obtainable open-circuit voltage for different currents flowing in the power line has increased by around three times, ranging from 0.7 to 3.5 V. Although the output voltage of the harvester has already been increased by series connecting three ferrite coils together, the induced voltage at some operating points, especially when the magnetic field is weak due to low current flows in the AC power line, is considerably low. As such, the magnetic energy harvester may not be able to drive the electronic output load. This low-output voltage generated from the magnetic energy harvester would pose a challenge on the design of the power management circuit. Another analysis carried out is on the power curve where maximum power is attainable at load resistance of 270 Ω. In the experiment, with the source current of 1 to 4 A flowing in the AC power line, the maximum electrical power available for harvesting ranges from 1 to 18 mW. The challenge here is that when the source current is low, say 1 A, the radiated magnetic field becomes weak and so the maximum power available for harvesting drops tremendously to around 1 mW or so and may not be sufficient to power the RF transmitter load continuously. Hence, a power management circuit designed to address the low-voltage and low-power challenges of the magnetic energy harvester has been proposed.

Power Management Circuit

Based on the analysis and characterization performed on the designed magnetic energy harvester, the concept of harvesting stray magnetic energy via an inductive coupled power transfer is found to be a viable solution for powering the low-power wireless sensor nodes. The block diagram in Fig. 11.15 illustrates the energy harvesting scheme and its application for wireless sensor nodes. Since the voltage source is inherently AC from the power supply along the power line, the induced voltage, V_{emf}, would appear as an alternative voltage source to the connected load. However, the wireless sensor node (i.e., AM RF transmitter) requires a DC source to operate; therefore, the induced voltage must be rectified to DC and regulated prior to powering up the device. This is achieved by using a voltage doubler instead of a standard diode-based full-wave rectifier, which is capable of rectifying and amplifying the low AC voltage to a higher DC voltage.

Referring to the experimental power curve, the amount of power that is generated across the designed ferrite core wound with copper

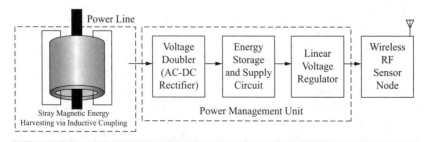

FIGURE 11.15 Block diagram of energy harvesting and wireless RF transmitter system.

wires is in a few milliwatts range. With the limited power generation level, it is not feasible for the magnetic energy harvester to power the wireless radio frequency (RF) transmitter continuously. To overcome that, an effective energy storage and supply circuit discussed by [75] is designed and inserted in between the energy source and the wireless load. This is to ensure that the electrical energy is stored in the capacitor and the energy stored is sufficient to sustain the operation of several RF transmissions. When the energy level of the storage capacitor in the power management unit is sufficient for operations, the RF transmitter would then start to transmit digital encoded information to the RF receiver located at some distance away. The amount of energy consumed by the transmitter is dependent on the amount of 12-bit digital encoded data to be transmitted.

The design specifications of the experiments to be carried out in the research work have been defined according to the practical field condition. The specifications are stated as follows: (1) source frequency of 50 Hz, which is the operating frequency in most of the countries; (2) electrical current flowing in the mainstream power line is set to be 4 A; and (3) number of turns in each winding is made to 500 turns. The advantage of the magnetic energy harvester is that it provides flexibility in the design parameters, i.e., N, ω, B involved, which can be designed accordingly to suit different operating conditions of the sensor node in certain specific applications.

Experimental Validation of a Developed Magnetic Energy Harvesting Wireless Sensor System

An experimental platform, which consists of a 220/230 Vac power supply connected to a bank of load resistances of 60 Ω, has been set up to emulate the electrical current of 1 to 4 A flowing in the power line. Since the primary-side power line is AC, the induced EMF would be AC voltage. This experimental setup is used as a testing platform to evaluate the performance of the magnetic energy harvesting system.

Figure 11.16 shows the waveforms of the induced AC voltage, V_{emf}; the output of the stray magnetic energy harvester; and the output DC voltage of the voltage doubler circuit. It can be observed that the induced voltage signal is a distorted sinusoidal wave rather than a smooth sinusoidal wave. The reason for this phenomenon may be due to the magnetic hysteresis effect and the magnetic saturation of the toroid core. Once the induced AC voltage, V_{emf}, of the stray magnetic energy harvester is inputted to the voltage doubler, the voltage doubler circuit outputs a DC and doubled voltage. As a result, the power management design would be simpler. The voltage from the secondary coil of the transformer is rectified in the voltage doubler and then electric charge is accumulated on the electrolytic storage capacitor of the power storage and supply circuit. The charging and discharging voltages of the storage capacitor are 4 and 6.72 V, respectively (as shown in Fig. 11.17). The amount of electrical

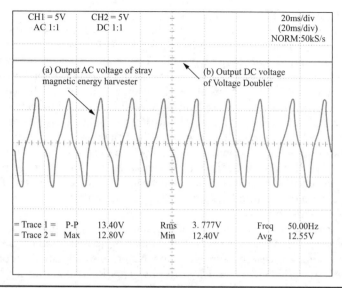

Figure 11.16 Waveforms of (a) output AC voltage of stray magnetic energy harvester and (b) output DC voltage of voltage doubler.

energy stored in the electrolytic storage capacitor with capacitance value of 47 µF is calculated to be 685 µJ.

For each packet of 12-bit digital data, the time taken for one transmission is 20 msec, i.e., 10 msec of active time and 10 msec of idle time. During the active transmission time, the supply voltage

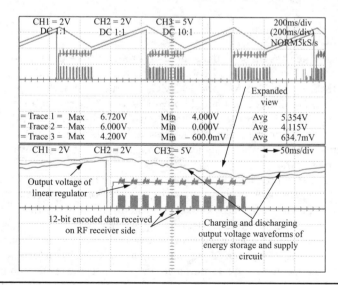

Figure 11.17 Waveforms collected at the RF receiver side to display the number of 12-bit encoded data packets received using the harvested energy.

and current of 3.3 V and 4 mA, respectively, are consumed by the RF transmitter load. As for the remaining time of 10 msec, the RF transmitter load is operating in idle mode, which means that a very minimal amount of energy would be consumed, so it is reasonable to exclude the power being consumed by the RF transmitter load during the idle time. Calculations show the average power, and hence the energy consumed by the RF transmitter load for one digital encoded data transmission, are 13.2 mW and 132 µJ, respectively. Using the harvested stray magnetic energy in the power lines via inductive coupling, the experimental results shown in Fig. 11.17 have verified that the RF transmitter is able to successfully transmit more than 10 packets of digital encoded data over to the receiver remotely. This is verified by the number of digitally encoded data packets received at the RF receiver side.

Although the amount of energy that the stray magnetic energy harvester can harness is relatively small as compared to other energy harvesting sources, i.e., solar and wind, nevertheless, the small amount of energy of 685 µJ is sufficient to power its RF transmitter load to transmit several packets of digitally encoded data in wireless transmission. The experimental results demonstrating the successful RF transmission using the harvested energy have been demonstrated. This implies that the magnetic energy harvester is able to meet its target objective of powering a wireless sensor node from the magnetic energy source generated from the power lines of the microgrid. The developed self-powered wireless RF transmitter working prototype is capable of transmitting 10 packets of 12 digital bits of information over a range of up to 70 meters in an open field with line of sight.

Conclusion

The chapter has presented a general approach of hierarchical control for MGs. The hierarchical control stems from ISA-95. A three-level control is applied to AC- and DC-MGs. On the one hand, the control of AC-MG mimics a large-scale power system AC grid, pointing out the similarities between both systems. On the other hand, the hierarchical control of DC-MGs presents novel features that can be useful in DPS applications such as telecom DC voltage networks, among others.

Consequently, flexible MGs are obtained that can be used for AC or DC interconnection with an AC or DC distribution system, controlling the power flow from the MG to these systems. In addition, these MGs are able to operate in both island or stiff-source connected modes, as well as to achieve a seamless transfer from one mode to the other.

By using the proposed approach, a multi-MG cluster can be performed, constituting a smart grid. In this sense, the tertiary control could provide high-level inertias to interconnect more MGs, thus acting as the primary control of the cluster. In this sense, MGs will behave like a voltage source with high inertias. Thus, a superior control level could send all the references to each cluster of the MG to restore the frequency and amplitude, i.e., the secondary control of the cluster level. Finally, the tertiary cluster control can fix the active and reactive power to be provided by this cluster, or it can act like a primary control to interconnect more clusters. As a result, we could scale the hierarchy of control as necessary.

Using this approach, the system turns more flexible and expandable, and consequently it could integrate more and more MGs without changing the local hierarchical control system associated with each MG.

With some prior knowledge of the operating conditions of the deployment ground in the multiple distributed microgrids, the wireless sensor nodes/network powered by ambient energy harvesting sources can be designed and developed. Kinetic energy from wind flow and magnetic energy from power lines are illustrated in this chapter. Experimental results have been given to demonstrate the technical feasibility of these self-powered wireless sensor nodes, and would be useful for future integration work with the microgrid application.

References

[1] H. Farhangi, "The path of the smart grid," IEEE Power and Energy Magazine, vol. 8, no. 1, Jan.-Feb. 2010, pp. 18–28.

[2] M. Mahmoodi, R. Noroozian, G. B. Gharehpetian, M. Abedi, "A Suitable Power Transfer Control System for Interconnection Converter of DC Microgrids," in Proc. ICREPQ Conf. , 2007.

[3] Y. Rebours and D. Kirschen, "A Survey of Definitions and Specifications of Reserve Services," Internal Report of the University of Manchester, Release 2, Oct. 2005.

[4] Y. Ito, Y. Zhongqing, and H. Akagi, "DC Micro-grid Based Distribution Power Generation System," Power Electronics and Motion Control Conference, 2004 (IPEMC 2004), vol. 3, pp. 1740–1745.

[5] H. Kakigano, Y. Miura, T. Ise, R. Uchida, "DC Micro-grid for Super High Quality Distribution—System Configuration and Control of Distributed Generations and Energy Storage Devices," IEEE IPEMC Power Electron Motion Control Conf., 2004 (3), pp. 1740–1745.

[6] H. Jiayi, J. Chuanwen, and X. Rong, "A review on distributed energy resources and MicroGrid," Renewable and Sustainable Energy Reviews, Elsevier, 12 (2008), pp. 2472–2483.

[7] "Technical paper—Definition of a set of requirements to generating units," UCTE 2008.

[8] H. Kakigano, Y. Miura, T. Ise, and R. Uchida, "DC Micro-grid for Super High Quality Distribution—System Configuration and Control of Distributed Generations and Energy Storage Devices," 37th IEEE Power Electronics Specialists Conference, 2006. PESC '06, June 2006, pp. 1–7.

[9] D. Salomonsson, L. Soder, and A. Sannino, "An Adaptive Control System for a dc Microgrid for Data Centers," in IEEE Proc. 42nd IAS Annual Meeting Industry Applications Conference, 2007, pp. 2414–2421.

[10] J. Bryan, R. Duke, and S. Round, "Decentralized generator scheduling in a nanogrid using dc bus signaling," in Proc. IEEE Power Engineering Society General Meeting, 2004, pp: 977–982.

[11] P. Viczel, "Power electronic converters in dc microgrid," in IEEE Proc. of 5th Int. Conf.-Workshop Power electronic converters in dc microgrid, CPE, 2007.

[12] P. Kundur, "Power System Stability and control," 1994, McGraw-Hill.

[13] E.C.W. de Jong and P.T.M. Vaessen, "DC power distribution for server farms," KEMA Consulting, September 2007.

[14] J. M. Guerrero, J. C. Vasquez, J. Matas, M. Castilla, L. G. de Vicuna, "Control Strategy for Flexible Microgrid Based on Parallel Line-Interactive UPS Systems," IEEE Trans. Ind. Electron., vol. 56, no. 3, March 2009, pp. 726–736.

[15] J. M. Guerrero, J. Matas, L. Garcia de Vicuna, M. Castilla, J. Miret, "Decentralized Control for Parallel Operation of Distributed Generation Inverters Using Resistive Output Impedance," IEEE Trans. Ind. Electron., vol. 54, no. 2, April 2007, pp. 994–1004.

[16] J. M. Guerrero, J. Matas, L. G. de Vicuna, M. Castilla, J. Miret, "Wireless-Control Strategy for Parallel Operation of Distributed-Generation Inverters," IEEE Trans. Ind. Electron., vol. 53, no. 5, Oct. 2006, pp. 1461–1470.

[17] J.M. Guerrero, L. Garcia de Vicuna, J. Matas, M. Castilla, J. Miret, "Output Impedance Design of Parallel-Connected UPS Inverters With Wireless Load-Sharing Control," IEEE Trans. Ind. Electronics, IEEE Transactions on, vol. 52, no. 4, Aug. 2005, pp. 1126–1135.

[18] J.M. Guerrero, L.G. de Vicuna, J. Matas, M. Castilla, J. Miret, "A wireless controller to enhance dynamic performance of parallel inverters in distributed generation systems," Power Electronics, IEEE Transactions on, vol. 19, no. 5, Sept. 2004, pp. 1205–1213.

[19] J. C. Vasquez, J. M. Guerrero, M. Liserre, A. Mastromauro, "Voltage Support Provided by a Droop-Controlled Multifunctional Inverter," IEEE Trans. Ind. Electron., vol. 56, no. 11, 2009, pp. 4510–4519.

[20] K. Alanne and A. Saari, "Distributed energy generation and sustainable development," Renewable & Sustainable Energy Reviews, vol. 10, pp. 539–558, 2006.

[21] R.H. Lasseter, A. Akhil, C. Marnay, J. Stevens, J. Dagle, R. Guttromson, A.S. Meliopoulous, R. Yinger, and J. Eto, "White paper on integration of distributed energy resources. The CERTS microgrid concept," in Consortium for Electric Reliability Technology Solutions, Apr. 2002, pp. 1–27.

[22] P. L. Villeneuve, "Concerns generated by islanding," IEEE Power & Energy Magazine, pp. 49–53., May/June 2004.

[23] M.C. Chandorkar and D. M. Divan, "Control of parallel operating inverters in standalone ac supply system," IEEE Transactions on Industrial Applications, vol. 29, pp. 136–143, 1993.

[24] C.C. Hua, K.A. Liao, and J.R. Lin, "Parallel operation of inverters for distributed photovoltaic power supply system," in Proc. IEEE PESC'02 Conf, 2002, pp. 1979–1983.

[25] Ambrosio, R, S.E. Widergren, "A Framework for Addressing Interoperability Issues," Proc. of 2007 IEEE PES General Meeting, Tampa, FL, June 2007.

[26] S.L. Hamilton, E.W. Gunther, Sr. Member, IEEE, R. V. Drummond, S.E. Widergren, "Interoperability – a Key Element for the Grid and DER of the Future,"

[27] A. A. Salam, A. Mohamed and M. A. Hannan, "Technical challenges on microgrids," ARPN Journal of Engineering and Applied Sciences vol. 3, no. 6, Dec. 2008, pp. 64–69.

[28] B, Kroposki, T. Basso, and R. DeBlasio, "Microgrid standards and technologies," in Proc. IEEE PES General Meeting, 2009, pp. 1–4.

[29] G. Suter and T.G. Werner, "The distributed control centre in a smartgrid," in Proc. CIRED'09, pp. 1–4, 2009.

[30] K. Visscher and S.W. de Haan, "Virtual synchronous machines (VSGs) for frequency stabilisation in future grids with significant share of decentralized generation," in Proc. CIRED '08, pp. 1–4, 2008.

[31] Q.-C. Zhong and G. Weiss, "Static Synchronous Generators for Distributed Generation and Renewable Energy," in Proc. IEEE Power Systems Conference and Exposition, IEEE/PES PSCE '09, 2009.

[32] Visscher, K.; De Haan, S.W.H., "Virtual synchronous machines (VSGs) for frequency stabilisation in future grids with a significant share of decentralized

generation," Smart Grids for Distribution, 2008. IET-CIRED. CIRED Seminar, 23–24 June 2008, pp. 1–4.

[33] A. Madureira, C. Moreira, and J. Peças Lopes, "Secondary Load-Frequency Control for MicroGrids in Islanded Operation," in Proc. International, Conference on Renewable Energy and Power Quality ICREPQ '05, Spain, 2005.

[34] J.P. Lopes, C. Moreira, and A.G. Madureira, "Defining control strategies for MicroGrids islanded operation," IEEE Transactions on Power Systems, May 2006, vol. 21, no. 2, pp. 916–924.

[35] B. Awad, J. Wu, N. Jenkins, "Control of distributed generation," Elektrotechnik & Informationstechnik (2008) 125/12, pp. 409–414.

[36] A. Mehrizi-Sanir and R. Iravani, "Secondary Control for Microgrids Using Potential Functions: Modeling Issues," Conf. Power Systems, CYGRE, 2009.

[37] K. Vanthournout, K. De Brabandere, E. Haesen, J. Van den Keybus, G.Deconinck, and R. Belmans, "Agora: distributed tertiary control of distributed resources," in Proc. 15th Power systems Computation Conference, Liege, Belgium, August 22–25, 2005.

[38] A.G. Madureira and J.A. Peças Lopes, "Voltage and Reactive Power Control in MV Networks integrating MicroGrids," Proceedings ICREPQ '07, 2007, Seville, Spain.

[39] T. Rigole, K. Vanthournout, G. Deconinck, "Resilience of Distributed Microgrid Control Systems to ICT Faults," 19th Int. Conf. and Exhibition on Electricity Distribution, CIRED-2007, Vienna, Austria.

[40] Z. Jiang, and X. Yu, "Hybrid DC- and AC-linked microgrids: towards integration of distributed energy resources," IEEE Conference on Global Sustainable Energy Infrastructure (Energy 2030), Atlanta, GA, Nov. 17–18, 2008.

[41] R. Nilsen and I. Sorfonn, "Hybrid power generation systems," EPE '09, 2009.

[42] A.D. Erdogan and M.T. Aydemir, "Use of input power information for load sharing in parallel connected boost converters," Electr. Eng., no. 91, pp. 229–250, 2009.

[43] Z. Ye, D. Boroyevich, K. Xing, and F. C. Lee, "Design of Parallel Sources in DC Distributed Power Systems using Gain-Scheduling Technique," in Proc. *IEEE PESC*, pp. 161–165, 1999.

[44] H. Kakigano, Y. Miura, T. Ise, and R. Uchida, "DC voltage control of the DC micro-grid for super high quality distribution," IEEJ Transactions on Industry Applications, vol. 127, no. 8, 2007, pp. 890–897.

[45] J. Schönberger, R. Duke, and S.D. Round, "DC-bus signaling: a distributed control strategy for a hybrid renewable nanogrid," IEEE Trans. Ind. Elecron., vol. 53, no. 5, Oct. 2006, pp. 1453–1460.

[46] T. Thringer, "Grid-friendly connecting of constant-speed wind turbines using external resistors," IEEE Trans. Energy Convers, 2002, vol. 17, no. 4, pp. 537–542.

[47] X. Sun, Y.-S. Lee, and D. Xu, "Modeling, Analysis, and Implementation of Parallel Multi-Inverter Systems With Instantaneous Average-Current-Sharing Scheme," IEEE Trans. Power Electron., vol. 18, no. 3, May 2003, pp. 844–856.

[48] J.M. Guerrero, L. Garcia de Vicuna, and J. Uceda, "Uninterruptible power supply systems provide protection," IEEE Ind. Electron. Magazine. vol 1. no. 1, May 2007, pp. 28–38.

[49] J.M. Guerrero, L. Hang, and J. Uceda, "Control of Distributed Uninterruptible Power Supply Systems," IEEE Trans. on Industrial Electronics, vol. 55, no. 8, pp. 2845–2859, August 2008.

[50] J.M. Guerrero and J. Uceda, "Guest Editorial," IEEE Trans. on Industrial Electronics, vol. 55, no. 8, pp. 2842–2844, August 2008.

[51] J.C. Vasquez, J. M. Guerrero, A. Luna, P. Rodriguez, and R. Teodorescu, "Adaptive Droop Control Applied to Voltage-Source Inverters Operating in Grid-Connected and Islanded Mode," IEEE Trans. on Industrial Electronics, vol. 56, no. 10, pp. 4088–4096, Oct. 2009.

[52] I.F. Akyildiz, W.L. Su, S. Yogesh, and C. Erdal, "A Survey on Sensor Networks," *IEEE Communications Magazine*, vol. 40, no. 8, pp. 102114, 2002.

[53] K. Sohrabi, J. Gao, V. Ailawadhi, and G. Pottie, "Protocols for self-organization of a wireless sensor network," *IEEE Personal Communications*, vol. 7, no. 5, 2000, p. 1627.

[54] Tsung-Hsien Lin, W.J. Kaiser, and G.J. Pottie, "Integrated low-power communication system design for wireless sensor networks," *IEEE Communications Magazine*, vol. 42, no. 12, pp. 142–150, 2004.

[55] V. Raghunathan, S. Ganeriwal, and M. Srivastava, "Emerging techniques for long lived wireless sensor networks," *IEEE Communications Magazine*, vol. 44, no. 4, pp. 108–114, 2006.

[56] D. Niyato, E. Hossain, M.M. Rashid, and V.K. Bhargava, "Wireless sensor networks with energy harvesting technologies: a game-theoretic approach to optimal energy management," *IEEE Wireless Communications*, vol. 14, no. 4, pp. 90–96, 2007.

[57] L. Doherty, B.A. Warneke, B.E. Boser, and K.S.J. Pister, "Energy and performance considerations for smart dust," *International Journal of Parallel and Distributed Systems and Networks*, vol. 4, no. 3, pp. 121–133, 2001.

[58] F.I. Simjee and P.H. Chou, "Efficient charging of supercapacitors for extended lifetime of wireless sensor nodes," *IEEE Transaction on Power Electronics*, vol. 23, no. 3, pp. 1526–1536, 2008.

[59] D. Dondi, A. Bertacchini, D. Brunelli, L. Larcher, and L. Benini, "Modeling and Optimization of a Solar Energy Harvester System for Self-Powered Wireless Sensor Networks," *IEEE Transaction on Industrial Electronics*, vol. 55, no. 7, pp. 2759–2766, 2008.

[60] C. Alippi and C. Galperti, "An Adaptive System for Optimal Solar Energy Harvesting in Wireless Sensor Network Nodes," *IEEE Transactions on Circuits and Systems I: Regular Papers*, vol. 55, no. 6, pp. 1742–1750, 2008.

[61] V. Raghunathan, S. Ganeriwal, and M. Srivastava, "Emerging techniques for long lived wireless sensor networks," *IEEE Communications Magazine*, vol. 44, no. 4, pp. 108–114, 2006.

[62] D. Niyato, E. Hossain, M.M. Rashid, and V.K. Bhargava, "Wireless sensor networks with energy harvesting technologies: a game-theoretic approach to optimal energy management," *IEEE Wireless Communications*, vol. 14, no. 4, pp. 90–96, 2007.

[63] Zhe Chen, J.M. Guerrero, and F. Blaabjerg, "A Review of the State of the Art of Power Electronics for Wind Turbines," *IEEE Transactions on Power Electronics*, vol. 24, no. 8, pp. 1859–1875, 2009.

[64] E. Koutroulis and K. Kalaitzakis, "Design of a maximum power tracking system for wind-energy-conversion applications," *IEEE Transactions on Industrial Electronics*, vol. 53, no. 2, pp. 486–494, 2006.

[65] Z. Chen and E. Spooner, "Grid Interface Options for Variable-Speed, Permanent-Magnet Generators," *IEE Proc. -Electr. Power Applications*, vol. 145, no. 4, pp. 273–283, 1998.

[66] Quincy Wang and Liuchen Chang, "An intelligent maximum power extraction algorithm for inverter-based variable speed wind turbine systems," *IEEE Transactions on Power Electronics*, vol. 19, no. 5, pp. 1242–1249, 2004.

[67] K. Khouzam and L. Khouzam, "Optimum matching of direct-coupled electromechanical loads to a photovoltaic generator," *IEEE Transaction on Energy Conversion*, vol. 8, issue 3, pp. 343–349, 1993.

[68] T. Paing, J. Shin, R. Zane, and Z. Popovic, "Resistor Emulation Approach to Low-Power RF Energy Harvesting," *IEEE Transaction on Power Electronics*, vol. 23, no. 3, pp. 1494–1501, 2008.

[69] R.W. Erickson and D. Maksimovic, "Fundamentals of Power Electronics," 2nd ed. New York: Springer, pp. 637–663, 2001.

[70] S. Heier (Author) and R. Waddington (Translator), *Grid Integration of Wind Energy Conversion Systems*, John Wiley & Sons Ltd, second edition, Chichester, West Sussex, England, 2006.

[71] Y.K. Tan and S.K. Panda, "Self-Autonomous Wireless Sensor Nodes with Wind Energy Harvesting for Remote Sensing of Wind-Driven Wildfire Spread," *IEEE Transactions on Instrumentation and Measurement*, 2011.

[72] V. Salas, E. Olias, A. Barrado, and A. Lazaro, "Review of the maximum power point tracking algorithms for stand-alone photovoltaic systems," *Solar Energy Materials and Solar Cells*, vol. 90, no. 11, pp. 1555–1578, 2006.

[73] Y.K. Tan and S.K. Panda, "A novel method of harvesting wind energy through piezoelectric vibration for low-power autonomous sensors," *nanoPower Forum (nPF '07)*, 2007.

CHAPTER 12

Wireless Sensor Networks for Consumer Applications in the Smart Grid

Hussein T. Mouftah, Melike Erol-Kantarci
School of Electrical Engineering and Computer Science, University of Ottawa, Ottawa, Ontario, Canada

Introduction

Wireless sensor networks (WSNs) consist of small, low-cost micro-electrical mechanical systems (MEMS) that have the ability to collect measurements from their surroundings by the help of several on-board sensors, process and store these measurements using their limited processing and storing resources, and transmit these data via their transceivers [1]. Sensor nodes perform these operations on their limited battery resources. Although they may be able to harvest energy from the environment as well, generally the harvested energy is relatively low [2].

WSNs are deployed in a wide range of environments and they are employed in various applications including military target tracking,

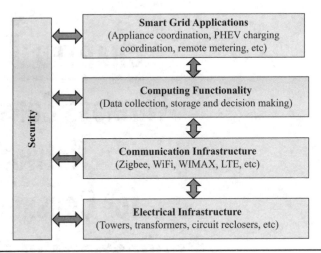

FIGURE 12.1 Layered architecture of the smart grid.

surveillance, health monitoring, disaster relief, seismic monitoring, wildlife monitoring, structural integrity verification, and hazardous environment exploration [3]. For instance, WSNs can be used in detecting forest fires or exploring volcanoes, which are both harsh operating environments for human operators [4]. Despite a wide range of WSN applications, the use of WSNs in the power grid has been recently explored, particularly after the invention of the smart grid.

Smart grid or the future electrical power grid integrates the information and communication technologies (ICTs) in its operation for the purpose of increasing reliability, security, and efficiency of the electrical services and reducing green house gas (GHG) emissions of the electricity production process [5, 6]. In Fig. 12.1, we present the layered smart grid architecture [7]. The bottom layer is the electrical infrastructure layer that contains the power delivery equipment of the traditional grid. On top of the electrical infrastructure, the smart grid adds the communication layer, which includes short-range and long-range communication standards such as Zigbee, Wi-Fi, WIMAX, or LTE [8–10]. Above the communication layer, the computing functionality layer performs data collection, storage, processing, and decision-making functions, which are key to the smart grid applications. The uppermost layer is the smart grid applications including remote metering, appliance coordination, smart grid monitoring, plug-in hybrid electric vehicle (PHEV) coordination, and so on. The security layer is associated with all of those four layers and security has to be handled carefully by each layer.

Initially, WSNs have been considered for smart grid asset and equipment monitoring [11]. WSNs provide opportunities for enhancing the reliability of the grid when adopted in the power grid assets. They have the potential to increase the efficiency of the grid by means of accurate monitoring of generation, transmission, distribution, and consumption of electricity. They are positioned as low-cost alternatives for the existing monitoring tools. Additionally, they can provide detailed diagnosis information that is not available with the current monitoring technologies. Remote system monitoring, equipment fault diagnostics, and wireless automatic meter reading are among the initial application areas that have been considered in [11].

Recently, WSNs have been penetrating residential premises, which has opened up further opportunities for integrating homes in the smart grid. In this context, the smart home concept [12] has regained importance with a focus on energy efficiency and smart grid functionality [13]. For example, Intel has developed the Home Energy Dashboard to provide a simple interface to consumers to monitor their monthly bills and the electricity consumption of the individual appliances [14]. On the other hand, combining smart grid and smart home concepts with the help of WSNs has been recently explored in [15–18] for the purpose of energy management in residential premises. These energy management schemes have been shown to provide savings for consumers and reduce peak load as well as reduce the carbon footprint of the household [19].

Consumer applications are among the significant smart grid applications taking advantage of two-way information and energy flow. Information flow is mainly governed by the smart meters. Smart meters deliver consumption information to the utilities for billing purposes while delivering time-differentiated pricing information to the consumers. The most commonly used time-differentiated pricing scheme is the time of use (TOU). TOU pricing offers different prices for peak, mid-peak, and off-peak hours. The time slices and the price associated with each time slice are based on the historical load/demand data collected over many years. TOU tariff is generally adjusted twice a year to reflect seasonal demand variations.

The main purpose of time-differentiated pricing is to discourage consumption during peak hours. In the power grid, managing peak hour consumption is important for healthy grid operation. It is also one of the factors affecting the price of electricity. In deregulated markets, price of electricity is determined by the market. Suppliers operating the base plants are able to accommodate an average load level. Generally, depending on the demand, they can be partially shut down or started within a daily cycle. During peak hours demand increases above the available capacity of the base plants, which makes peaker plants to be brought online to accommodate the peak load [20]. Peaker plants can respond to power demand in a short time but they generally use fossil fuels, which are expensive. Therefore, electricity is more expensive in peak hours. Instead of increasing the generation capacity, another approach would be to control the consumption. The efforts related with reduction of peak demand are known as demand response.

In the traditional power grid, demand response programs have been implemented for large-scale consumers such as industrial consumers or commercial buildings. Demand response is handled by basically calling consumers to reduce their consumption at peak hours. For instance, in commercial buildings, HVACs are cycled off to avoid penalties or to receive credits whenever the building operators receive a request from the utility. In industrial plants, equipment cooling can be cycled off during peak hours as long as the sensor readings are in safe operation margins [20]. On the other hand, demand response has not been implemented for residential customers in the traditional power grid due to scalability issues.

In the smart grid, it will be possible to communicate, monitor, and possibly control the power consumption of the consumers through WSNs without disrupting their business or comfort. For commercial customers, such as corporate buildings and shopping malls, occupancy, temperature, and air quality sensors can be integrated with smart grid operation. Furthermore, parking lots of offices and shopping malls bear novel opportunities for PHEV-related applications. For example, PHEVs can communicate with the other vehicles in the shopping mall, gather traffic and construction information along the route of the driver, and predict the required energy to complete the next trip. The charging duration or the discharging amount could be transmitted to the WSN of the parking lot where actors coordinate charging of the other vehicles according to the status of each vehicle. Furthermore, WSN-enabled applications for residential premises can enhance the controllability of consumer loads and provide fine-grained management capabilities.

In this chapter, we focus on WSN-enabled applications in the smart grid targeting residential consumers. These applications include interactive demand coordination for home appliances and coordination of PHEV charging/discharging cycle. Appliance coordination exploits communications among the appliances and a controller. The controller communicates with the consumer, appliances, smart meter, and storage units to determine a convenient time to accommodate the demand. It can suggest to shift the demand to a later time based on TOU pricing and consumer preferences. On the other hand, coordination of PHEV charging and discharging cycles aims to make optimum usage of PHEV batteries for energy storage and at the same time, it aims to find the optimum timeslot and optimum resource to charge the battery of the PHEV.

The rest of the chapter is organized as follows. In the next section, we first give an overview of the state-of-the-art in communication technologies for WSNs. WSNs can be implemented using Zigbee or alternatively they can use low-power Wi-Fi. We describe these communication technologies and give examples of WSNs using those technologies. In the third section, we introduce the WSN-enabled consumer applications in detail, including the interactive demand coordination for home appliances and coordination of PHEV charging/discharging cycles. Despite the advantages that are becoming available with WSNs, security and privacy of WSNs bear certain challenges for the smart grid. These challenges are discussed in the fourth section. Finally, the last section gives a summary of this chapter, and discusses the open issues related with the use of WSNs in consumer applications in the smart grid.

Communication Standards for WSNs

WSNs can employ various wireless communication technologies as long as they can maintain power efficiency. In general, short-range communication technologies are preferred due to dense deployment, cost, and energy concerns. WSNs may communicate via Zigbee or low-power Wi-Fi. In the following sections, we introduce and compare Zigbee and low-power Wi-Fi technologies.

Zigbee

Currently, there are various smart appliances and home automation tools that are able to communicate via Zigbee. Furthermore, several smart meter vendors have developed Zigbee-enabled smart meters which provide connectivity among the smart meter, home appliances, and home automation tools. For instance, Landis+Gyr, Itron, and Elster have advanced Zigbee-enabled smart meters. Landis+Gyr has also produced a home energy monitor that is able to communicate with the Landis+Gyr smart meters and report consumption data to the consumers [21]. ZigBee Alliance is also developing Smart Energy Profile (SEP) to support remote metering and Advanced Metering Infrastructure (AMI), and provide communication among utilities and household devices. Briefly, Zigbee is being widely implemented in various consumer electronic devices and smart meters. Using Zigbee for communication among sensor nodes increases the compatibility of the WSN with the other devices at the residential premises.

Zigbee is a convenient technology for WSNs since it is energy-efficient. Energy efficiency is maintained using a low duty-cycling mechanism. Zigbee is based on the IEEE 802.15.4 standard, which defines the physical (PHY) and medium access (MAC) layers. In the PHY, Zigbee utilizes different industrial scientific and medical (ISM) bands in different countries, except the 16 channels in the 2.4 GHz ISM band, which is used worldwide. Other than those channels, Zigbee uses 13 channels in the 915 MHz band in North America and one channel in the 868 MHz band in Europe. The supported data rates in Zigbee are 250 kbps, 100 kbps, 40 kbps, and 20 kbps. These are relatively low data rates; however, Zigbee was initially designed for monitoring and control applications which generally do not require high data rates. Current smart grid applications as well do not

Frequency Band	Region	Maximum Conductive Power/ Radiated Field Limit
Canada	2.4 GHz	1000 mW (with some limitations on installation location)
United States	2.4 GHz	1000 mW
United States	902–928 MHz	1000 mW
Europe (except France and Spain)	2.4 GHz	100 mW EIRP or 10 mW/MHz peak power density
Europe	868 MHz	25 mW
Japan	2.4 GHz	10 mW/MHz

TABLE 12.1   Maximum Transmit Power Levels in Zigbee ISM Bands

necessitate high data rates. Nevertheless, in the future, advanced applications demanding higher data rates may emerge.

The transmission power of a transmitter limits the communication range and according to the Federal Communications Commission (FCC) of the United States, IEEE 802.15.4 standard allows transmission powers up to 1 W. However, due to cost restrictions generally devices operate with lower transmit power. Maximum transmit power is regulated in different ways in each country. In Table 12.1, we provide the upper limits in various countries. Note that EIRP stands for effective isotropic radiated power.

Based on the limits on maximum transmit power, range of a Zigbee radio is approximately 30 m indoors. Zigbee's multi hop feature can expand the coverage of the WSN to cover a residential premise completely. Zigbee implements a multi hop routing protocol over the PHY and MAC layers of the IEEE 802.15.4 standard. A Zigbee network can support up to 64,000 nodes (devices) using two addressing modes: 16-bit and 64-bit addressing. Furthermore, according to recent studies on low-power personal area networks (PAN), it will be possible for Zigbee to adopt IPv6 addressing [22]. Recent amendments define packet fragmentation, reassembly, and header compression for PANs and allow IPv6 addressing over the short packet structure of Zigbee.

As we have mentioned earlier, Zigbee owes its energy-efficiency to a duty-cycling mechanism. Zigbee's duty-cycling mechanism relies on a PAN coordinator operating in beacon-enabled mode. In this mode, PAN coordinator synchronizes the nodes in the network via the superframe structure. Nodes communicate only in the active period which corresponds to a superframe duration (SD). They are inactive in the rest of the beacon interval (BI). SD is divided into contention access period (CAP) and contention free period (CFP). During CAP, nodes compete to achieve access to transmit their data by using the slotted carrier sense multiple access with collision avoidance (CSMA/CA) technique. On the other hand, CFP provides guaranteed time slots (GTS) which are reserved on the previous BI. The IEEE 802.15.4 standard defines the SD and BI as follows [23]:

$$SD = aBaseSuperframeDuration * 2^{SO}[symbols] \qquad (12.1)$$

$$BI = aBaseSuperframeDuration * 2^{BO}[symbols] \qquad (12.2)$$

where SO is the superframe order and BO is the beacon order. In the standard, the ranges of SO and BO are defined as $0 \leq SO \leq BO \leq 14$. PAN coordinator can also work in the beaconless mode. In this case, SO and BO are set to zero.

Zigbee's security builds on the mechanisms defined in the IEEE 802.15.4 standard. In the IEEE 802.15.4 standard, MAC layer is able to add an optional security subheader, which is denoted by the "S" flag in the base MAC header. The security subheader starts with a two-bit key identifier mode (KIM) field, followed by a three-bit security level (LVL) field, which provides integrity-checking capabilities as well as encryption functionalities. KIM defines the key property as follows [23]:

- 00: Key identified by source and destination
- 01: Key identified by macDefaultKeySource + one-byte key index
- 10: Key identified by four-byte key source + one-byte key index
- 11: Key identified by eight-byte key source + one-byte key index

Despite the advantages and wide acceptance of Zigbee, it has several drawbacks. As we have mentioned before, Zigbee is a low data rate technology. As smart grid applications become more complex and their bandwidth requirement increases, Zigbee's data rate may fall short for those applications. Additionally, Zigbee operates in the unlicensed spectrum, i.e., the ISM band, and its performance is affected negatively by interference from other wireless technologies using the same spectrum band, which are Wi-Fi, Bluetooth, and microwave appliances. This is known as the coexistence issue. The IEEE 802.15.4 standard uses O-QPSK PHY in the 2.4 GHz band. This quasi-orthogonal modulation scheme represents each symbol by one of 16 orthogonal (nearly) pseudo-random noise (PN) sequences. O-QPSK is a power-efficient modulation method that achieves low signal-to-noise ratio (SNR) and signal-to-interference ratio (SIR) [23]. In order to avoid coexistence issues, IEEE 802.15.4 devices perform dynamic channel selection, either at network initialization or in response to an outage. The device scans the set of channels specified by the ChannelList parameter. The channels of IEEE 802.11b and IEEE 802.15.4 around the 2.4 GHz ISM band are given in

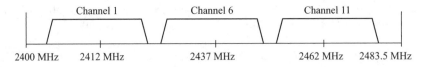

(a) IEEE 802.11 channels around 2.4 GHz ISM band.

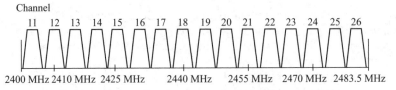

(b) IEEE 802.15.4 channels around 2.4 GHz ISM band.

FIGURE 12.2 Channel selection around 2.4 GHz ISM band for the IEEE 802.11 and IEEE 802.15.4 standards.

Figs. 12.2a and b, respectively. As seen in the figure some of the 802.15.4 channels overlap with the guard bands where interference is expected to be low.

Ultra Low-Power Wi-Fi

The traditional Wi-Fi technology is widely adopted in homes and commercial premises for providing relatively high data rate connectivity [24]. In the smart grid, Wi-Fi is targeting to be a part of home area networks (HANs) as well as, neighborhood area networks (NANs) and field area networks (FANs). Wi-Fi has longer range than Zigbee, i.e., approximately 500 m outdoors which positions it as a feasible solution for networks beyond distribution systems in the smart grid. Employing Wi-Fi in HANs, NANs, and FANs increases the interoperability; hence Wi-Fi is considered as a promising standard for the smart grid [25].

The IEEE 802.11 standard family defines the PHY and MAC layers of Wi-Fi. The IEEE 802.11 standard operates in the 2.4 GHz ISM band, and it utilizes two different physical layer specifications which are Frequency-Hopping Spread Spectrum (FHSS) and Direct Sequence Spread Spectrum (DSSS). FHSS separates the 2.4 GHz band into subchannels which are spaced by 1 MHz. The transmitter changes channels at least 2.5 times per second based on the predefined three sets of sequences in the standard. DSSS, which is a more advanced channel utilization technique, is also employed in CDMA. DSSS multiplies the data with a chip sequence and transmit this after employing either differential binary phase shift keying (DBPSK) or differential quadrature phase shift keying (DQPSK) modulation. To increase the data rates of DSSS, 8-chip complementary code keying (CCK) can be employed as well. This is called as high rate direct sequence spread spectrum (HR/DSSS) in the standard. In the MAC

	Low-Power Wi-Fi	Zigbee
Maximum data rate	2 Mbps	250 kbps
Rx energy (nJ/bit)	4	300
Tx energy (nJ/bit)	15 @ 18 dBm	280 @ 0 dBm
Sleep (uW)	15	5
Wake-up time (ms)	8–50	2

TABLE **12.2**   Comparison of Low-Power Wi-Fi and Zigbee

layer of IEEE 802.11, in addition to the data frame, request to send (RTS) and clear to send (CTS) control frames are employed. The standard also has advanced security and Quality of Service (QoS) settings. Furthermore Wi-Fi inherently supports IP addressing.

Despite the advantages of Wi-Fi, its high power consumption has remained as a drawback until the recently emerging ultra low-power Wi-Fi chips. With this advancement, Wi-Fi can be used in WSNs as well. Ultra low-power Wi-Fi is based on the IEEE 802.11b/g standard. It promises multiple years of operation similar to Zigbee, has data rates around 1 to 2 Mbps, and ranges of 10 to 70 m indoors [26,27]. Utilization of Wi-Fi based sensors in the smart grid has been studied in a recent work [28]. A comparison of low-power Wi-Fi and Zigbee is given in Table 12.2 [29].

Z-wave

Z-wave is a proprietary wireless communication protocol, particularly designed for home automation. It has been embedded into a large number of appliances by various vendors. The main goal of Z-wave is to provide wireless connectivity for HANs that include devices such as lamps, switches, thermostats, garage doors, etc. Therefore, Z-wave can be naturally adopted in the HAN segment of the smart grid.

Z-wave is a short-range, low-data rate radio frequency (RF) mesh networking standard operating in the 908 MHz ISM band in the Americas. It has data rates up to 40 kbps, and it uses BFSK modulation. The maximum range of a Z-wave radio is approximately 30 m indoors and around 100 m outdoors. Typical to all low power wireless communication technologies, poor signal propagation through walls and construction limits the communication range of Z-wave. Therefore it employs a mesh routing protocol. Although Z-wave does not require a central coordinator, it employs slave and master nodes.

Z-wave protocol stack consists of four layers [30]. The MAC layer controls access to RF media. A basic MAC frame contains preamble, start of frame, and end of frame fields for data encapsulation. MAC layer also employs a collision avoidance technique. The transport layer employs checksum for frame integrity check. Acknowledgment and retransmission is handled by the transport layer as well. The routing layer of Z-wave employs a table-based routing protocol. Both master and slave nodes are able to participate in routing. The Z-wave application layer decodes commands and executes them. Z-wave commands can be either protocol commands or application-specific commands. Protocol commands mostly specify ID assignment, and the application commands can be turning on/off devices or other home control-related commands.

The major drawback of Z-wave, similar to Zigbee, is its low data rate. Furthermore, Z-wave is able to support 232 devices, which is far fewer devices than Zigbee can support.

WirelessHART

WirelessHART is a wireless mesh network communications protocol, specifically designed for industrial automation and control applications. In this sense, it is suitable for WSN applications in power generation facilities of the smart grid. WirelessHART builds over IEEE 802.15.4 compatible radios, operating in the 2.4 GHz ISM band. Besides DSSS, time division multiple access (TDMA) technology is used where the nodes are provided with a schedule that is divided into 10 msec timeslots.

The range of a wirelessHART network can reach up to 200 m. In order to extend this range, in the mesh architecture each device is made capable of relaying packets of other nodes. In addition, the network manager determines the redundant routes based on latency, efficiency, and reliability and the WirelessHART gateway provides connectivity to the command center [31]. The protocol architecture is presented in Fig. 12.3.

Security of wireless communications is maintained with end-to-end sessions utilizing AES-128 bit encryption. Individual session keys as well as a common network encryption key is shared among all devices in order to facilitate broadcast activities.

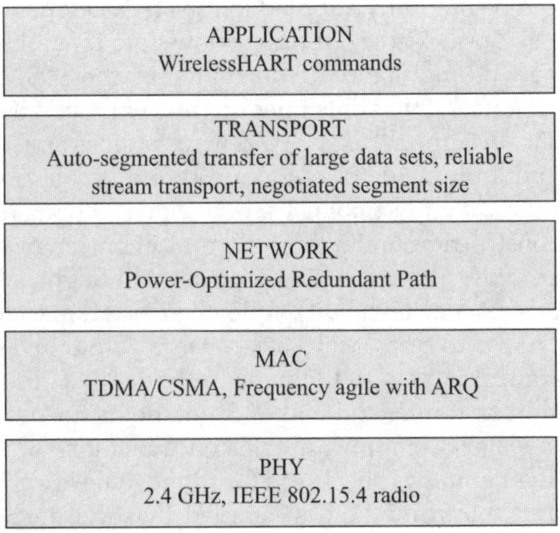

FIGURE **12.3**
WirelessHART
architecture [32].

APPLICATION
WirelessHART commands

TRANSPORT
Auto-segmented transfer of large data sets, reliable stream transport, negotiated segment size

NETWORK
Power-Optimized Redundant Path

MAC
TDMA/CSMA, Frequency agile with ARQ

PHY
2.4 GHz, IEEE 802.15.4 radio

ISA-100.11a

ISA-100.11a is an open standard for safety, monitoring, and control applications. It is developed by the ISA-100 committee, which was initiated in 2005 [33]. Similar to WirelessHART, it can be used in power generation facilities and safety-oriented utility personnel tracking applications in the smart grid domain. ISA-100.11a adopts IEEE 802.15.4 radios and the coexistence issue is handled by channel hopping.

ISA-100.11a allows mesh and star topologies using routing devices, nonrouting devices, backbone router, and a gateway, in addition to system and security managers. Having nonrouting devices aims at increasing lifetime of wireless nodes.

ISA-100.11a targets to support interoperability; therefore, it allows IP addressing. Furthermore, the application interface of ISA100.11a has been designed to support other protocols such as HART, FieldBus, etc.

In summary, WSNs can be implemented either with Zigbee, Z-wave, WirelessHART, ISA-100, or low-power Wi-Fi capabilities. In any case, they will have tremendous application areas in the smart grid. One natural area that WSNs can readily penetrate is the residential premises. In the following section, we summarize the use of WSNs for consumer applications in the smart grid.

WSN-Enabled Consumer Applications in the Smart Grid

WSNs are becoming an integral part of residential premises providing numerous opportunities for smart home functionalities to be available in a cost-effective way. In this section, we introduce two smart grid applications that use WSNs. We first focus on a demand management application that involves flexible appliances, and then we introduce another application that coordinates charging/discharging cycle of PHEV batteries.

WSN-Enabled Demand Management for Residential Consumers

Demand management in a smart grid refers to two separate functions. The first functionality is demand response, which is a well-known power grid method for reducing the consumption of the demand side. Second functionality is related with the energy generation, storage, and consumption of this self-generated energy in coordination with other loads in the smart grid, as well as selling energy back to the grid.

The aim of the WSN-based demand management application is to decrease the cost of energy usage at home and reduce peak load, while causing the least comfort degradation for the consumers. This application has been introduced in [18] and called in-Home Energy Management (iHEM) application. iHEM coordinates appliance usage times by taking into consideration both smart grid signals and consumer preferences. It assumes that several appliances have controllable loads, which means they can be scheduled to another time if the consumer agrees to do so.

iHEM employs a central controller that is in charge of communication with the smart meter as well as communicating with the appliances when they are turned on. In the iHEM application, consumers may turn on their appliances at any time, regardless of peak time concern. The central controller may suggest a delayed start time based on the condition of the smart grid. Unlike traditional optimization-based appliance scheduling techniques, iHEM processes the consumer demands in near real-time.

Appliance scheduling has been widely studied in the literature. Most of the approaches are either based on optimization, which requires initial knowledge of electricity price and appliance schedules [34, 35], or depend on game-theoretic approaches [36,37]. Although it may be possible to predict the schedule of a heating appliance, it may become very challenging to have an accurate estimate on when a washer or dryer will be turned on. Hence, optimization-based schemes have little applicability in practice. Nevertheless, as they can provide optimal solutions to the appliance scheduling problem, they can be used as a benchmark for other schemes. We introduce a simple optimization model that is called Optimization-based Residential Energy Management (OREM) in order to compare the performance of iHEM.

OREM scheme assumes that a day is divided into equal length consecutive timeslots, which employ varying prices for electricity consumption similar to TOU tariff. The objective function of OREM minimizes the total energy expenses by scheduling the appliances in the appropriate timeslots as given in Eq. 12.3. In the linear programming (LP) model, consumer requests are given as an input and an optimum schedule is achieved at the output.

$$\text{Minimize} \sum_{i=1}^{I}\sum_{j=1}^{J}\sum_{t=1}^{T}\sum_{k=1}^{K} E_i D_i U_t S_t^{ijk} \qquad (12.3)$$

The parameters used in the OREM model is given in Table 12.3. E_i is the average energy consumption of an appliance for a cycle. Appliances may have varying power consumption values within one cycle. For example, for a washer, heating consumes the highest amount of power. In the model, for the sake of simplicity, an average consumption value is assumed for the whole cycle.

The constraints of OREM are provided in Eqs. 12.4 through 12.7. Inequality Eq. 12.4 ensures that the total duration of the cycles of the scheduled appliances does not exceed the length of the timeslot that is assigned for them.

$$\sum_{k=1}^{K}\sum_{i=1}^{I} D_i S_t^{ijk} \le \Delta_t, \qquad \forall t \in T, \forall j \in J \qquad (12.4)$$

E_i	Energy consumption of appliance i
D_i	Length of the cycle of appliance i
U_t	Unit price for slot t
a_{ijk}	Timeslot of the arrival of request k of appliance i on day j
S_t^{ijk}	The ratio of time slot occupied by request k of appliance i on day j
Δ_t	Length of timeslot t
D_{max}	Maximum allowable delay
d_i	Delay of appliance i
I	Set of appliances
J	Set of days
T	Set of time slots
K	Set of requests for one day

TABLE 12.3 Parameters of OREM

A cycle may start at the end of one timeslot, and it will naturally continue in the consecutive timeslot. Equation 12.5 ensures that an appliance cycle is fully accommodated without experiencing any interruptions.

$$\sum_{t=1}^{T} D_i S_t^{ijk} = D_i, \qquad \forall i \in I, \forall j \in J, \forall k \in K \qquad (12.5)$$

OREM schedules the cycle of appliances to a convenient timeslot. As a result, appliances may start later than the time they are actually turned on, which creates a delay. On the other hand, to minimize the cost of energy usage, appliances could be scheduled to less expensive timeslots, and this may consequently generate bursts in the least expensive timeslots as well as increasing the waiting time (i.e., delay). Therefore, maximum delay, D_{max}, is limited to two timeslots. Equations 12.6 and 12.7 ensure that the maximum delay is limited by an upper bound as the request is either accommodated in the present or in the next timeslot. Hence, requests do not pile up in certain timeslots.

$$\sum_{t=1}^{m-1} S_t^{ijk} + \sum_{t=m+2}^{T} S_t^{ijk} = 0, \qquad \forall i \in I, \forall j \in J, \forall k \in K, m = a_{ijk} \qquad (12.6)$$

$$\sum_{t=m}^{m+1} S_t^{ijk} = 1, \qquad \forall i \in I, \forall j \in J, \forall k \in K, m = a_{ijk} \qquad (12.7)$$

The iHEM application works in a more interactive manner than OREM, where appliances communicate with the central controller

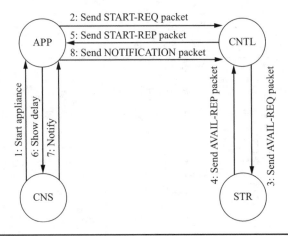

Figure 12.4   Packet flow of iHEM.

before they shift their cycles and an appliance cycle is shifted only if the consumer agrees to do so. The packet flow of iHEM is presented in Fig. 12.4. When a consumer (CNS) turns on an appliance (APP), the appliance generates a START-REQ packet and sends it to the controller (CNTL). Upon receiving a START-REQ packet, CNTL communicates with the storage unit (STR) of the local energy generator to retrieve the amount of the available energy. It also communicates with the smart meter to receive updated price information from the utility. This is not shown in Fig. 12.4 for the sake of readability. CNTL sends an availability request packet, namely AVAIL-REQ, to STR. Upon reception of AVAILREQ, STR replies with a AVAIL-REP packet where the amount of available energy is sent. After receiving the AVAIL-REP packet, CNTL determines the convenient starting time of the appliance by checking whether locally generated power is adequate for accommodating the demand. If this is the case, the appliance starts operating immediately; otherwise, the algorithm checks if the demand has arrived at a peak hour. If the demand corresponds to a peak hour, it is either shifted to off-peak hours or mid-peak hours as long as the waiting time does not exceed the maximum delay. The delay (d_i), which is the difference between the scheduled time and requested time, is sent to the APP in the START-REP packet as the waiting time. The consumer (CNS) decides whether to start the appliance right away or wait until the assigned timeslot. The decision of the consumer is sent back to the controller with a NOTIFICATION packet. This process is repeated every time a controllable appliance is turned on.

The packets of the iHEM application are relayed via the WSN, which is assumed to be present in the smart home for monitoring purposes. WSN employs a mixture of reduced function devices (RFD) and full function devices (FFD) where 5 FFDs are used for routing packets and 14 RFD devices, 4 of which are connected to the appliances, are employed. Sensor nodes communicate via Zigbee protocol utilizing the 2.4 GHz ISM band and the bandwidth is 250 kb/s. Deploying a dedicated WSN to relay the iHEM packets is costly, therefore, WSN of the smart home is used and the sensor nodes also relay the packets of other applications. We show the impact of the varying packet size of those applications on the overall performance of the network and iHEM by varying the packet sizes of the monitoring application between 32B and 128B. We assume the nodes generate packets at ten-minute intervals.

In Fig. 12.5a, we show that the packet delivery ratio of the WSN decreases as the packet size of the monitoring application increases. For packet size of 32B, the delivery ratio is almost 90%. On the other hand, for larger packets, delivery ratio reduces below 55%. As seen in Fig. 12.5b, end-to-end delay is around 0.75s for shorter packets, and it slightly increases to 0.77s as the packet size increases. Shorter packets decrease contention period; therefore, delivery ratio is higher and delay is less for those packets when compared to that of longer packets.

Performance of iHEM has been evaluated by a discrete event simulator and OREM has been solved in ILOG CPLEX optimization suite. Residential energy consumption may vary depending on a number of factors such as size of the house, number of occupants, location, and season. These parameters impact heating, cooling, lighting, and similar loads of the household. In [38], the authors have experimentally shown that consumption transition can be modeled by a Poisson process. Hence, to model the increasing demand during the peak hours a Poisson process with increasing arrival rate at peak hours has been used. The inter-arrival times between two requests is negatively exponentially distributed with a mean of 12 hours. During

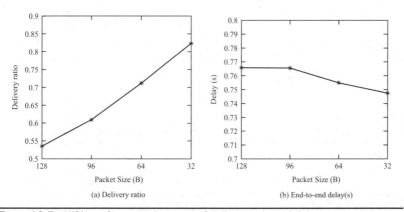

(a) Delivery ratio (b) End-to-end delay(s)

Figure 12.5 WSN performance in terms of delivery ratio and delay.

morning peak periods and evening peak periods the inter-arrival time is negatively exponentially distributed with a mean of 2 hours.

Four appliances have been considered, namely washer, dryer, dishwasher, and coffee maker. The duration and energy consumption of these appliances are vendor specific; however, reference values for average load per cycle has been given in [39]. Based on this study, washer, dryer, dishwasher, and coffer maker are assumed to consume 0.89 kWh, 2.46 kWh, 1.19 kWh, and 0.4 kWh, while the duration of the appliance cycles are taken as 30, 60, 90, and 10 minutes, respectively.

TOU rates used in the performance evaluation are based on the TOU tariff of an Ontario utility. The on-peak, mid-peak, and off-peak prices are taken as 9.3 cent/kWh, 8.0 cent/kWh, and 4.4 cent/kWh, respectively. For winter, on-peak hours are considered to be between 6 A.M. and 12 P.M. and 6 P.M. and 12 A.M. Mid-peak hours are from 12 P.M. to 6 P.M. On weekdays, between 12 A.M. and 6 A.M., and the weekends are the off-peak periods. We simulate iHEM and OREM for approximately 7 months.

Figure 12.6a presents the savings of the iHEM application, the optimal solution provided by OREM, and compares them to the case with no energy management. Note that total contribution of the appliances to the energy bill increases with increasing days because the bill is calculated cumulatively. As seen in Fig. 12.6a, the iHEM application decreases the contribution of the appliances to the energy bill and the savings of the iHEM application is are close to the optimal solution. After seven months, the iHEM scheme provides almost 30% reduction in the energy bill while the optimal solution reduces the bill by around 35%. These results show that residential energy management schemes are useful for decreasing energy bills. We also investigate the impact of the iHEM applications on the total load of the appliances. Figure 12.6b shows the contribution of the appliances on the average demand. When energy management is not employed, 30% of the load

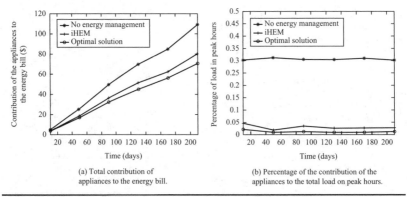

(a) Total contribution of appliances to the energy bill.

(b) Percentage of the contribution of the appliances to the total load on peak hours.

Figure 12.6 Performance comparison of iHEM and OREM with no energy management in terms of consumer expenses and load.

generated by the appliances takes place during peak periods while the iHEM application shifts those requests from peak times and only 5% of the total load is accommodated during peak hours. Therefore, the iHEM application is also capable of reducing the peak demand in the smart grid.

Throughout the previous evaluation, it has been assumed that TOU pricing is used. In the smart grid, it is also possible to have real-time (dynamic) pricing. Dynamic pricing reflects the actual price of the electricity in the market to the consumer bills. The market price of electricity is generally determined by the independent system operator where the day-ahead or hour-ahead prices are announced to the consumers. Raw market price of the electricity depends on several factors such as the load forecasts, supplier bids, and importer bids. The final price is determined after taxes, regulatory charges, transmission and distribution fees, and other service charges are added to the raw market price. Figure 12.7a presents the contribution of the appliances for the iHEM scheme with TOU (iHEM-TOU) and iHEM for real-time pricing. iHEM for realtime pricing still introduces savings when compared to the case without any energy management; however, iHEM-TOU performs better since the scheduling can be coordinated conveniently when the off-peak price stays fixed for a certain amount of time. The performance of scheduling under real-time pricing may be improved by demand prediction.

We also show the impact of iHEM on the carbon emissions resulting from the electricity consumption of the appliances during peak hours. Climate change and global warming are considered to be due to the amount of accumulating greenhouse gases (GHG) in the atmosphere; therefore, reducing carbon emissions is very significant

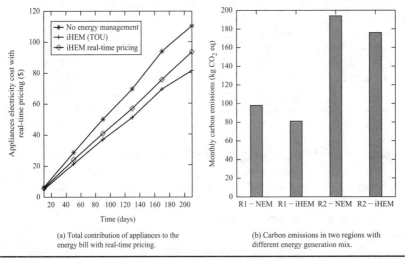

(a) Total contribution of appliances to the energy bill with real-time pricing.

(b) Carbon emissions in two regions with different energy generation mix.

FIGURE 12.7 Consumer expenses under real-time pricing and carbon emissions of iHEM.

Resource	Emissions (kgCO$_2$eq./kWh)
Nuclear	0.016
Coal and natural gas	0.760
Hydro,wind, and solar	0.048
Diesel and heavy oil	0.893
Region 1 base	0.21
Region 1 peak	0.37
Region 2 base	0.46
Region 2 peak	0.63

TABLE 12.4 Carbon Emissions of Energy Generation Resources and Two Regional Grids

for a sustainable habitat. Two different regional grids having different energy generation mixes are considered similar to [19, 40]. We assume that Region 1 is geographically rich in renewable resources where base generation mix is as follows: 50% nuclear, 25% coal and natural gas, 25% hydro, wind and solar. For peak generation mix we assume: 40% nuclear, 40% diesel and heavy oil, 20% hydro, wind and solar. For Region 2 we consider a grid where generation mostly depends on fossil fuels. We assume the base generation mix is 30% nuclear, 60% coal and natural gas, 10% hydro, wind and solar and peak generation mix is 25% nuclear, 70% diesel and heavy oil, 5% hydro, wind and solar. The carbon equivalent emissions of the generation resources are presented in Table 12.4, which are taken from [41]. The emissions related with Region 1 (R1) and Region 2 (R2) are calculated by using the previously mentioned mixture ratios.

In Fig. 12.7b, we show the carbon equivalent emissions for R1 and R2 for the two cases, namely No Energy Management (NEM) and iHEM. iHEM can provide 10% to 20% lower emissions depending on the regional characteristics where R1 represents a relatively optimistic scenario with higher penetration of renewable resources, and R2 represents a pessimistic scenario with less renewable energy generation penetration.

Coordination of PHEV Charging/Discharging Cycles

PHEVs is a hybrid electric vehicle that can drive on both gas and electrical power, which is stored in the lithium-ion battery of the vehicle. These batteries can be charged from the electrical power grid by plugging in the vehicle to a standard outlet or a charging station. For a full charging cycle, using a standard household outlet, a PHEV may have to remain plugged in for up to ten hours depending on the capacity of the battery. Charging from charging stations or fast chargers with higher power outputs takes shorter time. Fast chargers generally provide DC and their output can reach up to 90 kW. A fast charger can charge a PHEV in less than half an hour. Note that the exact energy consumption and charge duration differs from one PHEV brand to the other.

PHEVs are anticipated to be widely adopted as passenger vehicles in the following years. Since their batteries will be powered from the grid, the impact of their load needs to be evaluated as suggested in [42,43]. In [44,45], the authors have discussed the impact of PHEV charging loads on the distribution system, and presented that although utility transformers are designed to be capable of handling high loads for short time periods, as the number of PHEVs in the distribution system increases, transformer overloading may cause failures. Thus PHEVs may cause resilience problems in the smart grid unless charging is controlled and coordinated [46, 47]. Another aspect of the PHEV's integration to the smart grid is the use of PHEV batteries as a supply of electricity. When TOU is employed as a pricing policy, a PHEV can be coordinated to charge its battery when the price of electricity is low, and it can provide energy to home appliances during peak hours in order to reduce the consumer expenses while reducing the peak load.

In this section, we introduce a recently proposed residential energy management application where PHEV batteries are controlled by a home gateway/controller (HGC) and the appliances are scheduled based on the availability of PHEV battery and local energy generation [48]. The aim of this application is to reduce the amount of power acquired from the utility grid especially during peak hours. The gateway can receive price signals from the grid. It is also able to communicate with the rooftop solar power generator, the PHEV, and the registered appliances. The application that coordinates PHEV charging/discharging cycles determines the supply for each arriving request based on the status of the local renewable energy generation facility, electricity price, state of charge (SoC) of the PHEV battery, and the other PHEVs in the neighborhood. The HAN and NAN for this application is illustrated in Fig. 12.8.

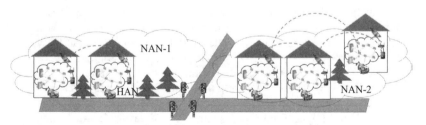

Figure 12.8   PHEV charging/discharging coordination for HANs and NANs of the smart grid.

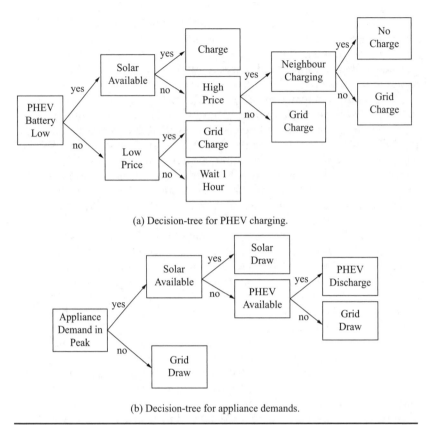

(a) Decision-tree for PHEV charging.

(b) Decision-tree for appliance demands.

FIGURE **12.9**   Coordination of PHEV charging/discharging cycles.

In this application, the gateway communicates with the PHEV and the appliances via an IEEE 802.15.4-based WSN similar to the demand management application of the previous section. The gateway uses predefined rules for coordinating charging/discharging cycles. Decision trees for charging and discharging are presented in Figs. 12.9a and 12.9b, respectively. The charging coordination works as follows. Driver of the PHEV sets a minimum battery level (B_{min}) and when SoC falls below this value, PHEV is charged from the grid. Otherwise, PHEV can be charged from the renewable resources if the stored solar energy is adequate for charging. In case of unavailability of solar energy and the price of electricity being higher than a preset threshold (P_{th}), which indicates that charging may incur resilience problems in the grid, PHEV charging in the neighborhood is checked. Charging is allowed only if neighbors are not simultaneously charging their PHEVs. Otherwise, charging is delayed for an hour and the PHEV communicates with the HGC at the end of one hour to repeat its demand. For PHEV discharging coordination, appliance demands are taken into consideration. PHEV battery can be used by the appliances during peak hours if renewable energy is not available. In this case, discharging is allowed until B_{min} is reached. When renewable energy and PHEV battery are not available, the HGC allows the use of electricity from the grid regardless of price.

To evaluate the performance of this application, B_{min} is set to 5 kWh and the maximum capacity of the battery is taken as 50 kWh [49]. The daily usage of the PHEV battery varies depending on the driving habits and the battery operation mode of the vehicle. We assume each day a random portion of the battery discharges while driving. The PHEV is assumed to be plugged in at random times in a day and stay plugged in from three hours to nine hours. Charging or discharging takes place during this period. The start time of charging is determined by the HGC. Therefore, plugging in the PHEV does not necessarily mean that the vehicle is charging. We assume PHEVs in the neighborhood are also plugged in at random times. Appliance demand arrivals and TOU prices are selected the same way as the previous section.

In Fig. 12.10a, we give the share of the resources in the smart home, in accommodating the demand on January, April, July, and October 2009. These months are selected to represent the variation of solar radiation in four seasons. During April and July, intuitively the solar radiation increases in the northern hemisphere, and as seen in Fig. 12.10a, the utilization of solar energy increases. Consequently, the utilization of PHEV battery and the power drawn from the grid decreases. Note that we only include controllable appliances in the simulations and the actual consumption of the household may be higher.

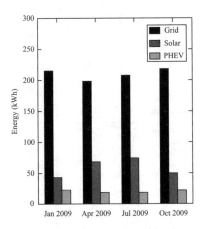

(a) Share of the resources in accommodating the demand on January, April, July and October 2009.

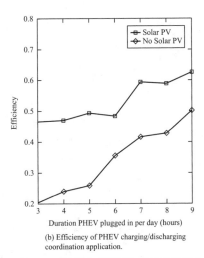

(b) Efficiency of PHEV charging/discharging coordination application.

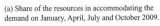

FIGURE 12.10 Performance evaluation of PHEV charging/discharging coordination application.

In Fig. 12.10b, we give the efficiency of the HGC in terms of consumer savings for two cases, i.e., a smart home with solar energy generation and a smart home with no local generation. Efficiency is defined as the ratio of the consumer savings to the electricity expenses. The efficiency is higher when renewable generation is available since it can be considered as a free energy source. Moreover, the efficiency increases as the duration of the PHEV being plugged in increases. PHEV provides storage of locally generated energy, besides, it can discharge to accommodate demand during peak hours. Also, PHEV charging can be delayed to skip peak hours when it is plugged in for longer periods. As a combination of these factors, PHEV coordination can help in reducing the consumer bills. Note that PHEV is assumed to be ready to charge as long as it is plugged in and its battery is below the maximum capacity. Charging stops when the PHEV is not plugged in or when the battery is fully charged.

In Fig. 12.11, we compare the ratio of the PHEV's battery used in one month for the smart home with solar energy generation and without local generation. When renewable energy is not available almost 10% of the PHEV battery is used to accommodate the demand of the appliances where for the case with solar energy, 8% of the PHEV battery is utilized. This means when possible, the HGC schedules the demands to utilize solar generation since solar energy is free. When solar energy is not available PHEV becomes an alternative source.

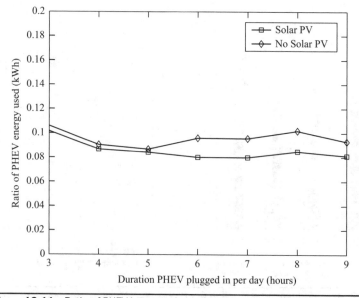

FIGURE 12.11 Ratio of PHEV battery used to accommodate household demand.

Security and Privacy of WSN-Based Consumer Applications

The power grid is becoming smarter by the integration of ICTs while the employment of ICTs may increase the vulnerability of the smart grid to cyberattacks. Particularly, employment of hundreds of small sensors, deployed in an ad hoc fashion, makes it very challenging to protect the smart grid data.

Security of a WSN refers to a combination of several criteria such as availability, authorization, authentication, integrity, and freshness [50]. Availability means that the network services are available without any interruptions even under attacks. Authorization controls the access of the sensors to the WSN where unauthorized sensors cannot send or receive data. Authentication ensures the authenticity of data, which prevents malicious nodes from sending bogus messages. Integrity means a message is not modified on its way to the destination. Finally, freshness ensures that old messages are not replayed by attackers. It is challenging to maintain those security measures in WSNs due to several reasons:

- WSNs use radio transmission.
- WSNs employ a large number of sensor nodes, deployed in unprotected environments.
- Sensor nodes have limited processing and storage capabilities.
- Sensor nodes have limited energy.

In WSNs, use of wireless medium for communications leaves the sensor nodes prone to eavesdropping or jamming attacks. Advanced signal processing techniques cannot be used since they increase cost and energy consumption. Furthermore, sensor nodes may have been deployed in an unprotected environment where tampering of a node or sensor theft may be relatively easier than it is in any other networks. Additionally, well-known cryptographic methods cannot be applied due to limited processing and storing capacity of the sensor nodes. Public key cryptography requires costly computations whereas symmetric key cryptography requires efficient key distribution techniques which are both challenging in WSNs. Last but not least, limited batteries of sensor nodes makes them prone to Denial of Service (DoS) attacks which can easily drain the batteries of sensor nodes. Briefly, WSN-related attacks may target various communication layers including physical, link, routing, or the transport layers. The defense mechanisms against those attacks generally increase the computation and communication overhead. However, they are essential for smart grid applications.

Since smart grids will be employing devices other than WSNs, their vulnerability to attacks and protection schemes is equally important to WSN security. In fact, smart grid security is a broad topic covering the consumer premises as well as the power grid assets [51-56]. In this section, we focus on the devices in residential premises, which are smart meters and home energy controllers. Smart meters deliver electricity consumption data of consumers to the utilities. Modification of these data by malicious attackers may result in incorrect billing, inaccurate load statistics, and false forecasting, as well as false pricing decisions. For instance, smart meters may be a target of Internet-based load alteration attacks which may be implemented by compromising the load control command signals of a utility [57]. These attacks may endanger grid stability if they are implemented from a large number of smart meters. Furthermore, misconfiguration of smart grid-integrated consumer devices may provide means for data modification attacks [58]. These type of attacks may work in two ways. Modified consumer data can be generated and transmitted to the utility or modified utility control signals may be sent to the consumers, both of which cause inconvenience for the customers and the grid operators. Devices on consumer premises are relatively easier to compromise and the utility has little or no control over these devices. Attackers may extract data from the memory of these devices including keys used for network authentication and insert malicious software, which could spread to other devices in the AMI [59].

Equally important as security, privacy needs to be carefully treated for WSN-based consumer applications in the smart grid. Privacy may refer to the privacy of a person or personal information, privacy of personal behavior, and privacy of personal communications. Privacy of personal information refers to any personal information such as physical, mental, economical condition that one does not wish to share with others. Privacy of personal behavior is related with one's activities and choices. Finally, privacy of personal communications is the right to communicate without being monitored or censored [60].

In the smart grid, consumer privacy may be violated if high-resolution electricity consumption data becomes available to malicious users. In [61], the authors have shown that it is possible to obtain a detailed information on the activities in a residential premise such as absence or presence of a person, sleep cycles, meal times, and shower times by accessing fine-grained electricity consumption data. Sophisticated attacks may further benefit from data leakage from consumer premises and reveal the properties of some consumer products to competitors as well. Particularly, information on electric vehicle performance can be sought after by manufacturers. Furthermore, mesh networking in the AMI may also raise privacy concerns as the data of smart meters may be routed by other smart meters.

In the future smart grid, communications are expected to be pervasive and the amount of information and associations may yield to more sophisticated information than anticipated now and can be subject to misuse. The smart grid data network will be handling a significant amount of personal and corporate information which needs to be secured. For these reasons, consumer applications require secure and privacy-preserving communications. Meanwhile, WSN security needs to be treated with care. Security mechanisms generally increase the cost of sensor nodes. Therefore, the cost-security tradeoff needs to be addressed in future studies.

Summary and Open Issues

WSNs are deployed in a wide range of environments for a broad type of applications including military, health, transport, and logistics. Most of these applications benefit from collective monitoring and acting capabilities of WSNs. WSNs can be deployed rapidly and self-organize to form an intelligent monitoring platform. For instance, low-cost sensor nodes can be spread over the field of operation of a utility and enhance utility asset monitoring capabilities. As the electricity generation, transmission, distribution, and consumption practices are undergoing a renovation, WSNs will become an integral part of the power grid functions.

WSNs can be employed in energy generation facilities such as nuclear, hydro, fossil fuelbased power plants, or renewable energy generation sites such as wind and solar farms. Apart from those centralized power generation facilities, in the smart grid, distributed generation is anticipated to be widely adopted. In order for distributed generation to be practical, storage of energy is essential. In the smart grid, it will be possible to coordinate distributed power generators and storage units thanks to fine-grained monitoring tools and distributed control algorithms. Moreover, transmission and distribution equipment can be monitored by WSNs. Last but not least, WSNs can be deployed in the consumer premises (demand-side) to coordinate supply and consumption. Especially, with the wide-scale wide scale adoption of PHEVs demand management will be critical for the stability of the smart grid and WSNs are promising tools for those applications.

In this chapter, we have introduced two WSN-based smart grid applications that target consumer premises. The first application is the iHEM application that aims to determine appliance schedules. The iHEM application has been shown to decrease the contribution of the appliances to the energy expenses of the consumers significantly. Additionally, iHEM application have decreased the load on the peak hours and the electricity-related carbon emissions as well. Furthermore, the performance of iHEM under real-time pricing has been evaluated, and iHEM has been shown to incur less expenses compared to the case of without energy management, even for dynamic pricing. The second application that we have introduced is the PHEV charging/discharging coordination application. Uncontrolled charging of PHEVs may endanger grid resilience, as well as incurring higher vehicle operating costs since the charging duration of a PHEV may coincide with peak hours. Communication among the PHEV, power generation units, appliances and the home controller, and communications among the home controllers in the NAN enable the PHEV to determine a convenient time of charging. Intercon necting PHEVs and renewable energy sources is also important for controlling the discharging profile of a PHEV. Time of discharging is coordinated so that the PHEV provides energy for appliances when the price of electricity is high, which consequently reduces the consumer expenses. It has been shown that this application makes efficient use of the local energy and the PHEV batteries. It also reduces the consumer bills and protects the grid from high PHEV loads due to simultaneous charging.

WSN-enabled smart grid applications rely on communications among sensor nodes. WSNs are generally implemented using Zigbee. Zigbee is based on the IEEE 802.15.4 standard, which designed for low power, low data rate, and short-range communications. Zigbee is used for machine-to-machine or appliance-to-appliance communications, as well as WSNs. Another emerging communication technology for WSNs is the low-power version of the well-known Wi-Fi technology. Wi-Fi is based on the IEEE 802.11 standard which has higher data rates and longer ranges than Zigbee. The traditional Wi-Fi consumes more power than Zigbee. Therefore, it has not been possible to implement Wi-Fi on sensor nodes with limited batteries and low-cost radios. Recently, low-power Wi-Fi has been developed promising several years of operation on battery.

WSNs are low-cost devices with limited processing, memory, and energy resources which position them as an ideal target for attacks. Besides, they are generally deployed in unprotected areas where they can be stolen or compromised. Limited processing and memory resources bear a challenge for computation-intensive public key cryptography mechanisms while symmetric keys are also challenging to implement due to their communication overhead. Meanwhile, limited battery resources makes WSNs an easy target for DoS attacks. Although Zigbee and Wi-Fi employ security mechanisms, they generally address security at physical and link layers. In the smart grid environment security vulnerabilities of the WSNs need to be addressed. Exploration of cost-effective security mechanisms is still an open issue.

Smart grid applications may be vulnerable to attacks generating not only from WSNs but also from smart meters or home controllers. Particularly, misconfiguration of consumer devices may provide opportunities for attackers to easily access a node and spread their attack. Considering that utilities will have little or no control over these devices, modified consumer data may even endanger grid stability. For this reason, it is essential to examine the grid response to the attacks and develop self-healing capabilities after attacks. Locating and isolating compromised equipment are significant concepts in improving the immunity of the smart grid.

The smart grid is a newly emerging field bearing numerous opportunities for novel consumer applications. Future applications are expected to involve learning techniques from the artificial intelligence (AI) field in order to increase consumer comfort and to be less intrusive. On the other hand, appliance technologies are continuously improving. In the future, appliances may tolerate interruptions, which may yield to applications with subcycle scheduling.

References

[1] I. F. Akyildiz, W. Su, Y. Sankarasubramaniam, E. Cayirci, "Wireless Sensor Networks: A Survey," Computer Networks (Elsevier) Journal, vol. 38, no. 4, pp. 393–422, March 2002.

[2] S. Sudevalayam, P. Kulkarni, "Energy Harvesting Sensor Nodes: Survey and Implications," IEEE Communications Surveys & Tutorials, vol. 13, no. 3. 2011, pp. 443–461.

[3] J. Yick, B. Mukherjee, D. Ghosal, "Wireless Sensor Network Survey," Computer Networks (Elsevier) Journal, vol. 52, 2008, pp. 2292–2330.

[4] Organisation for economic co-operation and development (OECD), "Smart Sensor Networks: Technologies and Applications for Green Growth," Technical report, December 2009.

[5] E. Santacana, G. Rackliffe, T. Le, X. Feng, "Getting Smart," IEEE Power and Energy Magazine, vol.8, no.2, March-April 2010, pp. 41–48.

[6] S.M. Amin, B.F. Wollenberg, "Toward a smart grid: power delivery for the 21st century," IEEE Power and Energy Magazine, vol. 3, no. 5, 2005, pp. 34–41.

[7] J. Gao, Y. Xiao, J. Liu, W. Liang, P. Chen, "A survey of communication/networking in Smart Grids," Future Generation Computer Systems (Elsevier), vol. 28, no. 2, 2012, pp. 391–404.

[8] V. Gungor, D. Sahin, T. Kocak, S. Ergut, C. Buccella, C. Cecati, G. Hancke, "Smart Grid Technologies: Communications Technologies and Standards," to appear in IEEE Transactions on Industrial Informatics, 2011.

[9] C. Lo, N. Ansari, "The Progressive Smart Grid System from Both Power and Communications Aspects," to appear in IEEE Communications Surveys and Tutorials, 2012.

[10] V.C. Gungor, F.C. Lambert, "A Survey on Communication Networks for Electric System Automation," Computer Networks Journal (Elsevier), vol. 50, pp. 877–897, May 2006.

[11] V.C. Gungor, B. Lu, G.P. Hancke, "Opportunities and Challenges of Wireless Sensor Networks in Smart Grid," IEEE Transactions on Industrial Electronics,vol.57, no.10, pp. 3557–3564, October 2010.

[12] A. Helal, W. Mann, H. Elzabadani, J. King, Y. Kaddourah and E. Jansen, "Gator Tech Smart House: A Programmable Pervasive Space," IEEE Computer magazine, pp. 64–74, March 2005.

[13] C.Warmer, K. Kok, S. Karnouskos, A. Weidlich, D. Nestle, P. Selzam, J. Ringelstein, A. Dimeas, and S. Drenkard, "Web services for integration of smart houses in the smart grid," In proc. of Grid-Interop Conference, Denver, CO, USA 2009.

[14] Intel Home Dashboard. Available [Online]http://www.intel.com/embedded/energy/homeenergy/demo. Last Accessed November 2011.

[15] M. Erol-Kantarci, H. T. Mouftah, "Using Wireless Sensor Networks for Energy-Aware Homes in Smart Grids," IEEE Symposium on Computers and Communications (ISCC), Riccione, Italy, June 22–25, 2010.

[16] M. Erol-Kantarci, H. T. Mouftah, "TOU-Aware Energy Management and Wireless Sensor Networks for Reducing Peak Load in Smart Grids," Green Wireless Communications and Networks Workshop (GreeNet) in IEEE VTC2010-Fall, Ottawa, ON, Canada, September 6–9, 2010.

[17] M. Erol-Kantarci, H. T. Mouftah, "Wireless Sensor Networks for Smart Grid Applications," in Proc of. International Electronics, Communications and Photonics, Conference (SIECPC) KSA, April 23–26, 2011.

[18] M. Erol-Kantarci, H. T. Mouftah, "Wireless Sensor Networks for Cost-Efficient Residential Energy Management in the Smart Grid," IEEE Transactions on Smart Grid, vol. 2, no. 2, pp. 314–325, June 2011.

[19] M. Erol-Kantarci, H. T. Mouftah, "The Impact of Smart Grid Residential Energy Management Schemes on the Carbon Footprint of the Household Electricity Consumption," IEEE Electrical Power and Energy Conference (EPEC), Halifax, NS, Canada, August 25–27, 2010.

[20] M. Erol-Kantarci, H. T. Mouftah, "Wireless Multimedia Sensor and Actor Networks for the Next-Generation Power Grid," Elsevier Ad Hoc Networks Journal, vol. 9 no. 4, pp. 542–511, 2011.

[21] Landis+Gyr. Available [Online] http://www.landisgyr.com/en/pub/home.cfm. Last accessed on September 2011.

[22] RFC4919: IPv6 over Low-Power Wireless Personal Area Networks (6LoWPANs). Available [Online] http://tools.ietf.org/html/rfc4919. Last accessed on September 2011.

[23] IEEE 802.15.4 standard. Available [Online] http://standards.ieee.org/about/get/802/802.15.html. Last accessed on November 2011.

[24] IEEE 802.11 standard. Available [Online] http://standards.ieee.org/about/get/802/802.11.html. Last accessed on April 2011.

[25] NIST Framework and Roadmap for Smart Grid Interoperability Standards, Release 1.0. Available [Online] http://www.nist.gov/public affairs/releases/upload/smartgrid interoperability final.pdf. Last accessed on September 2011.

[26] Ultra-low power wifi chips of Gainspan Inc.. Available [Online]http://www.gainspan.com/. Last accessed on October 2011.

[27] Ultra-low power wifi chips of Redpine Signals Inc.. Available [Online] http://www.redpinesignals.com/Renesas/index.html. Last accessed on September 2011.

[28] L.Li; X. Hu, C. Ke, K. He, "The applications of WiFi-based Wireless Sensor Network in Internet of Things and Smart Grid," 2011 6th IEEE Conference on Industrial Electronics and Applications (ICIEA), pp. 789–793, 21-23 June 2011.

[29] S. Tozlu, "Feasibility of Wi-Fi enabled sensors for Internet of Things," 7th International Wireless Communications and Mobile Computing Conference (IWCMC), pp. 291–296, 4-8 July 2011.

[30] M. T. Galeev, "Catching the Z-Wave," EE Times Design, Feb. 2006. Available [Online] http://www.eetimes.com/design/embedded/4025721/Catching-the-Z-Wave. Last accessed on September 2011.

[31] wirelessHART. Available [Online] http://www.hartcomm.org/. Last accessed November 2011.

[32] J. Song, S. Han, A.K. Mok, D. Chen, M. Lucas, M. Nixon, "WirelessHART: Applying Wireless Technology in Real-Time Industrial Process Control," IEEE Real-Time and Embedded Technology and Applications Symposium, pp. 377–386,April 2008.

[33] ISA-100.11a standard. Available [Online]http://www.isa100wci.org/. Last accessed November 2011.

[34] M. A. A. Pedrasa, T. D. Spooner, I. F. MacGill, "Coordinated Scheduling of Residential Distributed Energy Resources to Optimize Smart Home Energy Services," IEEE Transactions on Smart Grid, vol. 1, no. 2, pp. 134–143, 2010.

[35] A. Molderink, V. Bakker, M. Bosman Johann L. Hurink, Gerard J.M. Smit, "Management and control of domestic smart grid technology," IEEE Transactions on Smart Grid, vol. 1, no. 2, pp. 109–119, 2010.

[36] A.-H. Mohsenian-Rad, A. Leon-Garcia, "Optimal Residential Load Control with Price Prediction in Real-Time Electricity Pricing Environments," IEEE Transactions on Smart Grid, vol. 1, no. 2, pp. 120–133, 2010.

[37] A.-H. Mohsenian-Rad, V. W.S.Wong, J. Jatskevich, R. Schober, "Optimal and Autonomous Incentive-based Energy Consumption Scheduling Algorithm for Smart Grid," IEEE PES Innovative Smart Grid Technologies Conference, January 2010.

[38] S. W. Lai, G.G. Messier, H. Zareipour, C. H.Wai, "Wireless network performance for residential demand-side participation," IEEE PES Innovative Smart
Grid Technologies Conference Europe (ISGT Europe), Gothenburg, Sweden,
2010.

[39] R. Stamminger, "Synergy Potential of Smart Appliances," Deliverable 2.3 of
work package 2 from the Smart-A project, University of Bonn, March 2009.
[Online] http://www.smart-a.org. Last accessed November 2011.

[40] B. Kantarci, H. T. Mouftah, "Greening the Availability Design of Optical
WDM Networks," IEEE Globecom 2010 Workshop on Green Communications,
pp. 1447–1451, December 2010.

[41] Hydro Quebec, [Online] http://www.hydroquebec.com. Last accessed October
2011.

[42] K. Parks, P. Denholm, and T. Markel, "Costs and Emissions Associated with
Plug-In Hybrid Electric Vehicle Charging in the Xcel Energy Colorado Service
Territory," Technical Report NREL/TP-640-41410, May 2007.

[43] K. Mets, T. Verschueren, W. Haerick, C. Develder, F. De Turck, "Optimizing
smart energy control strategies for plug-in hybrid electric vehicle charging,"
IEEE/IFIP Network Operations and Management Symposium Workshops,
pp. 293–299, 19-23 April 2010.

[44] E. Sortomme, M.M. Hindi, S. D. J. MacPherson, S. S. Venkata, "Coordinated
Charging of Plug-In Hybrid Electric Vehicles to Minimize Distribution System
Losses," IEEE Transactions on Smart Grid, vol.2, no.1, pp. 198–205, March 2011.

[45] S. Shao, M. Pipattanasomporn, S. Rahman, "Challenges of PHEV penetration
to the residential distribution network," IEEE Power & Energy Society General
Meeting, pp. 1–8, 26-30 July 2009.

[46] M. Erol-Kantarci, J. H. Sarker, H. T. Mouftah, "Communication-based Plug-in
Hybrid Electrical Vehicle Load Management in the Smart Grid," IEEE
Symposium on Computers and Communications, Corfu, Greece, June 2011.

[47] M. Erol-Kantarci, J.H. Sarker, H.T. Mouftah, "Analysis of Plug-in Hybrid
Electrical Vehicle admission control in the smart grid," IEEE 16th International
Workshop on Computer Aided Modeling and Design of Communication
Links and Networks (CAMAD), pp. 56–60, 10-11 June 2011.

[48] M. Erol-Kantarci, H. T. Mouftah, "Management of PHEV Batteries in the
Smart Grid: Towards a Cyber-Physical Power Infrastructure," in Proc of.
Workshop on Design, Modeling and Evaluation of Cyber Physical Systems
(in IWCMC11), Istanbul, Turkey, July 5–8, 2011.

[49] G. Berdichevsky, K. Kelty, J.B. Straubel, E, Toomre, "The Tesla Roadster Battery
System," Report by Tesla Motors, August 2006.

[50] Y. Wang, G. Attebury, B. Ramamurthy, "A survey of security issues in wireless
sensor networks," IEEE Communications Surveys & Tutorials, vol.8, no.2,
pp. 2–23, Second Quarter 2006.

[51] M. Amin, "Securing the Electricity Grid," The Bridge, U.S. National Academy
of Engineering, vol. 40, no. 1, pp. 13–20, Spring 2010. Last accessed on October
2011.

[52] P. McDaniel, S. McLaughlin, "Security and Privacy Challenges in the Smart
Grid," IEEE Security & Privacy, vol. 7, no. 3, pp. 75–77, May-June 2009.

[53] The Smart Grid Interoperability Panel, Cyber Security Working Group,
"Guidelines for Smart Grid Cyber Security: Vol. 1, Smart Grid Cyber Security
Strategy, Architecture, and High-Level Requirements," August 2010. Available
[Online]http://csrc.nist.gov/publications/nistir/ir7628/nistir-7628 vol1.pdf.
Last accessed on October 2011.

[54] The Smart Grid Interoperability Panel, Cyber Security Working Group,
"Guidelines for Smart Grid Cyber Security: Vol. 3, Supportive Analyses and
References," August 2010. Available [Online]http://csrc.nist.gov/publica-
tions/nistir/ir7628/nistir-7628 vol3.pdf. Last accessed on October 2011.

[55] U.S. Department of Energy Office of Electricity Delivery and Energy
Reliability, "Study of Security Attributes of Smart Grid Systems - Current
Cyber Security Issues," April 2009. Available [Online] http://www.inl.gov/
scada/publications/d/securing_the_smart_grid_current_issues.pdf. Last
accessed on October 2011.

[56] M. Amin, "Toward A More Secure, Strong and Smart Electric Power Grid," IEEE Smart Grid Newsletter, January 2011.

[57] H. Mohsenian-Rad, A. Leon-Garcia, "Distributed Internet-Based Load Altering Attacks Against Smart Power Grids," to appear in IEEE Transactions on Smart Grid, 2011.

[58] Y. Simmhan, A.G. Kumbhare, B. Cao, V. Prasanna, "An Analysis of Security and Privacy Issues in Smart Grid Software Architectures on Clouds," IEEE International Conference on Cloud Computing (CLOUD), pp. 582–589, 4–9 July 2011.

[59] T. Goodspeed, D. R. Highfill, B. A. Singletary, "Low-level Design Vulnerabilities in Wireless Control Systems Hardware," Proceedings of the SCADA Security Scientific Symposium (S4), pp. 3–13–26, 21-22 January 2009.

[60] The Smart Grid Interoperability Panel, Cyber Security Working Group, "Guidelines for Smart Grid Cyber Security: Vol. 2, Privacy and the Smart Grid" August 2010. Available [Online]http://csrc.nist.gov/publications/nistir/ir7628/nistir-7628 vol2.pdf. Last accessed October 2011.

[61] M. A. Lisovich, D. K. Mulligan, S. B. Wicker, "Inferring Personal Information from Demand-Response Systems," IEEE Security & Privacy, vol. 8, no.1, pp. 11–20, Jan-Feb 2010.

CHAPTER 13

Zigbee-Based Wireless Monitoring and Control System for Smart Grids

Abiodun Iwayemi, Chi Zhou, Peizhong Yi

Illinois Institute of Technology

Introduction

The existing electric power system is undergoing a significant change due to numerous challenges. The typical efficiency of a fossil fuel plant is only 33% [1], while almost 8% of the generated power is lost during the transmission. In order to meet peak demand, 20% generation constructions are built, which only turn on 5% of the time [2]. In addition to that, the current power grid suffers from power outages and interruptions, which result in at least $150 billion in losses annually [3]. Thus, the new-generation power grid should address energy efficiency and reliability and be environment friendly.

The smart grid concept specifies the addition of intelligence and bidirectional digital communication to today's power grid in order to address the efficiency, stability, and flexibility issues that plague the grid. It facilitates several services, including wide-scale integration of renewable energy sources, rapid outage detection, real-time pricing feedback to customers, and demand–response programs involving residential and commercial customers.

All of these new capabilities emphasize the importance of the communication infrastructure and data management. These basic ingredients enable real-time data collection and analysis along with the control of electrical loads for peak-to-average ratio reduction and demand response.

The United States National Institute of Standards and Technology (NIST) has defined Zigbee and the Zigbee Smart Energy Profile (SEP) as two of the communication standards for use in the customer premise network domain of the smart grid [4]. Zigbee is a simple, low-cost, low-power, and low-data-rate [5] wireless technology based on the IEEE 802.15.4 standard. These features, along with its operation in the unlicensed industrial, scientific, and medical (ISM) spectrum band, make it very suitable for wireless smart grid applications. It has also been selected by a large number of utilities as the communications platform of choice for their smart metering devices as it provides a standardized platform for exchanging data between smart metering devices and appliances located on customer premises [2]. The SEP provides support for demand response, advanced metering support, real-time pricing, text messaging, load control, and other features.

However, by operating in the license-free ISM frequency band, Zigbee is subject to interference from various devices that also share this frequency band. These devices include IEEE 802.11 wireless local area networks (WLAN) or Wi-Fi networks, Bluetooth devices, baby monitors, and microwave ovens. Studies have shown that Wi-Fi is the most significant interferer for Zigbee within the 2.4 GHz ISM band [6–7]. With increasing adoption of Zigbee for smart grid applications within homes, campuses, and commercial buildings as shown in Fig. 13.1, their usage in environments with prevalent Wi-Fi networks

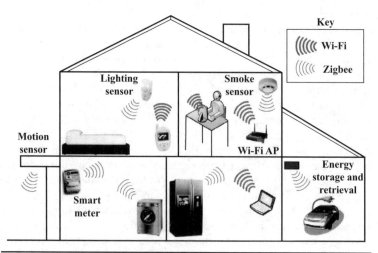

FIGURE 13.1 Zigbee and Wi-Fi device collocation.

introduces Zigbee and Wi-Fi coexistence problems, serving as the motivation for this work.

Zigbee-Based Building Energy Management Demonstration System

The Illinois Institute of Technology (IIT) Perfect Power project is a 5-year project sponsored by the U.S. Department of Energy (DOE), with the objective of implementing a smart grid in the IIT main campus. One of the primary research activities of IIT Perfect Power project is the evaluation of advanced wireless technologies for real-time system monitoring, load control and reduction, energy efficiency, and building automation. In line with NIST smart grid guidelines, the IIT Perfect Power project has adopted Zigbee as the wireless communications infrastructure for energy usage monitoring, net metering, and demand response.

In order to demonstrate the Perfect Power concept, a tabletop demonstration was created [1]. Two-way communication was used to transmit readings from Zigbee end nodes to a data collection and control center (DCCC), and to pass control messages from the DCCC to the end nodes. Each end node is able to relay the collected data to the DCCC via distributed Zigbee routing nodes. The test bed architecture is shown in Fig. 13.2. The Zigbee coordinator aggregates received data for display and processing, and transmit control signals to the end nodes according to the selected power management strategy.

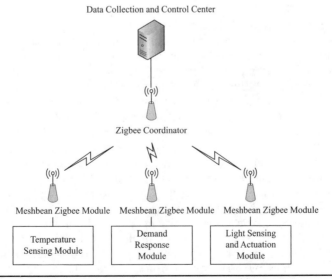

FIGURE 13.2 Perfect Power demonstration system architecture.

The DCCC serves as the system controller, integrating features of both the intelligent Perfect Power system controller and the building controller. It receives input from the various sensors along with power pricing, and manages the loads for energy efficiency, demand response, and cost savings. The DCCC functionality includes the display of received sensor data (temperature, light levels, room occupancy etc.); remote control of Zigbee modules; user configuration of timing; pricing and sensor data threshold values; control of externally connected loads on the basis of user-determined price thresholds; time of day and sensor readings; room occupancy-based lighting control; and other variables. All of these data enable DCCC to determine which loads to put in standby mode and which to switch off. This simple action can result in significant energy savings because loads operating in standby mode comprise up to 10% of all household electricity usage. It will also choose the power source based on simulated real-time price, selecting mains power or stored power resources (battery, solar, wind, etc.).

Zigbee/IEEE 802.15.4 and Wi-Fi/ IEEE 802.11b Overview

Zigbee/IEEE 802.15.4

IEEE 802.15.4 defines the physical (PHY) and medium access control (MAC) layer of the Zigbee protocol, while the Zigbee Alliance defines the network and application layers. The 802.15.4 standard specifies operation in the ISM 2.4 GHz, 915 MHz, and 868 MHz bands, along with two direct sequence spread spectrum (DSSS)-based PHY schemes. The basic channel access mode employs "carrier sense, multiple access with collision avoidance" (CSMA/CA). There are 16 Zigbee channels in the 2.4 GHz band, with each channel occupying 5 MHz of bandwidth. The maximum output power of the radios is generally 0 dBm (1 mw) and receiver sensitivities are –85 dBm for 2.4 GHz and –92 dBm for 868/915 MHz. Zigbee uses binary phase shift keying (BPSK) modulation for both 868 and 915 MHz bands, and offset quadrature phase-shift keying (OQPSK) modulation for 2.4 GHz band. Transmission range is between 1 and 100 m, and it is heavily dependent on the deployment environment [5] and the Zigbee radio transmission power capability. Frequency band and data rate information is summarized in Table 13.1(a).

Zigbee devices can be classified into two major categories, full function devices (FFDs) and reduced function devices (RFDs) [5]. FFDs can perform network establishment, routing, and management, while RFDs only support a subset of the Zigbee device functions, making them simple and low cost. As the root of the network and a bridge to other networks, the coordinator is responsible for network setup and management. Each Zigbee network contains only one

PHY (MHz)	Frequency Band (MHz)	Channel Number	Spreading Parameters		Data Parameters	
			Chip Rate (kchip/s)	Modulation	Bit Rate (Kb/s)	Symbol
868/ 915	868–868.6	0	300	BPSK	20	Binary
	902–928	1–10	600	BPSK	40	Binary
2450	2400–2483.5	11–26	2000	OQPSK	250	16-ary Orthogonal

TABLE 13.1(a)   Zigbee Frequency Bands and Data Rates [9]

Zigbee coordinator. Routers are used to connect between coordinators and other nodes. Routers and coordinators can communicate with all the devices on the network, and are typically powered by mains power since they cannot go to sleep without adversely affecting the ability to route traffic through the network. End devices communicate with routers, and are incapable of peer-to-peer communication. They tend to be battery-powered devices that spend most of their time in sleep mode. They periodically wake up, check for any messages buffered for them at their parent router, read their attached sensors, transmit the measured data, and return to sleep mode. Zigbee networks support three types of topologies: star, mesh, and cluster tree, which enable them to scale to support thousands of nodes.

The target market for Zigbee is general-purpose, inexpensive, self-organizing mesh networks for energy management, home automation, building automation, and industrial automation. The final goal of Zigbee Smart Energy is to stimulate social change in the way that people understand, manage, automate, and improve the energy efficiency dramatically [8].

Wi-Fi/IEEE 802.11b

The IEEE 802.11 standard specifies PHY and MAC layers for Wi-Fi. It defines 13 overlapping, 22 MHz–wide frequency channels in the ISM 2.4 GHz frequency band. There are only two groups of three nonoverlapping channels: channels 1, 6, and 11 in the United States, and channels 1, 7, and 13 in Europe. There are several versions of IEEE 802.11 with IEEE 802.11b being the most widely deployed version. IEEE 802.11b has a maximum transmission rate of 11 Mbps and uses the same CSMA/CA media access method defined in the original IEEE 802.11 standard. The 802.11b PHY layer incorporates DSSS modulation. Technically, the 802.11b standard uses Barker coding and complementary code keying (CCK) as its modulation technique. It is the amendment of CCK coding that enabled the dramatic increase in data rate compared to original standard. Typical indoor range is 100 ft at 11 Mbps and 300 ft at 1 Mbps. Different data rate specifications are shown in Table 13.1(b).

Data Rate	Code Length	Modulation	Symbol Rate	Bits/ Symbol	System
1 Mbit/s	11 (Barker C)	DBPSK	1	1	DSSS
2 Mbit/s	11 (Barker C)	DBPSK	1	2	DSSS
5.5 Mbit/s	4 (CCK)	DBPSK	1.375	4	HR/DSSS
11 Mbit/s	8 (CCK)	DBPSK	1.375	8	HR/DSSS

TABLE **13.1(b)** IEEE 802.11b Data Rate Specifications [11]

Main Interference Sources of IEEE 802.15.4

Due to its almost global availability, an increasingly diverse range of low-cost radio solutions utilize the 2.4 GHz ISM unlicensed band. This sharing of the spectrum among various wireless devices that can operate in the same environment leads to severe interference and results in significant performance degradation. The list of wireless technologies using the 2.4 GHz frequency band [27] includes:

- 802.11b networks
- 802.11g networks
- 802.11n networks
- Bluetooth Pico-Nets
- 802.15.4-based personal area networks (WPAN)
- Cordless phones
- Home monitoring cameras
- Microwave ovens
- Motorola canopy systems
- WiMax networks

Due to the variety of wireless technologies, different technologies using the 2.4 GHz ISM band will interfere with Zigbee in different ways. Most of these effects can be ignored, as only a few may cause severely impact Zigbee performance. Performance of Zigbee under Wi-Fi and Bluetooth is examined theoretically in [7] and the results show that WLAN is a much greater interferer than Bluetooth. The interaction between Zigbee and IEEE 802.11g is empirically evaluated in terms of throughput in [11], with results indicating that Zigbee does not affect IEEE 802.11g significantly; however, the effect on the throughput of Zigbee is significant when the spectrum of the chosen channels of operation coincide. The usage frequency is also a key factor in evaluating the differing degrees of interference. According to both [6, 7], most of the interference is caused by IEEE 802.11 transmitters due to their widespread usage in residential and public environments. Three Wi-Fi channels cover almost the entire spectrum of Zigbee, and even more importantly, Wi-Fi signals are almost 100 times stronger than the Zigbee signal.

In order to mitigate interference caused by Wi-Fi, the developers of the Zigbee standard added several functions in all four layers of the protocol. Direct sequence spread spectrum (DSSS) is a type of "spread spectrum" that broadens the signal to take up more bandwidth than the information signal modulated. Due to the wide bandwidth, it can coexist with narrow-band signals, which cause a slight reduction in the signal-to-noise ratio over the spectrum being used to spread spectrum. Carrier sense multiple access/collision avoidance (CSMA/CA) is used in the MAC sublayer. CSMA adopts a listen before you talk strategy, so users will not transmit until the channel is idle. The ad hoc on-demand distance vector (AODV) routing protocol is used in Zigbee. This is a pure on-demand route acquisition algorithm, and based on this routing algorithm, Zigbee can automatically construct a single cluster network or a potentially larger cluster tree network. The network is basically self-organized and supports network redundancy to attain a degree of fault resistance and self-healing.

Although there are so many interference mitigation techniques employed in Zigbee, the interference issue is still considered controversial. As a result, measuring the effect of interference and analyzing how the mitigation technology improves the coexistence of Wi-Fi and Zigbee is a worthwhile endeavor.

Performance Analysis of Zigbee under Wi-Fi

Bit error rate (BER) and packet error rate (PER) are two key parameters for evaluating how robust and reliable a digital communication technology is. The bit and packet error rate is defined as the rate at which errors occur in a transmission system, and is in direct proportion to signal-to-noise ratio. BER and PER are calculated in accordance with the analytical model in [9].

BER Analysis of Zigbee under Wi-Fi

The PHY of IEEE 802.15.4 at 2.4 GHz uses OQPSK modulation. For an additive white Gaussian noise (AWGN) channel, the BER can be calculated by the following equation [12]:

$$BER = Q\left(\sqrt{\frac{2E_b}{N_0}}\right) \tag{13.1}$$

where E_b/N_0 is the normalized signal-to-noise ratio (SNR) and Q(x) is the Q-function of Gaussian distribution:

$$Q(x) = \frac{1}{\sqrt{2\pi}} \int_x^\infty \exp\left(-\frac{u^2}{2}\right) du \tag{13.2}$$

When a Zigbee channel overlaps with a Wi-Fi channel, we can consider the Wi-Fi signal as partial band-jamming noise for the Zigbee signal [13] and the SNR is replaced by signal-to-interference-plus-noise ratio (SINR), which can be defined as:

$$SINR = \frac{P_{signal}}{P_{noise} + P_{interference}} \tag{13.3}$$

where P_{signal} is the power of the desired signal at the Zigbee receiver, P_{noise} is the noise power, and $P_{interference}$ is the received interference power from the Wi-Fi signal at the Zigbee receiver.

The path loss model represents the power loss between the transmitter and receiver, and can therefore be used in conjunction with the transmission power to enable the calculation of P_{signal} and $P_{interference}$. We define the maximum transmission power of Zigbee as 1 mw (0 dBm). Considering that Zigbee and Wi-Fi are most frequently deployed in indoor environments, an indoor path loss model is most suitable.

Considering that the power spectrum of IEEE 802.11b is 11 times wider than Zigbee and is not uniformly distributed, in-band interference power of IEEE 802.11 cannot be simply calculated by dividing 11 [14]. An amendment parameter of in-band power factor r is added to $P_{interference}$. Therefore, $P_{interference}$ is replaced by $r \cdot P_{interference}$.

To obtain the factor, the power spectral density of the IEEE 802.11b and offset frequency between the central frequency of Zigbee and Wi-Fi are considered. Since the power is concentrated around the central frequency, r increases as the offset frequency decreases.

PER Analysis of Zigbee under Wi-Fi Interference

The PER is calculated based on BER and collision time. The IEEE standards for both IEEE 802.11 [15] and 802.15.4 [16] specify three methods of clear channel assessment (CCA) to determine the channel occupancy [17]:

- CCA Mode 1: Energy detection. In this mode, the energy level within the channel is measured, and if the energy level is higher than a predefined threshold, the channel will be considered busy.

- CCA Mode 2: Carrier sensing. In this mode the channel is considered busy if the detected signal is compliant with the PHY of the device that is performing the CCA.

- CCA Mode 3: Carrier sensing with energy detection. If the detected energy level is above the threshold and the compliant carrier is detected, the channel is considered busy.

The default mode of operation of Wi-Fi is mode 2, in which a Wi-Fi node considers the channel free if no other Wi-Fi device is

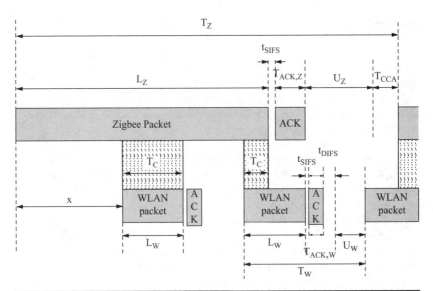

FIGURE 13.3 Interference model between IEEE 802.11b and IEEE 802.15.4 [14].

detected, even if some device other than Wi-Fi may be using the channel. In order to analyze the worst case, we assume that both Zigbee and Wi-Fi devices operate in CCA mode 2 here, meaning that they are essentially blind to each other's transmissions.

The collision time model is shown in Fig. 13.3. Based on the assumption of blind transmission, the contention window is not modified even when Zigbee and Wi-Fi coexist. Though both Zigbee and Wi-Fi adopt CSMA/CA, unlike Wi-Fi, Zigbee only detects the availability of a channel by CCA twice after its back-off time. The average collision time T_c can be calculated by means of Fig. 13.3, and the PER of Zigbee under Wi-Fi (IEEE 802.11b) interference is expressed as:

$$PER = 1 - \left[(1 - P_b)^{N_z - \lceil T_c/b \rceil} \times (1 - p_b^I)^{\lceil T_c/b \rceil}\right] \qquad (13.4)$$

where P_b is the BER without IEEE 802.11 interference, P_b^I is the BER with IEEE 802.11 interference, N_z is the number of the bits in a Zigbee packet, and b is duration of a bit transmission.

Frequency Agility–Based Interference Avoidance Scheme

According to the theoretical model, BER depends on noise level and interference power within the overlapping channel. Distance and offset frequency play a key role in interference power. If Zigbee devices can detect interference, find "safe channels," and migrate the entire PAN to a clear channel, their performance will be significantly improved. The solution should require minimal adjustments to the existing IEEE 802.15.4 standard, or it can be implemented via a software upgrade in order to facilitate easy adoption. In addition, any proposed solution must be simple and energy efficient. Considering these factors, a frequency agility algorithm is a perfect choice for IEEE 802.15.4 cluster-tree networks, which combine star and mesh topologies, to try to achieve both high level of reliability and scalability, and energy efficiency.

The primary elements of our scheme are interference detection and interference avoidance. Each sender node measures its PER periodically, and if the PER exceeds some threshold, the sender will transmit a link quality indicator (LQI) report to its parent router. If LQI is below a certain value, the coordinator instructs all the routers in the PAN to perform interference detection of the available channels. Interference detection is achieved by means of energy detection (ED) scans defined in the Zigbee protocol, and based on the feedback from all the ED scans, the coordinator selects a channel that has acceptable quality and is not used by other Zigbee PANs. The final step is the migration of all the PAN devices to this "safe" channel. We elaborate on the steps involved in the proposed frequency agility scheme in the following section.

Interference Detection

Zigbee end devices are battery-powered devices, with energy efficiency being a major feature of the Zigbee standard. Therefore it's essential that any interference detection scheme be energy efficient. Zigbee provides reliable service most of the time, so in order to extend device battery lifetime, interference mitigation functions should be only applied when necessary.

Interference detection studies for sensor networks include [18–20]. In [18], a radio interference detection protocol (RID) is presented to detect run-time radio interference among sensor nodes, while an interference detection scheme based on the ED scan results and received signal strength indication (RSSI) is proposed in [19]. In [20] the authors argue that RSSI is not an accurate measure of interference, as the RSSI values of Zigbee frames at a distance within 0.3 m can be very high. The author of [21] proposes an ACK/NACK-based interference detection scheme, which utilizes ACK/NACK reports to detect interference. The sender sends a beacon frame to the receiver and counts the number of NACKs. If the value exceeds the threshold, then interference is detected.

We propose a PER-LQI-based interference detection scheme in the Zigbee network to improve these schemes. Due to Zigbee's low duty cycle, which only requires a few milliseconds for packet transmission [17], a node can successfully deliver the majority of its packets by means of retransmission. To improve packet transmission and net-work battery life, regular packets are used to perform interference detection rather than dedicated signaling messages such as dedicated beacons or periodic packet transmissions. Each end device measures its PER over a transmission period of at least 20 packets [5]. When the PER exceeds 25%, the end device transmits an interference detection report to its parent router. The router checks the LQI between the router and end device; if the LQI [22] is smaller than 100 (which maps to PER 75%), it decides that the packet loss has occurred due to poor link quality rather than due to power outages or other problems at the end device. In this case, the router will perform ED scans on the current channel to ensure that interference is the actual cause of the degradation detected in link quality. Once the energy detection result RSSI exceeds a threshold of 35 (corresponding to a noise level between –65 and –51 dBm), it determines interference to have been detected. The node then makes an interference report to its router, which forwards the report to the coordinator. The coordinator then calls the corresponding interference avoidance scheme and initiates migration to a safe channel. The algorithm of interference detection is shown in the figure below:

Procedure 1: *Interference detection algorithm*

Begin:
  1. **PER = End devices report periodically**
  2. **If PER < threshold**
      Back to (1)
      Else end device sends messages to router
  3. **LQI = detected by router**
     If LQI > threshold
      Back to (1)
      Else RSSI = Energy detection on current channel
  4. **If RSSI < threshold**
      Back to (1)
      Else report to coordinator interference detected
End

For a specific case in which the interference is so severe that the end device can't successfully report it to the router, the router can still detect interference since it periodically monitors the link LQI between itself and all its child nodes. If the LQI is quite low over multiple cycles and the router doesn't receive any messages from its child nodes within the configured timeout period, the router automatically performs an energy detection scan and reports the results to the coordinator.

Interference Avoidance

Once interference is detected, some interference avoidance scheme needs to be applied to mitigate the effect. In [22], the authors consider a scenario in which multiple Zigbee PANs coexist, and suggest letting the PAN that experiences greater interference, or the PAN with the lower priority, change to another channel by means of beacon requests. The coordinator determines which channel they switch to based on the responses from the beacon requests that indicate free channels. In [14] a pseudorandom-based interference avoidance scheme is proposed whereby all devices move to the same channel based on a predefined pseudorandom sequence to avoid interference. This scheme doesn't consider factors such as the interference source and state of other channels; instead, channel selection is randomly performed and interference detection is repeated. An interference avoidance scheme that utilizes energy detection and active scans to determine which channel is appropriate for all the devices to change to is provided here.

In order to reduce the detection time and power consumption of our protocol, all Zigbee channels are divided into three classes based on the offset frequency. As shown in Fig. 13.4, class 1 (solid line) consists of channels 15, 20, 25, and 26 in which the offset frequency is larger than 12 MHz; class 2 (dashed line) is made up of channels 11, 14, 16, 19, 21, and 24 where the offset frequency is larger than 7 MHz and smaller than 12 MHz; while class 3 (dotted line) consists of channels 12, 13, 17, 18, 22, and 23, respectively, in which the offset frequency is smaller than 3 MHz. Class 1 has the highest priority while class 3 has the lowest. Upon receipt of an interference detection report, the coordinator sends an energy detection scan request to all routers in the PAN to check the status of channels from high priority to low priority until an available channel is found. The coordinator chooses the best channel by means of a weighted energy detection result in which each router is assigned a weight based on its priority, network topology, and location. Nodes that are near Wi-Fi access

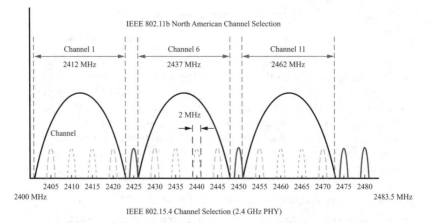

FIGURE 13.4   Zigbee and Wi-Fi channels in the 2.4 GHz band.

points (APs) or that possess a large number of child nodes are assigned larger weights, and the coordinator chooses the available channels based on how highly they score. In a cluster-tree Zigbee network, having all routers doing the energy detection helps to mitigate the effect of the hidden terminal problem. In comparison to having all the devices in the PAN performing an energy detection scan, our algorithm minimizes the complexity of the decision-making algorithm and is more energy efficient.

Upon completion of the energy detection scan, all routers in the PAN commence an active scan on the proposed migration channel selected by the coordinator. They send out a beacon request to determine if any other Zigbee or 802.15.4 PANs are currently active in that channel within hearing range of the radio and if a PAN ID conflict is detected, the coordinator selects a new channel and unique PAN. The decision algorithm is detailed in the figure below.

Procedure 2: *Interference avoidance algorithm*

Begin:
   1. $i = 1$;
   2. **If** $i \geq$ threshold
       back to (1)
       Else energy detection on channel i,
   3. **If** no available channel found
       i ++; back to (2)
       Else Q = available channel $i_1, i_2, \ldots i_k$
   4. **While** no safe channel find
       Active scan on channel $i_j \in Q$
       If $j == k$,
       Back to 2
       Else j++
   5. **Channel changed**
End

Simulation and Implementation Results

Simulation Results

Theoretical analysis and simulation of BER and PER are shown in Figs. 13.5(a) and 13.5(b), respectively. The solid line represents theoretical values while the dotted lines represent the values obtained via simulation. With the exception of a few channels that are far from the central Wi-Fi frequency, most of channels overlapping with Wi-Fi channels have 2 MHz, 3 MHz, 7 MHz, and 8 MHz offsets from the WLAN channel frequency. Simulations in these four scenarios are generated [23].

From both the simulation and theoretical results, we found that the BER and PER drop drastically as the offset frequency increases.

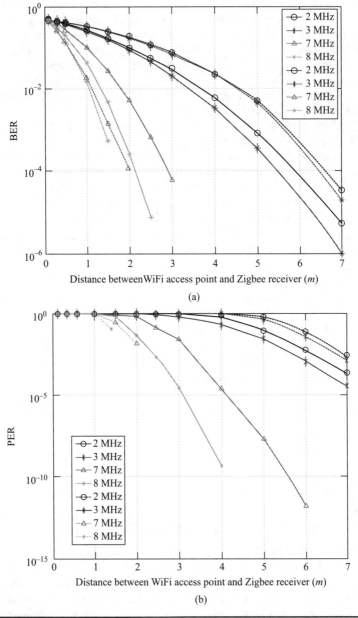

FIGURE 13.5 Theoretical and simulation (a) BER vs. distance, (b) PER vs. distance.

For the same offset frequency channel, the BER and PER decrease when the separation distance increases. BER and PER are higher when the offset frequency is 2 MHz and 3 MHz in the simulation compared to theoretic results; when the offset frequency is 7 MHz and 8 MHz, the BER is lower than the theoretical result. That is because the frequency band of the IEEE 802.11b simulation model is narrower than the theoretical one, with more power concentrated on the effective band frequency.

Both graphs prove that most interference power is congregated around the central frequency of Wi-Fi. "Safe distance" and "safe offset frequency" are two critical parameters that guide Zigbee network deployment in order to mitigate the Wi-Fi interference.

Implementation

The performance of Zigbee under Wi-Fi interference in real-world environments, as well as the performance of our scheme, is evaluated in terms of PER. In [6], Z-Wave reports that WLANs cause significant interference for 802.15.4 and Zigbee devices in the laboratory environment. This was refuted by a Zigbee Alliance report [24], which came to very different conclusions. Their report stated that "Zigbee contains a great many features that are designed to promote coexistence and robust operation in the face of interference. Even in the presence of a surprising amount of interference, Zigbee devices continue to communicate effectively. To understand these conflicting conclusions, we set up Zigbee-Wi-Fi coexistence test beds in residential and public environments to evaluate the performance of Zigbee.

Zigbee Performance under Wi-Fi Interference

Zigbee performance under Wi-Fi interference is examined in a typical residential high-rise environment, and the PER is calculated by the receiver board as follows:

$$PER = \left(\frac{\text{Number of Failed Messages}}{\text{Number of Attempted Measurements}}\right) * 100\% \qquad (13.5)$$

The performance in a residential environment was tested at Lake Meadows Apartments, a 22-story high-rise residential apartment building in Chicago with 15 units per floor. The majority of units each have a Wi-Fi router, which can cover multiple units in the physical communication range. The Internet connection is Digital Subscriber Line (DSL) with 768 kbps to 3 Mbps. Figure 13.6(a) shows the power spectrum measured within one unit. It is shown that a large number of Wi-Fi APs coexist with overlapping spectrum and various signal power strength. Since the DSL service in Lake Meadows Apartment doesn't support high data traffic, we use the Optical Wireless Integration Research Laboratory (OWIL) located in the basement of the Siegel Hall building on the IIT main campus to test the performance under high interference. The whole IIT campus is Wi-Fi enabled, with APs carefully deployed in a controlled manner to reduce the interference among multiple APs. Figure 13.6(b) shows the power spectrum measured in the lab, with the Zigbee signal clearly identified at 8 MHz offset from the Wi-Fi channel center frequency. Heavy

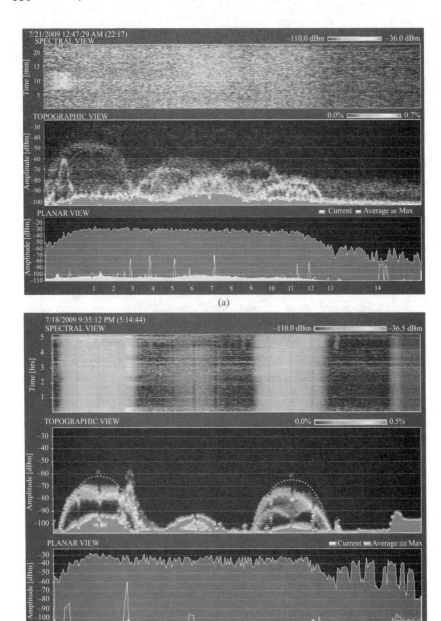

Figure 13.6 Power spectrum of Zigbee and Wi-Fi at (a) residential environment, (b) laboratory environment.

traffic (i.e., heavy Wi-Fi interference to Zigbee) is generated at the rate of 4.5 Mbps between two laptops through the router and varies the distance between the access point and the Zigbee receiver. We observe that the PER decreases as the frequency offset increases. However, in the residential environment although multiple Wi-Fi APs are present, the interference impact is not significant due to low Wi-Fi traffic. In comparison PER is much higher under the heavy interference in the laboratory environment [25].

Furthermore, impact from the uplink communication with the one from the downlink communication in Wi-Fi is compared. In this experiment, a pair of 2.4 GHz Meshnetics MeshBean full-function Zigbee modules is used to support more functions. Instead of using two laptops, one laptop and one PC desktop were used to connect to the Wi-Fi network so that we only have a single wireless link for either uplink or downlink communication, but not both.

From the test results shown in Table 13.2 we observe that the Wi-Fi downlink traffic causes more interference than the uplink traffic. This is due to the difference in the transmit power between the router and the laptop. In general, the transmit power of the Wi-Fi AP is higher than that of the laptop. The results show the performance of Zigbee can be improved dramatically when the offset frequency is larger than 8 MHz.

In order to verify the efficacy of our design guideline in multiple AP scenarios, we experimented with it in a two-AP Wi-Fi system. To measure the performance of Zigbee under the severe Wi-Fi interference, we generate streaming video traffic over Wi-Fi at 4.5 Mbps with two APs operating on the same channel. Due to the use of CSMA/CA, the maximum data rate per AP in the two-AP system is reduced to half, compared to a single AP case. Compared with Zigbee channel 12 at the same distance, the PER for Zigbee channel 14 drops dramatically as shown in Fig. 13.7. Thus, the results in a two-AP scenario are comparable to those in a single AP environment so our "safe distance" and "offset frequency" guidelines and interference avoidance scheme are applicable in multi-AP environments.

In summary, Zigbee provides good performance when Wi-Fi traffic is not heavy. With increasing traffic, Zigbee requires greater separation distance from Wi-Fi or more offset frequency to avoid strong interference from Wi-Fi. Our empirical results match our theoretical analysis and simulation results, confirming the accuracy of our "safe distance" and "safe offset frequency" guidelines.

Interference Detection

Obtaining accurate energy detection results within a short time is the key to guaranteeing the effectiveness of any interference avoidance scheme. We conducted a large number of tests on Zigbee nodes and found that each ED scan duration lasted 138 ms, which provided the best balance between the scan duration and accuracy. Tests show that

| Distance (meters) | Offset Frequency | | | | | | | |
| | 2 MHz | | 3 MHz | | 7 MHz | | 8 MHz | |
	Downlink	Uplink	Downlink	Uplink	Downlink	Uplink	Downlink	Uplink
1	662	657	598	594	499	489	8	0
2	597	579	586	556	414	278	6	0
3	558	529	525	504	373	207	1	0
4	495	454	470	446	353	77	0	0
5	415	375	372	358	216	4	0	0
6	407	350	358	339	69	0	0	0
7	356	306	321	283	6	0	0	0
8	304	272	295	241	3	0	0	0

TABLE 13.2 Number of Failed Packages (per 1000 Packages) vs. Distance (meter) at OWIL

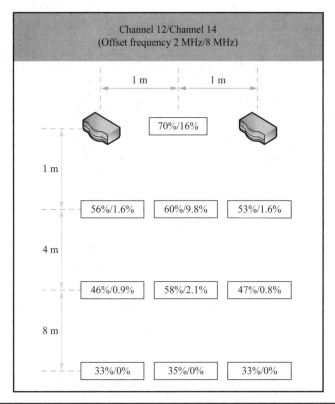

FIGURE 13.7   PER under two Wi-Fi APs.

100% of the quietest channels are in class 1 when we scan all 16 channels with a single Wi-Fi AP serving as the interferer. The implication is that a scan of only class 1 channels provides the same result as a complete scan of 16 channels. Therefore, scanning only class 1 channels provides savings of 75% on the time spent on energy detection.

The LQI is a parameter that indicates the strength or quality of the received packet. The range of LQI values is from 0 to 255, and the PER decreases as the LQI increases. The LQI measurement is performed for each received packet, and if a packet is lost, the transceiver sets LQI as 0. We analyze the LQI readings from 4600 packet transmissions for each channel. Results show that for Zigbee channels with a small offset frequency to the Wi-Fi central frequency, the link quality is bad and transmission packet strength is weak. When the offset frequency is larger than 8MHz, the LQI is larger than 220, which means PER is close to 0 [26].

Energy consumption is calculated based on the PER and battery life analysis [27]. Zigbee sleep duration between active event limits to 2 to 4000 s, while the battery life is around 5 years. We assume battery

Sleep Duration (ms)	Offset Frequency (MHz)	Active Energy Consumption (mAh/month)	Total Energy Consumption (mAh/month)	Battery Life (years)
2	2	73.69	77.27	0.54
	7	63.71	67.29	0.619
	8	19.95	23.53	1.77
	13	16.74	20.32	2.05
7.08	2	20.94	24.54	1.7
	7	18.11	21.69	1.92
	8	5.66	9.24	4.5
	13	4.75	8.33	5
31.3	2	4.75	8.33	5
	7	4.105	7.68	5.43
	8	1.28	4.86	8.57
	13	1.076	4.65	8.96

TABLE 13.3   Zigbee Power Consumption Data

capacity is 1000 mAH, battery efficiency is 50%, and retransmission times equal to 10. If the PAN operates under heavy interference, a high PER leads to a large amount of retransmission, which results in wasted energy throughout the sensor network. Table 13.3 shows that if we choose a cleaner channel, battery life can be prolonged by up to 2 to 3 years with the same sleep time between events.

Conclusion

Zigbee is a low-cost, short-range, low-energy wireless communication technology with wide applicability in many environments, especially in applications such as the smart grid. It can be used to interconnect, monitor, and remotely control data and appliances in homes, buildings, and factories. However, it shares the same license-free frequency band with Wi-Fi networks. In this chapter, Zigbee performance under Wi-Fi interference has been thoroughly evaluated. Using both numerical and empirical analysis, we demonstrate that Zigbee may be severely impacted by Wi-Fi, and that a "safe distance" and "safe offset frequency" can be identified to guide Zigbee deployment. In general, Zigbee provides satisfactory performance when the Wi-Fi interference is not significant. In the event of significant Wi-Fi interference, our frequency agility–based interference mitigation scheme can provide an effective and efficient means of providing reliable data service which enhances Zigbee performance to provide robust and reliable service in coexistence with Wi-Fi networks.

References

[1] A. Iwayemi, Peizhong Yi, Peng Liu, and Chi Zhou, "A Perfect Power demonstration system," in *Innovative Smart Grid Technologies (ISGT)*, 2010, pp. 1–7.

[2] H. Farhangi, "The path of the smart grid," *Power and Energy Magazine, IEEE,* vol. 8, no. 1, pp. 18 –28, Feb. 2010.

[3] United States Department of Energy White paper, "The smart grid: An introduction." 2008.

[4] Office of the National Coordinator for Smart Grid Interoperability, *NIST Framework and Roadmap for Smart Grid Interoperability Standards, Release 1.0.* National Institute of Standards and Technology, 2010.

[5] Zigbee Alliance, "Zigbee Specification: Zigbee Document 053474r17." Jan-2008.

[6] Zensys, *White Paper: WLAN interference to IEEE 802.15.4.* 2007.

[7] S.Y. Shin, H.S. Park, S. Choi, and W.H. Kwon, "Packet Error Rate Analysis of Zigbee Under WLAN and Bluetooth Interferences," *Wireless Communications, IEEE Transactions on*, vol. 6, no. 8, pp. 2825–2830, 2007.

[8] Zigbee Alliance, "Zigbee Smart Energy." Mar-2009.

[9] D.G. Yoon, S.Y. Shin, W.H. Kwon, and H.S. Park, "Packet Error Rate Analysis of IEEE 802.11b under IEEE 802.15.4 Interference," in *Vehicular Technology Conference, 2006. VTC 2006-Spring. IEEE 63rd*, 2006, vol. 3, pp. 1186–1190.

[10] J. Mikulka and S. Hanus, "Bluetooth and IEEE 802.11b/g Coexistence Simulation," *Radio Engineering*, vol. 17, no. 3, pp. 66–73, Sep. 2007.

[11] K. Shuaib, M. Boulmalf, F. Sallabi, and A. Lakas, "Co-existence of Zigbee and WLAN, A Performance Study," in *Wireless Telecommunications Symposium, WTS '06*, 2006, pp. 1–6.

[12] T. Rappaport, *Wireless Communications: Principles and Practice.* Upper Saddle River, NJ, USA: Prentice Hall PTR, 2001.

[13] R.L. Peterson, D.E. Borth, and R.E. Ziemer, *An Introduction to Spread-Spectrum Communications.* Upper Saddle River, NJ, USA: Prentice-Hall, Inc., 1995.

[14] S.Y. Shin, H.S. Park, and W.H. Kwon, "Mutual interference analysis of IEEE 802.15.4 and IEEE 802.11b," *Computer Networks*, vol. 51, no. 12, pp. 3338–3353, Aug. 2007.

[15] *IEEE Std 802.11 TM -2007 Part 11: Wireless LAN Medium Access Control (MAC) and Physical Layer (PHY) Specifications.* 2007.

[16] *IEEE Std 802.15.4a TM -2007: 802.15.4aPart 15.4: Wireless Medium Access Control (MAC) and Physical Layer (PHY) Specifications for Low-Rate Wireless Personal Area Networks (WPANs).* 2007.

[17] S. Farahani, *Zigbee Wireless Networks and Transceivers.* Newton, MA, USA: Newnes, 2008.

[18] G. Zhou, T. He, J.A. Stankovic, and T. Abdelzaber, "RID: radio interference detection in wireless sensor networks," presented at the Proceedings IEEE 24th Annual Joint Conference of the IEEE Computer and Communications Societies., Miami, FL, USA, 2010, pp. 891–901.

[19] C. Won, J-H Youn, H. Ali, H. Sharif, and J. Deogun, "Adaptive radio channel allocation for supporting coexistence of 802.15.4 and 802.11b," presented at the VTC-2005-Fall. 2005 IEEE 62nd Vehicular Technology Conference, 2005., Dallas, TX, USA, 2010, pp. 2522–2526.

[20] M. Kang, J. Chong, H. Hyun, S. Kim, B. Jung, and D. Sung, "Adaptive Interference-Aware Multi-Channel Clustering Algorithm in a Zigbee Network in the Presence of WLAN Interference," presented at the 2007 2nd International Symposium on Wireless Pervasive Computing, San Juan, PR, USA, 2007.

[21] S.M. Kim et al., "Experiments on Interference and Coexistence between Zigbee and WLAN Devices Operating in the 2.4 GHz ISM Band," *Proceedings of NGPC*, pp. 15–19, Nov. 2005.

[22] R.C. Shah and L. Nachman, "Interference Detection and Mitigation in IEEE 802.15.4 Networks," presented at the 2008 7th International Conference on Information Processing in Sensor Networks (IPSN), St. Louis, MO, USA, 2008, pp. 553–554.

[23] P. Yi, A. Iwayemi, and C. Zhou, "Frequency agility in a Zigbee network for smart grid application," presented at the Innovative Smart Grid Technologies (ISGT), 2010, 2010, pp. 1–6.

[24] G. Thonet and P. Allard-Jacquin, *Zigbee – WiFi Coexistence White Paper and Test Report*. Schneider Electric White Paper, 2008.

[25] P. Yi, A. Iwayemi, and C. Zhou, "Developing Zigbee Deployment Guideline Under WiFi Interference for Smart Grid Applications," *Smart Grid, IEEE Transactions on*, vol. 2, no. 1, pp. 110–120, 2011.

[26] Atmel Corp, "RF230: Low Power 2.4 GHz Transceiver for Zigbee, IEEE 802.15.4, 6LoWPAN, RF4CE and ISM Applications." Feb-2009.

[27] Zigbee Alliance, *Zigbee and Wireless Radio Frequency Coexistence*. 2007.

Index

A

AC-DC:
   microgrids, 251–253
   power conversion losses, 230
AC grid, 230–231
Active load:
   control, 183, 187–191
   communication-
      based, 188–191
   load following, 187
   without communication-
      based control, 191–195
   power flow equation, 192
   voltage-based active load,
      191
Adaptive voltage position, 252
Advanced metering
   infrastructure, 111, *see also*
   Automated metering
   infrastructure
AGC, *see* Automatic gain
   controller
AHM, *see* Asset management,
   health management
AMI, *see* Advanced metering
   infrastructure; Automated
   metering infrastructure
Ancillary services, 208–210
Asset management, 23, 44, 46
   best practices, 48
   health management, 44, 46
   maintenance practices, 49
   risk planning process, 51
   systems, 49–50
      availability, 50
      flexibility, 50
      reliability, 50

Automated metering
   infrastructure, 43,
   see also Advanced
   metering infrastructure
Automatic gain controller, 263
Automation technologies, 109
AVP, *see* Adaptive voltage position

B

Back to back:
   converter, 156
   voltage source converter, 158
      advantages, 159
      operation limits, 162
      PQ diagram, 162
      topology, 160
Battery:
   degradation, 207–208, 211,
      222–223
   virtual pool, 211–214
   wearing out, 207
BEV, *see* Electric vehicles,
   pure battery
BTB, *see* Back to back
Bypass switch, 255–264

C

CAP, *see* Contention access period
Carrier sense multiple access
   with collision avoidance,
   296, 324–325
Center of inertia:
   frequency domain, 69
   time domain, 68
Centralized generation, 111–131
CFP, *see* Contention free period

CIM, *see* Common information model
Coherency method for network partitioning, 63–67
COI, *see* Center of inertia
Common information model, 48
Conergy, 221
Conservation voltage reduction, 37
Contention access period, 296
Contention free period, 296
Control, 204–205
    automated, 205
    direct, 204
    distributed generation, 184
        central, 184
        communication-based, 184–185
        master slave, 184
        reversed droop, 185
        voltage-based droop, 185
        without communication, 185–187
    indirect, 205
    negative secondary, 223
    primary, 208
    secondary, 208–209, 223
    tertiary, 209
CSMA/CA, *see* Carrier sense multiple access with collision avoidance
Current sharing techniques, 251
CVR, *see* Conservation voltage reduction

══ **D** ══

DA, *see* Distribution automation
DALI, *see* Digital addressable lightning interface
Data collection and control center, 323–324
Data mining, 70, 78
    black-box predictors, 82–84
    catastrophe prediction, 78–79
    wide-area severity indices, 70–76
DC grid, 229–231
    microgrid, 229–231
DC link(s), 117–125
    base configuration, 169
    benefits of use in distribution networks, 166, 169–177

DC link(s) (*Cont.*):
    distributed generation penetration, 174
    economic assessment, 176–178
        pay back equation, 177
    increasing network load level, 172
    model, 167
    network loss reduction, 169–172
    optimal number of, 165–167
    radial distribution systems, 157
    voltage source converter-based, 157
DCCC, *see* Data collection and control center
Demand management, residential, 301–308
Demand response, 294
Demand-side developments, 3–4
Demand-side management, 2, 5, 9–10
    control methodology, 9, 10–11
    damage control, 10
    microgrid, 9
    virtual power plant, 9, 13
Denial of service, 313
Depth of discharge, 207
DFIG, *see* Wind turbines, double-fed induction generator
DG, *see* Distributed generation
DIALux, 240
Digital addressable lightning interface, 246
Digital relays, 47
Dispatch time, 214–216
Dispersed generation in medium voltage networks, 157
Distributed generation, 2, 8, 111–129
    control methodologies, 13–16
    networks constraints, 168
Distribution automation, 22–23, 111
    architecture, 23
    definition, 26
    IT and communications, 26
Distribution management systems, 43
DMS, *see* Distribution management systems
DoD, *see* Depth of discharge
DoS, *see* Denial of service

Droop control, 251–252
  paralleling inverters, 251
  P/V, *see* Control, distributed
    generation, reversed
    droop
  voltage-based, 183
DSM, *see* Demand-side
  management
DVA, *see* Dynamic vulnerability
  assessment
Dynamic vulnerability
  assessment, 64

━━━ **E** ━━━

Early warning system, 53–103
  design concept, 54–57
Electric Power Research
  Institute, 47
Electric vehicles, 201–226
  charging, 201–224, 308–312
    free, 223
    home, 205
    smart, 205–206
  grid-connected, 202
  plug-in hybrid, 202
  pure battery, 202
Electricity supply chain, 1
  demand side, 3–4, 6
  generation side, 3, 6
  trends, 6
EMerge Alliance, 231
EMU, *see* Energy management unit
Energy buffers, 4
Energy management unit, 54
Energy production tax
  credit (PTC), 108
Energy storage, 124–127, 130
  batteries, 114–127, 133
  electrochemical, 133
  hydrogen, 133
EPRI, *see* Electric Power Research
  Institute
EV, *see* Electric vehicles,
  grid-connected
EWS, *see* Early warning system

━━━ **F** ━━━

FACTS, *see* Flexible AC
  transmission systems
Fault detection, isolation, and
  restoration, 27–30
  decentralized, 31, 32

Fault detection, isolation, and
  restoration (*Cont.*):
  loop control scheme, 30
  peer-to-peer messaging, 30
  system architecture, 29
FCMdd, *see* Fuzzy c-meoids
  algorithm
FDIR, *see* Fault detection, isolation,
  and restoration
Federal Energy Regulatory
  Commission, 118
Feed-in tariffs, 221–222
FERC, *see* Federal Energy
  Regulatory Commission
FFD, *see* ZigBee, full function
  device
Flexible AC transmission
  systems, 109–110
Fuzzy c-meoids algorithm, 64

━━━ **G** ━━━

Generation storage, 7–8, 17
Grid management
  time, 202–203
Grid parity, 222
GTS, *see* Guaranteed time slot
Guaranteed time slot, 297

━━━ **H** ━━━

High-voltage direct current,
  109–111, 126–127
Home energy dashboard, 293
HQRI, *see* Hydro-Québec
  Research Institute
HVDC, *see* High-voltage
  direct current
Hydro-Québec Research
  Institute, 57

━━━ **I** ━━━

IEC 61851, 206
IEDs, *see* Intelligent electronic
  devices
IEEE 82.11b, 325
IEEE 82.15.4, 295–298, 300,
  322, 324–325
  interference sources, 326–327
IGBT, *see* Insulated-gate bipolar
  transistor
IGCT, *see* Integrated gate
  commutated thyristor
iHEM, *see* In-home energy
  management

In-home energy
management, 301–308
Independent system
operators, 119
Insulated-gate bipolar
transistor, 124–125
Intel, 293
Intelligent electronic devices, 27
Intelligent volt/var optimization
scheme, 39
International standards
organization, 48
ISA-100.11a, 301
ISO, *see* Independent system
operators, International
standards organization

■■■ **L** ■■■

LED, *see* Lighting, light-emitting
diodes
LFC, *see* Load frequency control
Lighting, 227–248
light-emitting diodes, 228, 233
simulation, 240–243
solid-state:
Roland Haitz law, 234
sources, 228, 233
system implementation,
243–247
systems, 232–234
sources, 233
Load frequency control, 263
Load models, 37
kI load, 38
kP load, 38
kZ load, 37
Load shifting, 202–204
Low-voltage direct current,
231–232
emerging standards,
231–232
LVDC, *see* Low-voltage direct
current

■■■ **M** ■■■

Market, 217–218
regular, 223–224
regulation reserve, 208–210,
222
Maximum power point tracking,
275, 277–278
Medium voltage networks, 155

Metering:
off-board, 206
onboard, 206
MG, *see* Microgrid
Microgenerators, 4
Microgrid, 12, 142–145,
181–188, 195
advantages, 143–144
architecture and design of a
microgrid, 145–150
barriers, 150–152
control:
distributed, 145
hierarchical, 254–272
inner loop, 255
primary 194, 257–260
secondary194, 260–264
simulation results,
265–272
tertiary control,
264–265
definition, 249
energy neutral, 12
expectations of, 182
grid connected, 181
grid synchronization,
252–264
hybrid DC-AC, 253
island mode operation, 12, 142,
181–183
pricing aspects, 189–191
critical peak pricing, 190
real-time pricing, 190
tariff metering, 191
time-of-use, 189
volt-var variability, 145
Micro-wind energy harvesting,
274
MPPT, *see* Maximum power point
tracking
Multiband modal analysis,
89–91

■■■ **N** ■■■

NASPI, *see* North American
SynchronoPhasor Initiative
NIST, *see* United States National
Institute of Standards and
Technology
North American SynchronoPhasor
Initiative, 54
NPC, *see* Power converters,
neutral point connected

O

Offshore wind, 111–122
  high voltage:
    dc interconnection, 124
    subsea cables, 122
  medium voltage subsea
      cables, 122
  onshore substation, 122–123
  subsea interconnection,
      121–124
  underwater connectivity, 115
Oscillatory stability problem, 88

P

P/V drop control, *see* Control,
    distributed generation,
    reversed droop
PAN, *see* Personal area networks
Paralleling inverters, 251
Parking, 203–204
  home, 204
  leisure, 204
  public, 204
  work, 204
PDC, *see* Phasor data concentrator
PE, *see* Power electronics
Peaker plants, 293
PEBB, *see* Power converters,
    power electronics building
    block
Perfect power project, 323–324
Performance assessment, 84–88
Personal area networks, 296
Phasor data concentrator, 54
Phasor measurement unit,
    54, 57–63
  performance under
      voltage dip, 61
  requirements, 57–59
  vulnerability assessment, 57
PHEV, *see* Electric vehicles,
    plug-in hybrid
Philips, 240
Photovoltaic, 129–138
  cell, 130
  cell equivalent circuit, 133–135
  cells strings, 137
  energy production, 135
  plants connected to the grid,
      138
  private generation, 224
  voltage-current characteristics,
      133–135

PIR, *see* Sensor, pyroelectric
    infrared
PM, *see* Wind turbines, variable
    speed permanent magnet
PMU, *see* Phasor
    measurement unit
Power converters:
  dynamic var support, 120–123
  four-quadrant, three-level
      converter, 117
  integrated gate commutated
      thyristor, 117
  neutral point connected, 117
  power electronics building
      block, 117
  power factor, 120
  power quality, 120
Power electronics, 109
Power flow controllers, 156
Power system stabilizers, 54
Price:
  critical peak pricing, 190
  microgrid pricing
      aspects, 189–191
  real-time pricing, 190
  signals, 204
  spread, 220–222
  tariff metering, 191
  time-of-use, 189
PSS, *see* Power system stabilizers
PTC, *see* Energy production tax
    credit
Pulse-width modulation, 111, 124
PV, *see* photovoltaic
PWM, *see* Pulse-width modulation

R

Radio frequency identification,
    272
RC, *see* Reliability coordinator
Reactive power, 118
  calculation, 34–35
Rectifiers, 230
Regional transmission operator,
    54
Regional transmission
    organizations, 119
Regulation capacity:
  negative, 211
  positive, 210–211
Reliability coordinator, 55
Renewable energy, 108–110, 127
Renewable portfolio standard, 108

RFD, *see* ZigBee, reduced function device
RFID, *see* Radio frequency identification
Ringdown analysis, 96
RTO, *see* Regional transmission operator
RTOs, *see* Regional transmission organizations

━━ **S** ━━

SAIDI, *see* System average interruption duration index
SCADA, *see* Supervisory control and data acquisition
Secondary control, 253
Self-powered smart wireless sensor, 272–286, *see also* Wireless sensor networks, energy harvesting
Sensor nodes, 235–238
low-power wireless, 273
pyroelectric infrared, 245
SEP, *see* Smart energy profile
SG, *see* Smart grid
SIMO, *see* Single input multiple output
SimpliciTI, 244
Single input multiple output, 91
SMA, 221
Smart building, 234–235, *see also* Smart home
case study, 238–243
Smart devices, 4
Smart energy profile, 295, 322
Smart grid, 7–8, 24, 250
architecture, 292
barriers, 218
capabilities, 109
communications, 8
consumer applications, 301–312
security and privacy, 313–315
definition, 7
earning potential, 220
monitoring and forecasting, 7
power flows, 8
sensing and measuring, 7
smart technologies, 108
ZigBee-based control system, 321–340
Smart home, 293, *see also* Smart building

Smart-induced green building, 234–235, *see also* Smart building
case study, 238–243
SOC, *see* State of charge
Solar power, 128–139
power plants, 132–133
energy storage systems, 133
intermittency power production, 132
irradiation, 128
permanent grid connected, 131
State modal analysis, 91–95
deterministic SIMO-ERA, 92–93
linear SIMO signal identification, 91
stochastic SIMO-SSSID, 93–94
State of charge, 202
Static var compensators, 109–110
with energy storage, 127
Substation automation, 108
Supervisory control and data acquisition, 109–111
SVC, *see* Static var compensators
System average interruption duration index, 27

━━ **T** ━━

Texas Instruments, 244
TOU, *see* Price, time-of-use
Transmission and distribution infrastructure, 44, 45

━━ **U** ━━

U.S. Department of Energy, 251, 323
Uninterruptible power systems, 251
United States National Institute of Standards and Technology, 322
UPS, *see* Uninterruptible power systems
U.S. Advanced Battery Consortium, 207
USABC, *see* U.S. Advanced Battery Consortium
Usage pattern, irregular, 238

V

V2G, *see* Vehicle-to-grid
VAR, *see* Reactive power
Vehicle-to-grid, 202, 216–217,
    222–223
Virtual output impedance, 252,
    259–260
Virtual output resistance, 252
VLU, *see* Wind turbines, voltage
    limiter unit
Volt/var optimization, 111
Voltage regulators, 34, 36
Voltage source converters,
    124, 126–127, 158
  advantages, 159
  operation limits, 162
  PQ diagram, 162
  topology, 160
Voltage-based droop control, 183
Voltage support, 123
VSC, *see* Voltage source converters
Vulnerability assessments,
    57, 63, 76
VVO, *see* Volt/var optimization
VVV, *see* Microgrid, volt-var
    variability

W

WAMS, *see* Wide-area
    measurement system
WASAS, *see* Wide-area situational
    awareness system
WASI, *see* Wide-area severity index
Wide-area measurement
    system, 53
Wide-area monitoring system, 57
Wide-area severity index, 67, 70–76
Wide-area situational awareness
    system, 54
Wi-Fi, 325–327
  ultra low-power, 298–299
Wind energy power converters,
    116–120
Wind power:
  backup generation, 114
  backup storage, *see* Wind power,
    backup generation
  grid reinforcement, 112
  integration, 110
  interconnection, 121
  predictability, 113–115

Wind turbines, 110, 126–127
  AC substation, 121
  brake chopper, 119
  double-fed induction
    generator, 110, 116
  low-voltage ride-through
    capability, 111, 118, 119
  medium voltage
    switchgear, 120–121
  power coefficient factor, 276
  power dynamic power
    control, 125–127
  variable speed permanent
    magnet, 116
  voltage dip, 119–120
  voltage limiter unit, 119
Wireless sensor devices, 250
Wireless sensor networks,
    235–238, 272, 274–280
  communication standards,
    295–301
  comparison, 299
  energy harvesting, 273–283
    magnetic, 281–284
    wind, 274–280
  sensor node, 272–273
WirelessHART, 300
WSN, *see* Wireless sensor
    networks

Z

ZigBee, 295–298, 324–325
  Alliance, 295
  coordinator, 325
  full function device, 324
  interference avoidance
    scheme, 329–340
    detection, 330–331
    implementation, 335
    performance results,
      335–340
    simulation, 333–335
  reduced function device, 324
  security, 297
  test bed architecture, 323
  Wi-Fi interference, 327–329
    bit error rate analysis,
      327–328
    packet error rate analysis,
      328–329
Z-wave, 299–300